T0328912

Liutex and Its Applications in Turbulence Research

Liutex and Its Applications in Turbulence Research

Chaoqun Liu

Center for Numerical Simulation and Modeling
Department of Mathematics
University of Texas at Arlington
Arlington, Texas, United States

Hongyi Xu

Fudan University
Shanghai, China

Xiaoshu Cai

University of Shanghai for Science and Technology,
Shanghai, China

Yisheng Gao

Nanjing University of Aeronautics and Astronautics,
Nanjing, China

ACADEMIC PRESS

An imprint of Elsevier

ELSEVIER

Academic Press is an imprint of Elsevier
125 London Wall, London EC2Y 5AS, United Kingdom
525 B Street, Suite 1650, San Diego, CA 92101, United States
50 Hampshire Street, 5th Floor, Cambridge, MA 02139, United States
The Boulevard, Langford Lane, Kidlington, Oxford OX5 1GB, United Kingdom

Copyright © 2021 Elsevier Inc. All rights reserved.

No part of this publication may be reproduced or transmitted in any form or by any means, electronic or mechanical, including photocopying, recording, or any information storage and retrieval system, without permission in writing from the publisher. Details on how to seek permission, further information about the Publisher's permissions policies and our arrangements with organizations such as the Copyright Clearance Center and the Copyright Licensing Agency, can be found at our website: www.elsevier.com/permissions.

This book and the individual contributions contained in it are protected under copyright by the Publisher (other than as may be noted herein).

Notices
Knowledge and best practice in this field are constantly changing. As new research and experience broaden our understanding, changes in research methods, professional practices, or medical treatment may become necessary.

Practitioners and researchers must always rely on their own experience and knowledge in evaluating and using any information, methods, compounds, or experiments described herein. In using such information or methods they should be mindful of their own safety and the safety of others, including parties for whom they have a professional responsibility.

To the fullest extent of the law, neither the Publisher nor the authors, contributors, or editors, assume any liability for any injury and/or damage to persons or property as a matter of products liability, negligence or otherwise, or from any use or operation of any methods, products, instructions, or ideas contained in the material herein.

Library of Congress Cataloging-in-Publication Data
A catalog record for this book is available from the Library of Congress

British Library Cataloguing-in-Publication Data
A catalogue record for this book is available from the British Library

ISBN: 978-0-12-819023-4

For information on all Academic Press publications visit our website at
https://www.elsevier.com/books-and-journals

Publisher: Matthew Deans
Acquisitions Editor: Brian Gueri
Editorial Project Manager: Mariana Kuhl
Production Project Manager: Sreejith Viswanathan
Cover Designer: Alan Studholme

Typeset by TNQ Technologies

Working together
to grow libraries in
developing countries

www.elsevier.com • www.bookaid.org

Contents

Biography of Authors

Dr. Chaoqun Liu received both BS (1968) and MS (1981) from Tsinghua University, Beijing, China, and PhD (1989) from University of Colorado at Denver, USA. He is currently the Tenured and Distinguished Professor and the Director of Center for Numerical Simulation and Modeling at University of Texas at Arlington, Arlington, Texas, USA. He has worked on high-order direct numerical simulation (DNS) and large eddy simulation (LES) for flow transition and turbulence for almost 30 years since 1990. As PI, he has been awarded by NASA, US Air Force, and US Navy with 50 federal research grants of over 5.7×10^6 US dollars in the United States. He has published 13 professional books, 124 journal papers, and 157 conference papers. He is the founder and major contributor of Liutex and the third generation of vortex definition and identification methods including the Omega, Liutex/Rortex, Modified Liutex Omega, Liutex Core Line methods, RS vorticity decomposition, and UTA R-NR velocity gradient tensor decomposition. His email is cliu@uta.edu.

Dr. Hongyi Xu got his BS (1985) and MS (1988) from Shanghai University of Technology, Shanghai, China, and received PhD (1998) from Queens University at Kingston, Canada. He is currently a Professor at Fudan University, Shanghai, China. He has conducted professional research in the fixed-wing and rotary-wing aerodynamics and thermal turbulence in high-temperature turbine of aero-engine. He has 15 years of working experience as a senior research officer in the Institute for Aerospace Research, National Research Council (IAR/NRC) of Canada. He was the rotary-wing group leader in the aerodynamic lab of IAR/NRC and was appointed by Department of National Defense as a chief expert representative of Canada participating in The Technology Collaboration Program (TTCP) activities involving United States, British, Canada, Australia, and New Zealand. He was leading and participating in a number of R&D research programs from both North American major aero-manufacturers and government organizations, such as Canadian Bombardier Aircraft Company, Pratt–Whitney Aero-Engine Company, U.S. Bell Helicopter Incorporation, and Defence Research and Development Canada, Defence Research Establishment, etc. He joined the Department of Aeronautics and Astronautics, Fudan University, as a senior professor in 2013. He has published many research papers in both internationally prestigious journals and conferences. He is currently leading a number of research programs from Chinese Natural Science Foundation and Shanghai Aeroengine Corporation. Academically, he is a well-known expert in direct numerical simulation of turbulence and established the wall-turbulence big databank at Fudan University. Based on these data, he pushed the current turbulent boundary layer theory to the forefront and developed the complete law-of-the-wall formulations for wall-bounded turbulence. Moreover, he applied the modern neuro-network techniques to explore the innovative turbulence closure modeling.

Dr. Xiaoshu Cai received his doctor degree at Shanghai Institute of Mechanical Engineering, 1991. He has been employed as a professor in University of Shanghai for Science and Technology since 1996 and the director of the Institute of Particle and Two-phase Flow Measurement. His research areas are as follows: particle sizing, measurement techniques for two-phase flow, turbulence, combustion, environment and emission monitoring, turbomachinery, and life science. Prof. Cai was granted more than 10 million RMB and half million USD research funding for aforementioned research in recent years. Until now, Prof. Cai has published more than 200 papers in international and Chinese journals and conferences and some books and obtained more than 30 patents. He has been elected as the vice chairman of the Chinese Society of Particuology, the chairman of Shanghai Society of Particuology, and the member of executive board of some national academic organizations and has also been invited as the member of editorial boards of 4 international journals and 4 Chinese journals.

Dr. Yisheng Gao received his Bachelor of Engineering degree in Aircraft Design and Engineering (2007), Master of Engineering degree in Fluid Mechanics (2009), and PhD degree in Fluid Mechanics (2016) from Nanjing University of Aeronautics and Astronautics University, China. He was a postdoctoral fellow at the University of Texas at Arlington, Texas, USA, in 2017—19 and is currently a lecturer at Nanjing University of Aeronautics and Astronautics University, Nanjing, China. His research is concerned with computational fluid dynamics, design optimization methods, vortex identification, and turbulence. He has published 1 professional book and 24 journal papers.

Preface

First edition

Liutex is a new physical quantity discovered by Professor Chaoqun Liu at University of Texas at Arlington (UTA) in 2018 to represent fluid rotation, which is similarly important as velocity, pressure, temperature, and vorticity for fluid dynamics. The discovery of Liutex, we believe, is probably one of the most important breakthroughs in modern fluid dynamics especially for vortex science and turbulence research. Vortex is ubiquitous in nature and viewed as the building blocks or more vividly the muscles and sinews of turbulent flows. However, vortex had no mathematical definition before the introduction of Liutex, which was the bottleneck of modern fluid dynamics and caused countless confusion in vortex and turbulence research.

According to Liu et al. (2019), there are three generations of vortex identification methods in history. In 1858, Helmholtz first defined vortex as vortex tube composed of the so-called vortex filaments, which were really infinitesimal vorticity tubes. It is classified as the first generation of vortex identification and vortex is defined as vorticity tube. Science and engineering applications have shown that the correlation between vortex and vorticity is very weak, especially in the near-wall region. During the past three decades, many vortex identification criteria, Q, Δ, λ_2, and λ_{ci} methods, for example, have been developed, which are classified as the second generation of vortex identification. They are all based on the eigenvalues of the velocity gradient tensor. However, they are all scalars and thus strongly dependent on the factitious and arbitrary threshold, when plotting the isosurface to represent the vortical structures. In addition, they are all obviously contaminated by stretching (or compression) and shearing. Liutex as the third generation of vortex definition and identification was developed by Liu and his students at UTA. Liutex is defined as a vector that uses the real eigenvector of velocity gradient tensor as its direction and twice the local angular speed of the rigid rotation as its magnitude. The major idea of Liutex is to extract the rigid rotation part from fluid motion to represent vortex. After almost 200 years of efforts, human beings, for the first time, found an accurate physical quantity to represent fluid rotation or vortex.

After that, a number of vortex identification methods have been developed by Liu and his UTA Team including Liutex vector, Liutex vector lines, Liutex tubes, Liutex isosurface, Liutex Omega methods, Objective Liutex, and, more recently, Liutex Core Line methods, which can more accurately visualize the vortical structures in turbulent flows, proved by countless users and research papers. Liutex Core Line, which is defined as a special Liutex line, where the gradient of Liutex magnitude is parallel to Liutex vector, is unique and threshold-free.

In addition, the existence, uniqueness, stability, and Galilean invariance of Liutex have been proved by the UTA Team. Fitting for local and global rotation has been proved in addition to applicability for both compressible and incompressible flows.

A so-called "Principal Coordinate" based on the velocity gradient tensor is defined and the fundamental vector and tensor decompositions are made in the Principal Coordinate by Liu and his students. A new vorticity decomposition to Liutex and shear, namely RS decomposition of vorticity, is proposed by Liu to reveal non-dissipative rigid rotation and dissipative shear for decomposition of fluid motion to replace Helmholtz velocity decomposition. A new Liutex-based UTA R-NR tensor decomposition is also proposed by Liu to replace the traditional Cauchy−Stokes (Helmholtz) decomposition. The Liutex-based UTA R-NR tensor decomposition is unique and Galilean invariant while Cauchy−Stokes is not, and R-part is the rigid fluid rotation while the antisymmetric part is not. The Liutex definition, Principal Coordinate, Principal Decomposition, vorticity RS decomposition, velocity gradient tensor UTA R-NR decomposition, proposed by Liu, integrate the Liutex theory and pave the foundation for new vortex science, new turbulence research, and new fluid dynamics.

Liutex (rigid rotation) spectrum similarity theory of power of −5/3 in transitional and turbulent boundary layer is discovered by Liu and his students while Kolmogorov's −5/3 law is not matched well with DNS or experiments especially in low Reynolds number turbulent boundary layers. Liutex dynamics and modified Navier−Stokes equations, which govern both laminar and turbulent flows without models, are under development by the UTA team and their collaborators.

Vortex identification has six core elements including (1) absolute strength, (2) relative strength, (3) local rotational axis, (4) vortex rotation axis, (5) vortex core size, (6) vortex boundary, which are touchstones that the vortex identification methods should be tested against. It is confirmed with illustrative examples that only the Liutex system is able to give precise information of all six core elements in contrast to the failure of the first- and second-generation methods in vortex identification except for rough estimates of vortex boundaries.

Prof. Chaoqun Liu, as the founder and major contributor of Liutex, has worked on direct numerical simulation (DNS) for flow transition and turbulence for 30 years since 1990. He was very skeptical about the existing vortex definition and turbulence theories. There are several questions that bothered him much for a long time. First, we never had a mathematical definition of vortex that has been viewed as the building blocks of turbulence. Liu did not understand why our textbook teaches student that vortex tube can never break down and cannot end inside flow field, but can only end on either a solid surface or boundary at infinity. Later in the turbulence part of the same book, the textbook teaches student that turbulence is generated by "vortex breakdown." This kind of self-contradiction in our textbooks made people very strayed. In fact, all natural vortices observed by DNS, LES, or experiments end inside the flow field and none of them end on the solid surface, which is just the opposite of what our textbooks teach. The other turbulence theories are also very confusing. Richardson (1928) gave the eddy cascade and vortex breakdown theory on turbulence, which was supported by Kolmogorov (1941) with three famous hypotheses. The most famous one is the similarity law of turbulence energy spectrum or the −5/3 law. However, as the computers and experiment instruments are so advanced nowadays, there is no body who reported they find the vortex cascade caused by large vortex breakdown to two or three smaller vortices, and no DNS results can match the −5/3 law very well as the assumptions of infinitely large Reynolds number and isotropic turbulence do not exist in boundary layers. In addition, Helmholtz's three theories of vortex may work for inviscid flow but definitely fail in viscous flow or real flow. How can we use Helmholtz theories for turbulence, which has no possibility to be inviscid? Although Robinson (1990) and Adrian (2007) provided some new scenarios on turbulence, researchers were still not awakened since the Helmholtz's and Kolmogorov's theories still dominate the turbulence research and the turbulence community still had no mathematical definition for vortex. After Liu gave several times of presentation on his concerns in the AIAA conferences, where he met strong resistance from different sides, Liu then published a paper titled "Physics of turbulence generation and sustenance in boundary layer" in Computers and Fluids in 2014 with his students, Yonghua Yan and Ping Lu, which clearly shows his opinion that turbulence is not generated by large vortex breakdown, but by the transformation of shear to rotation or, in other words, by the transformation from shear to Liutex. "Shear layer instability is the mother of turbulence."

Two events happened in 2014. One was that Mr. Yiqian Wang, who is now an Associate Professor in Soochow University of China, joined the UTA team as a visiting student, and the other is Liu met Prof. Xiaoshu Cai of University of Shanghai for Science and Technology in China through the introduction of Prof. Jiyuan Tu, who is a professor of RMIT in Australia. Through Prof. Cai, Liu later met Prof. Hongyi Xu, who is a professor of Fudan University of China. They then formed a small research group to focus on turbulence research.

In 2015, as the Principal Lecturer, Prof. Liu was invited to hold a short course in Beijing, China, titled as "New Theory on Turbulence Generation and Sustenance," at Tsinghua University organized by Prof. Song Fu and sponsored by 24 major universities and institutes in China. The short course received unprecedented enthusiasm from Chinese turbulence research community as 240 professional people registered for the short course. The organizers originally prepared a lecture room that can host around 200 audiences, but on the opening day, they had to remove many desks from the back half room and move in many chairs instead to save the space. As more people registered and came to the lecture, the organizers had to ask all local teachers and students of Tsinghua University not to enter the room and leave more space for the guest participants. Even though, there were still many people who had to stand in the room without seat. This event gave a shock to the Chinese turbulence research community that a systematical turbulence theory was lectured by a visiting professor from University of Texas at Arlington, Texas, USA.

In 2016, Liu published a paper "New Omega Vortex Identification Method" in Science China: Physics, Mechanics & Astronomy with Yiqian Wang, Yong Yang, and Zhiwei Duan. Omega method represents the relative strength of vortex and is insensitive to threshold selection. In addition, the Omega method can visualize both strong and weak vortices simultaneously. In the Omega paper, Liu also proposed to decompose vorticity to a rotational part and a nonrotational part (RS decomposition of vorticity). Later, Yiqian Wang helped Liu find that the vorticity is smaller inside vortex, but larger outside vortex, which is an observation directly opposing to the traditional opinion that vortex is a concentration of vorticity magnitude. The UTA team then decided to focus their research on vortex and its definition and identification ever since.

During the period of 2016–19, several visiting students and post doctors joined the UTA team including Xiangrui Dong, Shuling Tian, Yisheng Gao, Panpan Yan, Jianming Liu, and Wenqian Xu. Their participation greatly accelerated the pace in finding the rotational part of vorticity or Liutex, which represents vortex.

In December of 2017, Prof. Xiaoshu Cai invited Liu to visit his university. They then decided to organize a conference called "New Development and Key Issues of Vortex and Turbulence Research" or "First Vortex Conference," which attracted more than 200 participants although they faced some boycotts. The conference lasted for 2 days. Liu gave lectures in the whole morning on the first day about the new definition of vortex with the assistances from his students. There were several other speakers in the first afternoon and second morning. Liu answered many questions and distributed the UTA's computational codes to participants in the second afternoon. During the conference, participants had very hot discussions over a variety of questions on vortex and turbulence research. Many people were excited with the new definition of vortex as they heard that "vortex" is not "vorticity" for the first time. During the discussion, several professors suggested that the new definition of vortex is not a traditional concept of vortex and it should be given a new name. Liu accepted one suggestion to name the new definition of vortex as "Rortex." Unfortunately, there was a concern that the name of "Rortex" was not given by Liu. The UTA team including many collaborators decided to rename the new vortex definition as "Liutex" in late 2018. Actually, the Shanghai Conference is the place where Liu for the first time announced a new vortex definition in December 2017. In March 2018, Liu found that the direction of the new vortex definition or Liuex direction is the real eigenvector of the velocity gradient tensor. Therefore, a new and accurate mathematical definition of fluid rotation or vortex was born in University of Texas at Arlington in 2018. Later, Liu and his students published two papers in PoF on Liutex in 2018.

We then had three more conferences on vortex and turbulence research. The second one was held in Hangzhou with about 200 participants and organized by Profs. Weifang Chen and Lihua Chen of Zhejiang University of China in June 2018, the third one was held in Beijing with over 100 participants and organized by Prof. Yuning Zhang of Northern China Electric Power University in December 2018, and the fourth one was again held in Shanghai with about 200 participants and organized by Prof. Xiaoshu Cai in June 2019. Liu was the Principal Speaker for all four conferences. An alliance of vortex research was announced in 2018, which attracted over 100 members.

One important event in 2019 is that Prof. Lindi Zhou, the Chief Executive Editor of Journal of Hydrodynamics (JHD), found our work very interesting and decided to give the UTA team a full support. Prof. Zhou also proposed his six key elements of vortex identification. He believes only Liutex and third generation of vortex identification methods can answer all the six key issues, but the first generation and second generation of vortex identification methods fail.

It is worth to be mentioned that Yiqian Wang gave an explicit formula for Liutex magnitude, Xiangrui Dong gave an empirical epsilon formula for Omega and Liutex Omega, Yisheng Gao proposed a real Schur decomposition to find a fast way to look for the rigid rotation part. He also developed all computational codes to apply these Liutex methods, which have been published in the UTA website at https://www.uta.edu/math/cnsm/public_html/cnsm/cnsm.html. In addition, Jianming Liu developed objective Omega and objective Liutex, Xiaoshu Cai conducted a lot of experiments to support Liutex, and Hongyi Xu developed an automatic Liutex Core Line algorithm, which was successfully implemented into a computing code after Liu gave the Liutex Core Line definition and manual generation method in 2019. The automatic generation of Liutex core lines, for the first time, presented the unique and complete vortical structures in their entirety to the world in 2019 JHD featured paper.

After the 2017 Shanghai Conference, the UTA team has published 13 papers in Physics of Fluids (PoF) and more than 15 papers in Journal of Hydrodynamics. The UTA Team's papers were ranked as Top 2, 6, 7 of the best 2019 PoF featured letters, Top 2 and 5 of the 2019 most cited PoF papers, and Top 3 and 15 of the 2019 best PoF articles.

This book is a new book and the first book to systematically introduce Liutex theory, its mathematical definition, its theoretical foundation, and its applications in science and engineering. Liutex has been applied by countless scientists and engineers to identify vortex, to visualize new vortex structure, to study new physics, and more important, to uncover the turbulence mysterious veil. We believe that as more and more scientists apply Liutex and the third generation of vortex methods, people will reveal more and more secrets of turbulence. The key idea of Liutex is to extract the rigid rotation from the fluid motion and use it as the exact quantity to represent the fluid rotation or vortex. Liutex will open a new gate for quantified vortex and turbulence research to replace traditional qualitative turbulence research, which is in general based on observations, graphics, visualizations, approximations, assumptions, and hypotheses.

As vortex exists everywhere in the universe, a mathematical definition of vortex or Liutex will play a critical role in scientific research. There is almost no place without vortex in fluid dynamics. As a projection, the Liutex theory will be critical to the investigations of the vortex dynamics applying to hydrodynamics, aerodynamics, thermodynamics, oceanography, meteorology, metallurgy, civil engineering, astronomy, biology, etc. and to the research of the generation, sustenance, modeling and controlling of turbulence.

The book was finished mostly during the pandemic period of COVID-19, which is one of the most difficult times in human history. Liu is grateful to his coauthors, Prof. Hongyi Xu, Prof. Xiaoshu Cai, and Dr. Yisheng Gao, for their hard working and full cooperations during the pandemic period. Liu is also deeply thankful to his wife, Weilan Jin, for her full support by taking care of all housework. Liu also would like to thank his daughter, Haiyan Liu, and his son, Haifeng Liu, for their full support.

As the second author of this book, Prof. Hongyi Xu is very much enjoying the collaborations with Profs Liu and Cai since we all believe in the value of academic integrity and share the curiosity of vortex and turbulence. In particular, he is grateful to Prof. Cai who let him be one of the first few of the practitioners lucky to have the opportunity to appreciate the beauty of Liutex theory from the very beginning of its birth. As presented in Chapters 7 and 10 of the book, Xu fully believes that Liutex does lift and uncover the mask covering vortex, which has puzzled our science community for so many centuries. Specifically, the Liutex core lines limpidly bring out the skeleton of vortex structures and for the first time, vividly exhibit these structures to our visual world, which, from Xu's experience, is so far the unique representation of vortical structures with the true, only true, nothing else but the true mathematical essences of vortex physics in entirety. Another important and beauty part of the Liutex theory is the R-S decomposition given by Liu, which clearly explains the current misunderstandings of vortex based on the vorticity and provides a clearer roadmap to apply the Helmholtz equation to rigorously study the dynamics of vortices in the transition leading to turbulence. Within the context, the findings and practices elucidated in Chapters 7 and 10 open the door and start an era to truly and rigorously interrogate vortex dynamic details after the DNS database in entirety is available to the science community. With that being said, the DNS big data will keep on growing in future at the website of http://simplefluids.fudan.edu.cn, which is going to constantly bring out the surprising and enlightening findings enabled by the Liutex theory. Finally, Xu would like to express his thanks to his wife, Danhua Wang, working at Canadian Science Publishing, who provided him very professional and constructive advice toward writing and publishing the book.

Prof. Xiaoshu Cai is very much enjoying the collaborations with Prof. Liu and Prof. Xu in the work presented in this book. His main research areas are the development of measurement techniques and experimental study on fluid dynamics, two-phase flow and particles. Around 2014, he and his students did some experiments with the new measurement technique on small submerged water jet and found some interesting phenomena on vortex and turbulence. Prof. Liu was interested in these phenomena very much. He has been engaged in the DNS for flow transition and turbulence for the last 30 years since 1990. It is a very successful and fruitful fortune that the three researchers collaborated to attack the centennial challenges of vortex and turbulence since the research expertise of Prof. Hongyi Xu, as Cai's friend, is in the fluid dynamics and turbulence, as well as Prof. Liu. However, with the various research backgrounds in mathematics, physics, and computations as well as experiments, three of us combined our strengths together to make a historical research of vortex and turbulence by working closely with each other in the field since our fortunate meeting. Cai along with his colleagues and students made some significant contributions to the experimental work presented in Chapter 13, and here Cai would like to deeply express his thanks to them.

Dr. Yisheng Gao really appreciates this opportunity to collaborate with Prof. Liu, Prof. Xu, and Prof. Cai on this excited topic. During 2017—19, Dr. Gao joined Prof. Liu's group as a postdoctoral fellow to develop the Liutex theory system from the very beginning. The first idea, proposed and extensively discussed in late 2017, already showed the significant advantage over the existing vortex identification methods. With Prof. Liu's deep insight, the further development during the next 2 years makes Liutex a complete theoretical framework for vortex and turbulence research, not merely a vortex identification method. Especially the discovery of

Liutex similarity marks an extremely important contribution to the turbulence research, though the underlying mechanism is not yet fully understood. Currently, Dr. Gao is working with Prof. Xu to improve Liutex core lines for scientific and engineering applications. In our opinion, it is a real breakthrough to reveal the "secrets" under the sophisticated vortical structures usually hidden by the isosurfaces of the current vortex identification methods. Prof. Cai provides many beautiful experimental evidences to support our work. Gao is also grateful to Prof. Cai for his support for Liutex from the very beginning since Prof. Cai has already organized several conferences on the topic of Liutex.

We appreciate that the tremendous help and support in the editorial work were provided by Profs. Yiqian Wang, Xiangrui Dong, Jianming Liu, Mr. Heng Li, and Mr. Duo Wang toward the book publication. The authors would like to give an acknowledgment here to express a special gratitude to them.

The authors would like to thank the Department of Mathematics of University of Texas at Arlington where the UTA Team is housed and where the birthplace of Liutex is. The authors are grateful to Texas Advanced Computational Center (TACC) for providing computation hours. Prof. Xu would like to acknowledge the financial supports from the Fudan University to build the DNS big data at http://simplefluids.fudan.edu.cn.

Chaoqun Liu
Distinguished Professor and Director of
Center for Numerical Simulation and Modeling
Department of Mathematics
University of Texas at Arlington, Arlington, Texas, United States

Hongyi Xu
Professor
Department of Aeronautics and Astronautics
Fudan University
Shanghai, China

Xiaoshu Cai
Professor
Institute of Particle and Two-phase Flow Measurement
University of Shanghai for Science and Technology
Shanghai, China

Ysheng Gao
Lecturer
College of Aerospace Engineering
Nanjing University of Aeronautics and Astronautics
Nanjing, China

August 31, 2020

Short review of three generations of vortex identification methods

1.1 Short review

Vortices are ubiquitous in nature, such as tornado, hurricane, turbulence, and galaxy as shown in Fig. 1.1. In transitional and turbulent flows, there exist countless vortices with variety of sizes and strength. These vortical structures, also formally referred to

Tornado

Hurricane

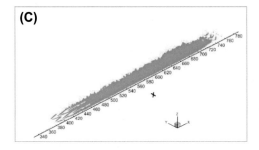

Flow transition

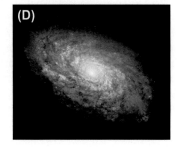

Galaxy

FIGURE 1.1

Vortex examples in nature.

Reproduced from [Liu, C., et al, "Third generation of vortex identification methods: Omega and Liutex/Rortex based systems", Journal of Hydrodynamics, 31(2), 2019], with the permission of Journal of Hydrodynamics

Liutex and Its Applications in Turbulence Research. https://doi.org/10.1016/B978-0-12-819023-4.00009-4
Copyright © 2021 Elsevier Inc. All rights reserved.

as coherent turbulent structures (Hussain, 1986; Sirovich, 1987; Robinson, 1991; Haller, 2015), are regarded as muscles and building blocks of turbulent flows and play a crucial role in turbulence generation and sustenance. Nowadays, several important coherent structures, including hairpin vortices in wall-bounded turbulence (Theodorsen, 1952; Liu et al., 1995; Adrian, 2007; Wu et al., 2009; Liu et al., 2014), quasi-streamwise vortices (Brooke and Hanratty, 1993), and vortex braids in turbulent shear layers (Rogers and Moser, 1993; Martin and Meiburg, 1991), have been identified. Naturally, an unambiguous and rigorous definition of vortex is indispensable for the comprehensive and thorough investigation of these sophisticated phenomena. Unfortunately, although vortices can be intuitively recognized as the rotational/swirling motion of fluids and have been intensively studied, a universally accepted mathematical definition is yet to be achieved ever since the beginning of the subject (Chong et al., 1990; Wu et al., 2006; Epps, 2017), which is probably one of the major obstacles to understanding and visualizing vortices (Chashechkin, 1993; Lugt, 1983; Green, 1995; Liu et al., 2014).

According to Liu et al. (2019), most of the currently available Eulerian vortex identification methods belong to three generations. For a long time in history, vortices are generally associated with vorticity, which has a rigorous mathematical definition, i.e., the curl of velocity. Helmholtz (1858) first defined the concept of vortex tube and vortex filament. Since then, vorticity-based methods have been adopted in the vast majority of research papers and textbooks and vorticity dynamics has been developed to the study of the generation and evolution of vorticity and vortical-flow stability (Saffman, 1992; Majda and Bertozzi, 2001). These methods are based on the concept that vortices are defined as vorticity tubes and vortex strength is measured by the magnitude of vorticity and can be classified as the first generation of vortex identification methods. However, in viscous flows, especially in turbulent flows, the use of vorticity will run into severe difficulties. Vorticity cannot distinguish between a shear layer region and a real rotational region, and the correlation between vortex and vorticity is very weak in science and engineering applications. For example, Robinson (1991) found that the association between regions of strong vorticity and actual vortices can be rather weak. To overcome the problems associated with the first-generation vortex identification methods, many local Eulerian vortex identification criteria have been proposed, including the Q criterion (Hunt et al., 1988), the λ_2 criterion (Jeong et al., 1995), the Δ criterion (Chong et al., 1990; Dallmann, 1983; Vollmers et al., 1983), the λ_{ci} criterion (Zhou et al., 1999; Chakraborty et al., 2005). While these criteria are widely used, these criteria have several shortcomings. (1) The physical meaning is not very clear. (2) A user-defined threshold is required, which is case-related and empirical. (3) These methods provide different values as the measure of vortex strength, but these values are usually inconsistent with each other. (4) These methods are uniquely determined by the eigenvalues of the velocity gradient tensor with different formula, which implies that they could be contaminated by stretching and shear in different levels. (5) As an absolute strength indicator, they cannot capture both strong and weak vortices simultaneously due to the threshold selection. More seriously, these criteria can only give a scalar value without any information about the rotation axis or orientation. These eigenvalue-based criteria are classified as the second generation of vortex identification methods. To address the threshold issue, Liu et al. (2016) propose a new vortex identification method called "Omega" method, based on the idea that a vortex is a region where the vorticity overtakes the deformation. The Omega method has several advantages: (1) simple implementation, (2) clear physical meaning, (3) nondimensional and normalized, (4) robust to moderate threshold change (from 0.52 to 0.6), (5) capable to capture both strong and weak vortices simultaneously. The Omega method has been proved to be efficient and easy to use without the threshold adjustment in various applications. But it still only provides a scalar and is based on the Cauchy—Stokes decomposition, which makes the Omega method to be among the second-generation methods. It is clearly pointed out by Liu et al. (2016) that vortex cannot be represented by vorticity and vorticity must be decomposed to a rotational part and a shear part, i.e., $\nabla \times \vec{v} = \vec{R} + \vec{S}$. However, a rigorous mathematical definition of $\vec{R}$ and $\vec{S}$ and the relationship between vortex and vorticity remained elusive at that time.

Recently, a mathematical definition called vortex vector or "Rortex" is introduced by Liu and his team (Liu et al., 2018a; Gao et al., 2018) to identify the local rigid rotation of the fluid motion. The name was later changed to "Liutex." Liutex is a mathematical definition of the rigid rotation part of fluid motion or vortex without stretching and/or shear contamination. Since vortex cores trend to the rigid rotation, vortex cores can be naturally represented by Liutex. Based on the definition of Liutex, the accurate mathematical relation between vorticity and Liutex, namely $\nabla \times \vec{v} = \vec{R} + \vec{S}$ (vorticity consists of Liutex and the antisymmetric shear), is finally obtained. In addition, the velocity gradient tensor can be decomposed to a rotational part (R) and a non-rotational part (NR). The non-rotational part can be further decomposed to a pure shear part (PS) and a stretching or compression part (SC), which leads to a new velocity gradient tensor decomposition written as $\nabla \vec{V} = R + PS + SC$, based on a so-called "Principal Coordinate" system using the direction of Liutex as the Z-axis (see Chapter 2). All three parts including Liutex (rigid rotation), pure shear (PS), stretching and compression (SC) can clearly represent fluid motion, which are uniquely and mathematically defined for further turbulence research. The new velocity gradient decomposition has more clear physical meaning for fluid motion in comparison with the traditional Cauchy—Stokes decomposition (Batchelor, 2000), which decomposes the velocity gradient tensor to a symmetric part and an antisymmetric part. Since the antisymmetric part cannot represent the flow rotation or vorticity cannot represent vortex, the Cauchy—Stokes decomposition is not appropriate to represent the decomposition of the fluid motion. This new definition of Liutex and the associated velocity gradient tensor decomposition are expected to be an important breakthrough in modern fluid dynamics and extremely important for turbulence research. Since the introduction of Liutex, considerable developments have been made in the vortex definition and identification. Multiple

Liutex-based methods to visualize vortices from different perspectives have been proposed, including the Liutex vectors and lines, Liutex iso-surfaces, Liutex Omega method (Dong et al., 2019b; Liu et al., 2019d), Liutex core line method (Gao et al., 2019b; Xu et al., 2019), objective Liutex method (Liu et al., 2019b). More importantly, it has been demonstrated by Xu et al. (2019b) that in a low Reynolds number turbulent boundary layer, the frequency and wavenumber spectrum of the Liutex magnitude follow the $-5/3$ law very well, while the spectrum of vorticity and parameters of other vortex identification methods deviate from the power law spectrum of any order. In contrast, the energy spectrum of the same flow only marginally matches the $-5/3$ law shown by K41 theory (Kolmogorov, 1941a,b,c) in a much smaller frequency range. This discovery implies Liutex similarity law and reveals that the Liutex vector might be directly connected to turbulence generation and sustenance. These series of works based on the concept of Liutex are collectively classified as the third generation of the vortex definition and identification methods, which not only visualize vortices from different perspectives but also provide a quantified way to study vortices and turbulence.

In Liu et al. (2019), six core elements of vortex identification methods, including (1) absolute strength, (2) relative strength, (3) local rotational axis, (4) vortex rotation axes, (5) vortex core size, and (6) vortex boundary, are brought up to examine vortex definition and visualization methods. The original concept of Liutex has a systematic definition for vortex including a scalar for the rotation strength, a vector for the local rotation axis, and a tensor for the fluid motion decomposition. Accordingly, the magnitude and the direction of the Liutex vector can answer the first and third core elements, respectively. The recently developed Liutex Omega method (Dong et al., 2019b; Liu et al., 2019d) combines the Liutex method and the idea of the Omega method to use a proportion to represent the rigidity of fluid motion and thus can describe the relative strength and the vortex boundary. In addition, Liutex core line method has been proposed (Gao et al., 2019b; Xu et al., 2019) to uniquely capture all vortex core center lines to represent a skeleton of the vortical structure and these Liutex core lines can serve as vortex rotation axes. Therefore, the Liutex concept and the recent development can establish a complete theoretical system, which answers all the six core elements of vortex identification methods. On the other hand, the first generation of vortex identification methods fails to answer any of the core elements. The scalar-valued second generation of vortex identification methods can only present the roughly vortex boundary based on the somewhat arbitrary threshold selection. These breakthroughs in the development of vortex science are critical for modern fluid dynamics and turbulence research. The Liutex theoretical system and methods developed by the UTA Team led by Prof. Chaoqun Liu, classified as the third generation of vortex definition and identification methods, may open a door for deeper and quantified turbulence research instead of graphics, movies, hypotheses, observations, assumptions, and other qualitative methods.

Although substantial progresses have been made by the Liutex theoretical system, there still remain many important questions to be answered. For example, the Liutex (vortex) dynamics, the universal governing equations for laminar and turbulent flows, the mechanism of turbulence generation and sustenance, turbulence control, and so on are all needed for further research. The current chapter has no intention to conduct a comprehensive review of the historical development of vortex identification methods, but only brief introduce several traditional vortex identification methods and the third-generation methods. For an overview of the existing vortex identification methods, one can refer to recently published review papers (Epps, 2017; Zhang et al., 2018a,c).

This chapter is organized as follows. Section 1.2 is an introduction of the vorticity-based first generation of vortex identification methods. Section 1.3 describes the second generation of vortex identification methods including some currently popular vortex identification methods and their shortcomings. In Section 1.4, the Omega vortex identification method is introduced and its advantages are elaborated. The Liutex method is shortly introduced in Section 1.5.

1.2 First generation: vorticity-based vortex identification methods

Since the concepts of vorticity tube and filament were proposed by Helmholtz in 1858, it is generally believed that vortices consist of vorticity tubes and vortex strength is measured by the magnitude of vorticity, i.e., $\nabla \times \vec{v}$. In his original paper, Helmholtz provided the following definitions: "by vortex lines I denote lines drawn through the fluid mass so that their direction at every point coincides with the direction of the momentary axis of rotation of the water particles lying on it ...; " "by vortex filaments I denote portions of the fluid mass cut out from it by way of constructing corresponding vortex lines through all points of the circumference of an infinitely small surface element." According to these definitions, three vortex theorems can be obtained as follows: (1) **Helmholtz's first theorem:** the strength of a vortex filament is constant along its length; (2) **Helmholtz's second theorem:** a vortex filament cannot end in a fluid; it must extend to the boundaries of the fluid or form a closed path; (3) **Helmholtz's third theorem:** in the absence of rotational external forces, a fluid that is initially irrotational remains irrotational (Helmholtz, 1858) (it should be noted that the order may be different in different books). These theorems serve as the foundation of vorticity dynamics. Lamb (1932) further points out that: "If through every point of a small closed curve we draw the corresponding vortex-line, we mark out a tube, which we call a 'vortex-tube.' The fluid contained within such a tube constitutes what is called a 'vortex-filament,' or simply a 'vortex.'" Saffman (1992) defines vortex as a finite volume of vorticity immersed in irrotational fluid in his monograph. As described in Nitsche (2006), "A vortex is commonly associated with the rotational motion of fluid around a common centerline. It is defined by the vorticity in the fluid, which measures the rate of local fluid rotation." Wu et al. (2006) also define vortex as "a connected fluid region with high concentration of vorticity compared with its surrounding." Although the vorticity is widely adopted to detect vortices, one common counter-example is immediately raised that in the laminar boundary layer the average shear force generated by the no-slip wall is so strong that an extremely large amount of vorticity exists but no rotation motions (vortices) are observed in the near-wall regions, which implies that vortex cannot be represented by vorticity. Similarly, in the laminar channel flow, which has an analytic solution: $U = (1+y)(1-y)$, $V = 0$; $\omega_z = \frac{\partial V}{\partial x} - \frac{\partial U}{\partial y} = -2y$ as shown in Fig. 1.2, the magnitude of vorticity is very large near the wall. If a close line circle is picked to do the velocity integration to find the circulation near the wall, the circulation and then vorticity are very large according to the Stokes theory, but there is no fluid rotation near the wall, namely, no vortex. Therefore, the concept that vorticity is equivalent to the vortex is a mistake originated by Helmholtz due to the limit of the scientific research level at that time. Robinson (1990a,b) has found that "the association between regions of strong vorticity and actual vortices can be rather weak in the turbulent boundary layer, especially in the near wall region." Actually, it is not uncommon that the maximum magnitude of vorticity occurs outside the central region of vortical structures. As shown by Wang, Y., et al. (2017), the magnitude of vorticity can be substantially reduced along vorticity lines entering the vortex core region near the solid wall in a flat plate boundary layer (Fig. 1.3).

For a transitional flow over a flat plate, Figs. 1.4 and 1.5 clearly indicate that in the near-wall region of the boundary layer, the local vorticity vector can deviate from the direction of vortical structures and a vortex can appear in the area where the vorticity is smaller than the surrounding area where the vorticity is larger than the vorticity inside the vortex. These results demonstrate that vorticity cannot be used to represent vortex.

Helmholtz's vortex definition and three theorems are still the foundation of modern vortex (vorticity) dynamics. However, they may be useful for inviscid flows but cannot be used for viscous flow and especially are not appropriate for turbulent flows. One may argue that Helmholtz theories have a few preconditions such as for inviscid, barotropic flows with conservative body force and nonrotating force.

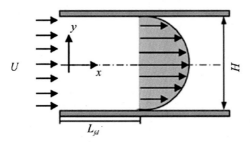

FIGURE 1.2

Laminar channel flow.

Reproduced from [Liu, J., et al, "Mathematical Foundation of Turbulence Generation-Symmetric to Asymmetric Liutex/Rortex", Journal of Hydrodynamics, 31(3), 2019], with the permission of Journal of Hydrodynamics

vortices with relatively smaller vorticity magnitude

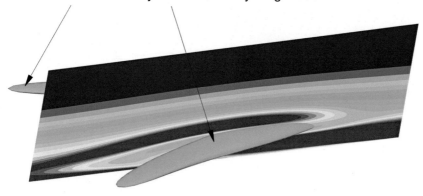

|vorticity|: 0.00 0.03 0.07 0.10 0.13 0.17 0.20 0.23 0.27 0.30 0.33 0.37 0.40 0.43 0.47 0.50

FIGURE 1.3

Vortex appears in the area where vorticity is relatively smaller.

Reproduced from [Liu, C., et al, "Third generation of vortex identification methods: Omega and Liutex/Rortex based systems", Journal of Hydrodynamics, 31(2), 2019], with the permission of Journal of Hydrodynamics

(A) |vorticity|: 0.2 0.3 0.4 0.5 0.6 0.7 0.8 0.9 1   **(B)** |vorticity|: 0.2 0.3 0.4 0.5 0.6 0.7 0.8 0.9 1

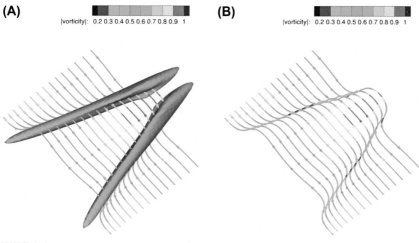

FIGURE 1.4

Vorticity and vortex are not aligned and not correlated in vortex legs.

Reproduced from [Liu, C., et al, "Third generation of vortex identification methods: Omega and Liutex/Rortex based systems", Journal of Hydrodynamics, 31(2), 2019], with the permission of Journal of Hydrodynamics

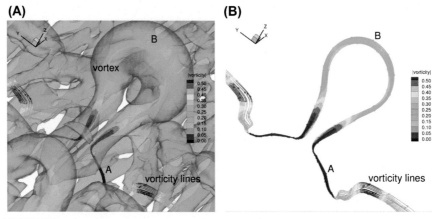

FIGURE 1.5

Vorticity tube and vortex tube are not aligned and not correlated in vortex rings.

Reproduced from [Liu, C., et al, "Third generation of vortex identification methods: Omega and Liutex/Rortex based systems", Journal of Hydrodynamics, 31(2), 2019], with the permission of Journal of Hydrodynamics

However, these conditions never exist in real flows, especially for turbulence. The real outcome of teaching students Helmholtz vortex definition and three theorems is that almost all students believe vortices are vorticity tubes and vortex strength is measured by the magnitude of vorticity, and Helmholtz's three theorems are widely used to study vortices and turbulence. This is really a tragedy to science and education. Vortex is a natural phenomenon, but vorticity is a mathematical definition. There is no reason to say that vortex is vorticity. Instead, "vortex is not vorticity." They are two different concepts. Anyone who insists "vortex is vorticity" is hopeless and can never be successful. One more question is that if vortex cannot be ended inside the flow field, how turbulence is generated by vortex breakdown? Vortex can break down, which means vortex is not vorticity tubes. The correlation analysis between vorticity and vortex (fluid rotation) clearly shows there exists no correlation between vortex and vorticity in general.

1.3 Second generation: eigenvalue-based vortex identification methods

Another obvious candidate for vortex identification would be the one based on closed or spiraling streamlines. Robinson et al. (1989) has proposed a definition: "A vortex exist when instantaneous streamlines mapped onto a plane normal to the vortex core exhibit a roughly circular or spiral pattern, when view from a reference frame moving with the center of the vortex core." Though it seems intuitively viable, the determination of the vortex core is a chicken and egg problem. What's worse, streamlines are not Galilean invariant. Lugt (1979) tried to use pathlines to define a vortex: "A vortex is the rotation motion of a multitude of material particles around a common center." However, the pathlines of material particles are not Galilean invariant either. And if the pathlines are used to describe vortices, how to determine the common center is also vague.

Eulerian velocity-gradient-based criteria then emerge to overcome the problem for vortex identification. These criteria can discriminate against vorticity due to shear and are Galilean invariant, thus are widely used in research and engineering. Most of these criteria are based on the analysis of the velocity gradient tensor $\nabla \vec{v}$. More specifically, these methods are exclusively determined by the eigenvalues of the velocity gradient tensor or the related invariants. If λ_1, λ_2, and λ_3 are used to denote three eigenvalues, the characteristic equation can be written as

$$\lambda^3 + P\lambda^2 + Q\lambda + R = 0 \tag{1.1}$$

where

$$P = -(\lambda_1 + \lambda_2 + \lambda_3) = -tr\left(\nabla \vec{v}\right) \tag{1.2}$$

$$Q = \lambda_1\lambda_2 + \lambda_2\lambda_3 + \lambda_3\lambda_1 = -\frac{1}{2}\left(tr\left(\nabla \vec{v}^2\right) - tr\left(\nabla \vec{v}\right)^2\right) \tag{1.3}$$

$$R = -\lambda_1\lambda_2\lambda_3 = -\det\left(\nabla \vec{v}\right) \tag{1.4}$$

P, Q, and R are three invariants, respectively. For incompressible flow, from the continuous equation, $P = 0$. In the following, four representatives of eigenvalue-based criteria, namely the Q-criterion, Δ-criterion, λ_{ci}-criterion, and λ_2-criterion are briefly introduced.

1.3.1 Q criterion

The Q criterion, proposed by Hunt et al. (1988), is one of the most currently popular vortex identification methods. Q is a measure of the local rotation rate in excess of the strain rate. It also requires the pressure in the vortical regions to be lower than the ambient pressure. But this condition is often omitted in practice. Therefore, Q is expressed as a residual of the vorticity tensor norm square subtracted from the strain-rate tensor norm square,

$$Q = \frac{1}{2}\left(\|B\|_F^2 - \|A\|_F^2\right) \tag{1.5}$$

where A and B are the symmetric and antisymmetric parts of the velocity gradient tensor, respectively:

$$A = \frac{1}{2}\left(\nabla\vec{v} + \nabla\vec{v}^{\mathrm{T}}\right) = \begin{bmatrix} \frac{\partial u}{\partial x} & \frac{1}{2}\left(\frac{\partial u}{\partial y} + \frac{\partial v}{\partial x}\right) & \frac{1}{2}\left(\frac{\partial u}{\partial z} + \frac{\partial w}{\partial x}\right) \\ \frac{1}{2}\left(\frac{\partial v}{\partial x} + \frac{\partial u}{\partial y}\right) & \frac{\partial v}{\partial y} & \frac{1}{2}\left(\frac{\partial v}{\partial z} + \frac{\partial w}{\partial y}\right) \\ \frac{1}{2}\left(\frac{\partial w}{\partial x} + \frac{\partial u}{\partial z}\right) & \frac{1}{2}\left(\frac{\partial w}{\partial y} + \frac{\partial v}{\partial z}\right) & \frac{\partial w}{\partial z} \end{bmatrix}$$

(1.6)

$$B = \frac{1}{2}\left(\nabla\vec{v} - \nabla\vec{v}^{\mathrm{T}}\right) = \begin{bmatrix} 0 & \frac{1}{2}\left(\frac{\partial u}{\partial y} - \frac{\partial v}{\partial x}\right) & \frac{1}{2}\left(\frac{\partial u}{\partial z} - \frac{\partial w}{\partial x}\right) \\ \frac{1}{2}\left(\frac{\partial v}{\partial x} - \frac{\partial u}{\partial y}\right) & 0 & \frac{1}{2}\left(\frac{\partial v}{\partial z} - \frac{\partial w}{\partial y}\right) \\ \frac{1}{2}\left(\frac{\partial w}{\partial x} - \frac{\partial u}{\partial z}\right) & \frac{1}{2}\left(\frac{\partial w}{\partial y} - \frac{\partial v}{\partial z}\right) & 0 \end{bmatrix}$$

(1.7)

and $\|\cdot\|_F^2$ represents the Frobenius norm. Theoretically, $Q > 0$ can identify the vortex boundary. But in practice, a threshold $Q_{threshold}$ must be specified to define the regions where $Q > Q_{threshold}$. The Q-criterion shows the symmetric tensor has a role to balance the antisymmetric tensor. It also means the existence of a vortex requires not only vorticity, but also the capacity of vorticity to overcome the fluid deformation:

$$a = \|A\|_F^2 = \left(\frac{\partial u}{\partial x}\right)^2 + \left(\frac{\partial v}{\partial y}\right)^2 + \left(\frac{\partial w}{\partial z}\right)^2 + \frac{1}{2}\left(\frac{\partial u}{\partial y} + \frac{\partial v}{\partial x}\right)^2$$

$$+ \frac{1}{2}\left(\frac{\partial u}{\partial z} + \frac{\partial w}{\partial x}\right)^2 + \frac{1}{2}\left(\frac{\partial v}{\partial z} + \frac{\partial w}{\partial y}\right)^2$$

$$= \left(\frac{\partial u}{\partial x}\right)^2 + \left(\frac{\partial v}{\partial y}\right)^2 + \left(\frac{\partial w}{\partial z}\right)^2 + \frac{1}{2}\left(\frac{\partial u}{\partial y}\right)^2 + \frac{1}{2}\left(\frac{\partial u}{\partial z}\right)^2 + \frac{1}{2}\left(\frac{\partial v}{\partial x}\right)^2 + \frac{1}{2}\left(\frac{\partial v}{\partial z}\right)^2$$

$$+ \frac{1}{2}\left(\frac{\partial w}{\partial x}\right)^2 + \frac{1}{2}\left(\frac{\partial w}{\partial y}\right)^2 + \frac{\partial u}{\partial y}\frac{\partial v}{\partial x} + \frac{\partial u}{\partial z}\frac{\partial w}{\partial x} + \frac{\partial v}{\partial z}\frac{\partial w}{\partial y}$$

(1.8)

$$b = ||\mathbf{B}||_F^2 = \frac{1}{2}\left(\frac{\partial u}{\partial y} - \frac{\partial v}{\partial x}\right)^2 + \frac{1}{2}\left(\frac{\partial u}{\partial z} - \frac{\partial w}{\partial x}\right)^2 + \frac{1}{2}\left(\frac{\partial v}{\partial z} - \frac{\partial w}{\partial y}\right)^2$$

$$= \frac{1}{2}\left(\frac{\partial u}{\partial y}\right)^2 + \frac{1}{2}\left(\frac{\partial u}{\partial z}\right)^2 + \frac{1}{2}\left(\frac{\partial v}{\partial x}\right)^2 + \frac{1}{2}\left(\frac{\partial v}{\partial z}\right)^2 + \frac{1}{2}\left(\frac{\partial w}{\partial x}\right)^2$$

$$+ \frac{1}{2}\left(\frac{\partial w}{\partial y}\right)^2 - \frac{\partial u}{\partial y}\frac{\partial v}{\partial x} - \frac{\partial u}{\partial z}\frac{\partial w}{\partial x} - \frac{\partial v}{\partial z}\frac{\partial w}{\partial y} \tag{1.9}$$

$$||\mathbf{A}||_F^2 = ||\mathbf{B}||_F^2 + \left(\frac{\partial u}{\partial x}\right)^2 + \left(\frac{\partial v}{\partial y}\right)^2 + \left(\frac{\partial w}{\partial z}\right)^2 + 2\frac{\partial u}{\partial y}\frac{\partial v}{\partial x} + 2\frac{\partial u}{\partial z}\frac{\partial w}{\partial x} + 2\frac{\partial v}{\partial z}\frac{\partial w}{\partial y}$$

$$Q = \frac{1}{2}\left(||\mathbf{B}||_F^2 - ||\mathbf{A}||_F^2\right)$$

$$= -\frac{1}{2}\left[\left(\frac{\partial u}{\partial x}\right)^2 + \left(\frac{\partial v}{\partial y}\right)^2 + \left(\frac{\partial w}{\partial z}\right)^2 + 2\frac{\partial u}{\partial y}\frac{\partial v}{\partial x} + 2\frac{\partial u}{\partial z}\frac{\partial w}{\partial x} + 2\frac{\partial v}{\partial z}\frac{\partial w}{\partial y}\right]$$

$$\tag{1.10}$$

In the principal coordinate, the aforementioned formula shows the first three items representing the effects of stretching or compression, but Q could mistreat them as part of vortex no matter they are stretching $\left(\frac{\partial u}{\partial x} > 0\right)$ or compression $\left(\frac{\partial u}{\partial x} < 0\right)$. The rest three items on the right-hand side contain shears or, in other words, Q treats shear as part of vortex. Accordingly, Q is contaminated by stretching (or compression) and shear. In addition, a vortex has a fluid rotation axis. But Q is a scalar and cannot present the direction of the fluid rotation axis. Actually, $Q > Q_{threshold}$ is a subset of $Q > 0$ and different $Q_{threshold}$ would give different subsets, which is not unique and dependent on the threshold selection and pretty much arbitrary. There is really no proper or improper threshold and no one is able to give a proper threshold. This is the problem associated with most of second generation of vortex identification methods.

1.3.2 Δ criterion

According to critical point theory, in a nonrotating reference frame translating with a fluid particle, the instantaneous streamline pattern presents closed or spiraling where $\nabla\vec{v}$ has one real and two conjugate complex eigenvalues (Chong et al., 1990). The instantaneous streamline pattern (obtained from Taylor series expansion of the local velocity to a linear order) is governed by the eigenvalues of $\nabla\vec{v}$. The discriminant for the characteristic equation for the velocity gradient tensor $\nabla\vec{v}$ is

$$\Delta = \left(\frac{\tilde{Q}}{3}\right)^3 + \left(\frac{\tilde{R}}{2}\right)^2, \qquad \tilde{Q} = Q - \tfrac{1}{3}P^2, \qquad \tilde{R} = R + \tfrac{2}{27}P^3 - \tfrac{1}{3}PQ \qquad ,$$

$R = -\lambda_1\lambda_2\lambda_3 = -\det\left(\nabla\vec{v}\right)$ where det stands for determinant. If $\Delta \leq 0$, the three eigenvalues of $\nabla\vec{v}$ are real; if $\Delta > 0$, there exist one real eigenvalue and two conjugate complex eigenvalues and then the point is located inside the vortex. Note that for incompressible flow, $P = 0$, so the following results can be obtained: $\tilde{Q} = Q$, $\tilde{R} = R$, and

$$\Delta = (Q/3)^3 + (R/2)^2. \tag{1.11}$$

If $Q > 0$, $\Delta > 0$, the point is inside the vortex. However, If $Q < 0$, since $\left(\frac{R}{2}\right)^2 > 0$, there is still a possibility $\Delta > 0$. Therefore, a point with $Q < 0$ is still possible to locate inside the vortex. This implies the inconsistence between Q and Δ. It should be pointed out that $\Delta > 0$ is the sufficient and necessary condition for a point inside the vortex zone, which is a mathematical criterion and cannot be violated. However, Δ doesn't have a clear physical meaning, and it is in general cannot be applied to measure the strength of vortex.

1.3.3 λ_{ci} criterion

The λ_{ci} criterion (Zhou et al., 1999) can be regarded as an extension of the Δ criterion and is consistent with the Δ criterion in that the theoretical vortex boundary is identical when zero threshold is applied. When the velocity gradient tensor $\nabla\vec{v}$ has two conjugate complex eigenvalues, the local time-frozen streamlines exhibit a swirling flow pattern. In this case, the eigendecomposition of $\nabla\vec{v}$ can be written as

$$\nabla\vec{v} = \left[\vec{v}_r \ \vec{v}_{cr} \ \vec{v}_{ci}\right]\begin{bmatrix} \lambda_r & 0 & 0 \\ 0 & \lambda_{cr} & \lambda_{ci} \\ 0 & -\lambda_{ci} & \lambda_{cr} \end{bmatrix}\left[\vec{v}_r \ \vec{v}_{cr} \ \vec{v}_{ci}\right]^{-1} \tag{1.12}$$

Here, $\left(\lambda_r, \vec{v}_r\right)$ is the real eigenpair and $\left(\lambda_{cr} \pm i\lambda_{ci}, \vec{v}_{cr} \pm i\vec{v}_{ci}\right)$ the complex conjugate eigenpairs. In the local curvilinear coordinate system (c_1, c_2, c_3) spanned by the eigenvector $\left(\vec{v}_r, \vec{v}_{cr}, \vec{v}_{ci}\right)$, the instantaneous streamlines are the same as pathlines and can be expressed as

$$\begin{aligned} c_1(t) &= c_1(0)e^{\lambda_r t} \\ c_2(t) &= [c_2(0)\cos(\lambda_{ci}t) + c_3(0)\sin(\lambda_{ci}t)]e^{\lambda_{cr}t} \\ c_3(t) &= [c_3(0)\cos(\lambda_{ci}t) - c_2(0)\sin(\lambda_{ci}t)]e^{\lambda_{cr}t} \end{aligned} \tag{1.13}$$

where t represents the time-like parameter and the constants $c_1(0)$, $c_2(0)$, and $c_3(0)$ are determined by the initial conditions. From Eq. (1.13), the period of orbit of a fluid particle is $2\pi/\lambda_{ci}$, so the imaginary part of the complex value λ_{ci} is called swirling strength. However, Eq. (1.12) is a similar transformation, but not orthogonal transformation, which is corresponding to a nonorthogonal coordinate system and thus the antisymmetric term represented by λ_{ci} in general cannot be used as a precise measure for fluid rotation. Again, λ_{ci} is a scalar and would be contaminated by shear. It should be also noted that vorticity is Galilean invariant in an orthogonal transform system, but does not keep unchanged in a similar transformation.

1.3.4 λ_2 criterion

The underlying idea of the λ_2 criterion is rooted in the observation that the concept of a local pressure minimum in a plane fails to identify vortices under strong unsteady and viscous effects. By neglecting these unsteady and viscous effects, the symmetric part of the gradient of the incompressible Navier–Stokes equation can be expressed as $\boldsymbol{A}^2 + \boldsymbol{B}^2 = -\frac{1}{\rho}\nabla(\nabla p)$ where p is the pressure. To identify the region of local pressure minimum in a plane, Jeong and Hussain (1995) define the vortex core as a connected region with two negative eigenvalues of the symmetric tensor $\boldsymbol{A}^2 + \boldsymbol{B}^2$. If the eigenvalues of the symmetric tensor $\boldsymbol{A}^2 + \boldsymbol{B}^2$ are ordered as $\lambda_1 \geq \lambda_2 \geq \lambda_3$, this definition is equivalent to the requirement that $\lambda_2 < 0$. Note that, generally, λ_2 cannot be expressed in terms of the eigenvalues of the velocity gradient tensor. However, in the special case when the eigenvectors are orthonormal, λ_2 can be exclusively determined by the eigenvalues. Since the λ_2 criterion ignores the unsteady and viscous effects and has square order of the fluid angular speed dimension, it cannot be used to measure the fluid rotation (vortex) strength. Table 1.1 shows these vortex identification methods with their authors and publication year.

1.3.5 Three problems: threshold, rotation direction, and strength of vortex

Without exception, the aforementioned criteria require user-specified thresholds. Since different thresholds will indicate different vortical structures, it is critical to determine an appropriate threshold. When a large threshold for the Q criterion is used, "vortex breakdown" (see Fig. 1.6) will be observed in the late boundary layer transition. But if a small threshold is applied, no "vortex breakdown" (see Fig. 1.7) will be exposed, which means that an appropriate threshold is vital to these vortex identification methods. Many computational results have revealed that the threshold is case-related, empirical, sensitive, time-step-related, and hard to adjust, case by

Table 1.1 Major eigenvalue-based vortex identification methods.

Name	Authors	Year
Q criterion	Hunt, Wray, Moin	1988
Δ criterion	Chong, Perry	1990
λ_2 criterion	Jeong, Hussain	1995
λ_{ci} criterion	Zhou, Adrian, Balachandar	1999

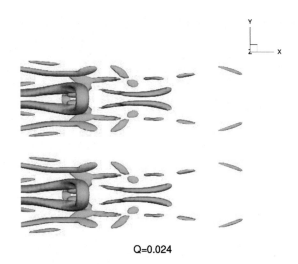

Q=0.024

FIGURE 1.6

Vortex breakdown with a large threshold of $Q=0.024$ (same DNS data set).

Reproduced from [Liu, C., et al, "Third generation of vortex identification methods: Omega and Liutex/Rortex based systems", Journal of Hydrodynamics, 31(2), 2019], with the permission of Journal of Hydrodynamics

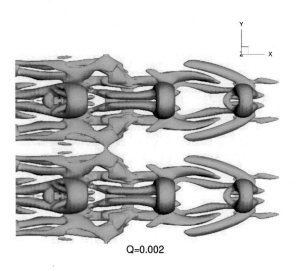

Q=0.002

FIGURE 1.7

No vortex breakdown with a small threshold of $Q=0.002$ (same DNS data set).

Reproduced from [Liu, C., et al, "Third generation of vortex identification methods: Omega and Liutex/Rortex based systems", Journal of Hydrodynamics, 31(2), 2019], with the permission of Journal of Hydrodynamics

case and even time step by time step. Furthermore, no one knows whether the specified threshold is proper or improper. Actually there may be no single proper threshold especially if strong and weak vortices coexist. If the threshold is too small, weak vortices may be captured, but strong vortices could be smeared and become vague. If the threshold is too large, weak vortices will be wiped out. The other disadvantage of these vortex identification methods is that these vortex identification methods can only provide iso-surfaces without any information about the rotation axis or vortex direction. A more serious question is raised: if the iso-surface can represent the rotation strength? The answer is no, since they are different to each other (not unique) in quantity and dimension, contaminated by shear and stretching or compression in different degrees, and fail to represent the rigid rotation of fluid motion.

1.4 Omega vortex identification method

Since 2014, the UTA team led by Chaoqun Liu started to develop new vortex identification methods to address the shortcomings of the aforementioned traditional vortex identification methods. The first breakthrough was the Omega vortex identification method proposed by Liu et al. (2016). This method was further modified by Dong et al. (2018c) to better determine the parameter ε.

1.4.1 Definition of Omega

The Omega method was originated from an important physical understanding that vortex is a region where vorticity overtakes deformation. Vorticity cannot directly represent the fluid rotation, although the rigid body rotation must possess vorticity. Therefore, in fluid motion, the vorticity could be small with a strong rotation and could be large without rotation, such as the laminar channel flow shown in Fig. 1.2. Deformation is also an important factor while vortical motion exists. Therefore, it is reasonable to consider the ratio of vorticity and deformation when defining a vortex. As given in the original paper (Liu et al., 2016), Ω is defined as a ratio of vorticity tensor norm squared over the sum of vorticity tensor norm squared and strain rate tensor norm squared,

$$\Omega = \frac{\|B\|_F^2}{\|A\|_F^2 + \|B\|_F^2} = \frac{b}{a+b} \tag{1.14}$$

To avoid division by zero or an extremely small number, a small parameter ε is applied, so Eq. (1.14) becomes

$$\Omega = \frac{b}{a+b+\varepsilon} \tag{1.15}$$

where a and b are given in Eq. (1.8) and Eq. (1.9) respectively.

1.4.2 Determination of ε

The introduction of a small positive parameter ε is to avoid division by zero or an extremely small number in the original definition. According to Eq. (1.14), Ω is nondimensional and normalized as $\Omega \in [0, 1]$. But some serious noises or clouds may appear inside the flow domain if both terms a and b are close to zero. These noises can be reduced or even removed by a small positive parameter ε. However, the proper value of ε is still case-related and dimension-related. Apparently, ε is strongly related to the case and time steps especially if the dimensional governing equations are applied. Even for the nondimensional governing equations, the reference length and reference speed could be very different in different computation cases. A proper selection of ε is critical. Dong et al. (2018c) propose a linear expression to relate ε and the maximum of $b - a$. ε is defined as a function of $(b-a)_{\max}$, which is a fixed parameter at each time step in each case. In Dong's paper, ε is proposed as follows:

$$\varepsilon = 0.001 * (b - a)_{\max} = 0.002 * Q_{\max} \tag{1.16}$$

It should be noted that Eq. (1.16) is an empirical formula based on a large number of test results from different cases. The term $(b - a)_{\max}$ represents the maximum of the difference of vorticity tensor squared and strain rate tensor squared and is easy to obtain as a fixed number at each time step in a certain case. The adjustment of ε in many cases is thus not necessary after ε is determined by Eq. (1.16). However, the users may still need to adjust the small number ε for their own computations.

1.4.3 Advantages of the Omega method

The Omega method is originated from the physical concept that vortex is a connected region where vorticity overtakes deformation, which easily shows $\Omega > 0.5$ inside vortex. In practice, it is common to choose $\Omega = 0.51$ or $\Omega = 0.52$ as the fixed threshold. According to several computational results including DNS for flow transition and LES for shock and boundary layer interaction, Omega is insensitive to the threshold change and the vortex structures are very similar when $\Omega = 0.52-0.6$ is applied. $\Omega = 0.52$ shows the vortex boundaries for many different cases. This conclusion has been validated by many users. This means that one of the major shortcomings of the traditional vortex identification methods, the determination of the proper threshold, is avoided by the Omega method. This is one of the distinguished advantages of the Omega method over all other first and second generations of vortex identification methods. Actually, the moderate change of the threshold of Ω will only change the vortex shape to be fatter or thinner.

There are many examples that can show that, with the same threshold, the Omega method can capture both strong vortices and weak vortices while other vortex identification methods fail. The following two examples are provided by other research groups in Behang University and Tsinghua University through courtesy personal contact. Fig. 1.8 shows the vortex structures around turbine tip identified by Ω

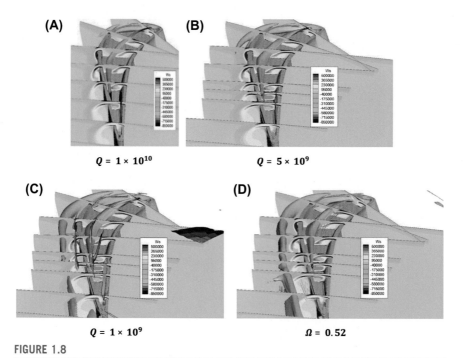

FIGURE 1.8

Different iso-surfaces of vortex structures around the turbine tip.

Reproduced from [Liu, C., et al, "Third generation of vortex identification methods: Omega and Liutex/Rortex based systems", Journal of Hydrodynamics, 31(2), 2019], with the permission of Journal of Hydrodynamics

and Q criteria conducted by Zou et al. in Beihang University (provided by Courtesy Personal Contact of Zou et al.). As can be seen from Fig. 1.8A and B, the weak vortex structures located on the left fail to be captured by a larger Q iso-surface, although the main strong vortex in the central area is similar with the one captured by $\Omega = 0.52$; however, when adjusting Q to a smaller value (Fig. 1.8C), the weak vortices are basically captured but strong vortices are smeared by several structures, which may be contaminated. Therefore, it is obvious that the Omega method can successfully capture both strong and weak vortex structures simultaneously.

Fig. 1.9 shows the vortex structure in swirling flow given by Dr. Gui et al. (Liu et al., 2019) in Tsinghua University. Note that the swirling flow is characterized by the motion of fluid swirl imparted onto a directional jet flow or without directional jet flow. The direct numerical simulation data of swirling jets flows in a rectangular container were utilized here, and the comparison of several vortex identification methods including Liutex, Q, λ_2, Ω_L (same as the Ω method) has been performed to assess the performance and capacity of these vortex criteria. The conclusion is that Ω_L and Liutex have the ability to identify more additional secondary weak vortices immediately after the vortex breakdown (VB) and in far downstream, while such strongly kinked weak vortices cannot be clearly seen by λ_2 and Q.

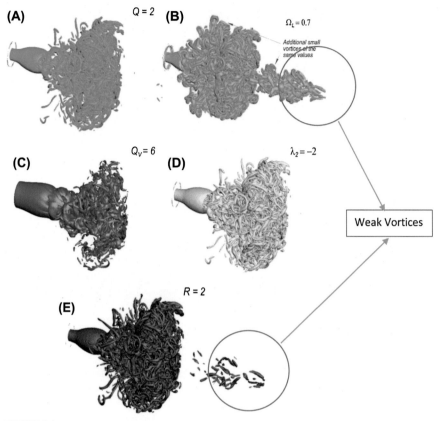

FIGURE 1.9

Vortex structures of Sn = 0.36 at t = 20, visualized by Q-criterion (A), Ω_L (B), Vorticity (C), λ_2 (D), and Liutex (E), respectively.

Reproduced from [Gui, N., et al, "Comparative assessment and analysis of Rorticity by Rortex in swirling jets", Journal of Hydrodynamics, 31(3), 2019], with the permission of Journal of Hydrodynamics

1.5 Liutex and third generation of vortex definition and identification

During the past three decades, many vortex identification criteria, such as Q, Δ, λ_2 and λ_{ci} methods, have been developed, which are classified as the second generation of vortex identification methods. They are all based on the eigenvalues of the velocity gradient tensor. They are all scalars and then strongly dependent on so-called threshold, which is man-made and arbitrary, to show the iso-surface as the vertical structure. In addition, they are all contaminated by stretching, compression, and shearing. Liutex and Liutex-based methods, as the third generation of vortex definition and identification methods, are developed by Liu and his team at UTA. Liutex is defined as a vector that uses the real eigenvector of velocity gradient tensor as its direction and twice local fluid angular speed as its magnitude. The major idea of Liutex is to extract the rigid rotation part from fluid motion to represent vortex. After almost 200 years of struggle, human beings for the first time find a physical quantity to represent fluid rotation or vortex. Liutex is defined as $\vec{R} = R\vec{r}$ and $\vec{\omega} \cdot \vec{r} > 0$ where:

$$R = \vec{\omega} \cdot \vec{r} - \sqrt{\left(\vec{\omega} \cdot \vec{r}\right)^2 - 4\lambda_{ci}^2}, \qquad (1.17)$$

$\vec{\omega}$ is the vorticity, λ_{ci} is the imaginary part of the eigenvalue, and $\vec{r}$ is the real eigenvector of $\nabla \vec{v}$.

Liutex, which is the exact quantity of fluid rotation, may open a door for quantified vortex and turbulence research to replace the traditional qualitative turbulence research, which is in general based on observations, graphics, visualizations, approximations, assumptions, hypotheses, and other qualitative tools.

1.6 Summary

Vortex is a universal form of fluid rotational motion in nature. A rigorous and universally accepted mathematical definition is critical to vortex science and turbulence research. But it remained an open issue before Liutex was discoverd by Liu et al. at UTA. To properly define and visualize a vortex, there are six core elements to be answered. Unfortunately, the first generation of vortex identification methods, or vorticity-based methods, cannot provide any answers to the six elements since vortex is not vorticity tube, which completely disagrees with most textbooks. The second generation of vortex identification methods cannot answer any elements either except for the approximate vortex boundaries, which are strongly dependent on the arbitrary threshold selection. The main reason of the failure of the first and second generations of vortex identification methods is that they are all contaminated by stretching, compression, and/or shear in addition to that all second generations of vortex identification methods are scalars while vortex is a vector field. As the third generation of vortex definition and identification, Liutex, which is an extraction of rigid rotation of fluid motion, can successfully answer all the six elements of vortex identification. These will be presented in the following chapters.

Principal coordinate system based on velocity gradient tensor

In the Eulerian description of fluid motion, the change of velocity $d\vec{v}$ corresponding to a small vector $d\vec{l}$ can be approximated by a linear map $d\vec{v} = \nabla\vec{v} \cdot d\vec{l}$. Here, $\nabla\vec{v}$ denotes the velocity gradient tensor. From critical point theory, the velocity gradient tensor $\nabla\vec{v}$ can provide a complete description of the local flow pattern in a reference frame moving with the fluid particle. Therefore, the velocity gradient tensor becomes an important topic for vortex identification. In this chapter, a specific coordinate system, called principal coordinate system, will be first presented in two-dimensional case to demonstrate the concept of fluid rotation, based on the analysis of the velocity gradient tensor. Then the three-dimensional principal coordinate system is obtained by the real eigenvector. These principal coordinates will serve as the foundation for the concept of Liutex described in the next chapter.

2.1 Matrix representation of velocity gradient tensor and Cauchy–Stokes decomposition

A velocity gradient tensor in three-dimensional case can be represented by a 3×3 matrix:

$$\nabla\vec{v} = \begin{bmatrix} \dfrac{\partial u}{\partial x} & \dfrac{\partial u}{\partial y} & \dfrac{\partial u}{\partial z} \\[2ex] \dfrac{\partial v}{\partial x} & \dfrac{\partial v}{\partial y} & \dfrac{\partial v}{\partial z} \\[2ex] \dfrac{\partial w}{\partial x} & \dfrac{\partial w}{\partial y} & \dfrac{\partial w}{\partial z} \end{bmatrix}, \tag{2.1}$$

where u, v, w represent the components of the velocity vector $\vec{v}$ respectively and x, y, z the axes of the original Cartesian coordinate respectively.

The Cauchy–Stokes decomposition decomposes $\nabla\vec{v}$ into a symmetric part and an antisymmetric part, analogous to the Helmholtz velocity decomposition, which decomposes a smooth, rapidly decaying vector field into an irrotational (curl-free) vector field and a solenoidal (divergence-free) vector field:

Liutex and Its Applications in Turbulence Research. https://doi.org/10.1016/B978-0-12-819023-4.00016-1
Copyright © 2021 Elsevier Inc. All rights reserved.

$$\nabla \vec{v} = \begin{bmatrix} \dfrac{\partial u}{\partial x} & \dfrac{\partial u}{\partial y} & \dfrac{\partial u}{\partial z} \\[2mm] \dfrac{\partial v}{\partial x} & \dfrac{\partial v}{\partial y} & \dfrac{\partial v}{\partial z} \\[2mm] \dfrac{\partial w}{\partial x} & \dfrac{\partial w}{\partial y} & \dfrac{\partial w}{\partial z} \end{bmatrix} = A + B \tag{2.2}$$

$$A = \frac{1}{2}\left(\nabla \vec{v} + \nabla \vec{v}^{T}\right) = \begin{bmatrix} \dfrac{\partial u}{\partial x} & \dfrac{1}{2}\left(\dfrac{\partial u}{\partial y}+\dfrac{\partial v}{\partial x}\right) & \dfrac{1}{2}\left(\dfrac{\partial u}{\partial z}+\dfrac{\partial w}{\partial x}\right) \\[3mm] \dfrac{1}{2}\left(\dfrac{\partial v}{\partial x}+\dfrac{\partial u}{\partial y}\right) & \dfrac{\partial v}{\partial y} & \dfrac{1}{2}\left(\dfrac{\partial v}{\partial z}+\dfrac{\partial w}{\partial y}\right) \\[3mm] \dfrac{1}{2}\left(\dfrac{\partial w}{\partial x}+\dfrac{\partial u}{\partial z}\right) & \dfrac{1}{2}\left(\dfrac{\partial w}{\partial y}+\dfrac{\partial v}{\partial z}\right) & \dfrac{\partial w}{\partial z} \end{bmatrix}$$

$$B = \frac{1}{2}\left(\nabla \vec{v} - \nabla \vec{v}^{T}\right) = \begin{bmatrix} 0 & \dfrac{1}{2}\left(\dfrac{\partial u}{\partial y}-\dfrac{\partial v}{\partial x}\right) & \dfrac{1}{2}\left(\dfrac{\partial u}{\partial z}-\dfrac{\partial w}{\partial x}\right) \\[3mm] -\dfrac{1}{2}\left(\dfrac{\partial u}{\partial y}-\dfrac{\partial v}{\partial x}\right) & 0 & \dfrac{1}{2}\left(\dfrac{\partial v}{\partial z}-\dfrac{\partial w}{\partial y}\right) \\[3mm] -\dfrac{1}{2}\left(\dfrac{\partial u}{\partial z}-\dfrac{\partial w}{\partial x}\right) & -\dfrac{1}{2}\left(\dfrac{\partial v}{\partial z}-\dfrac{\partial w}{\partial y}\right) & 0 \end{bmatrix}$$

and

$$\nabla \vec{v} \cdot d\vec{l} = A \cdot d\vec{l} + B \cdot d\vec{l} = A \cdot d\vec{l} + \frac{1}{2}\nabla \times \vec{v} \times d\vec{l}. \tag{2.3}$$

The right-hand side of Eq. (2.3) uses a property of 3×3 antisymmetric matrix, which can represent cross-products as matrix multiplications. The symmetric part A, called strain rate tensor, is considered as the average deformation. The antisymmetric part B, called angular rotation rate tensor, spin tensor, or vorticity tensor, is considered as the average rotation. But as will be shown in the following chapters, the antisymmetric part, namely vorticity tensor, cannot represent fluid rotation and must be decomposed to a rigid rotation part and an antisymmetric shear part.

2.2 2D principal coordinate system

2.2.1 2D coordinate rotation

Assume two Cartesian coordinate systems have the same origin, but the x axis of the new (rotated) coordinate system is obtained by rotating the x axis of the original coordinate system counter-clockwise through an angle θ, as shown in Fig. 2.1. For one

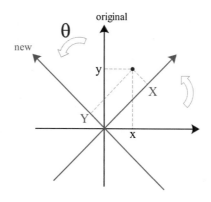

FIGURE 2.1

2D Coordinate rotation.

point with the coordinates $\begin{bmatrix} x \\ y \end{bmatrix}$ in the original coordinates, the coordinates $\begin{bmatrix} X \\ Y \end{bmatrix}$ in the new (rotated) coordinates can be expressed as

$$\begin{bmatrix} X \\ Y \end{bmatrix} = P \begin{bmatrix} x \\ y \end{bmatrix} \tag{2.4}$$

where $P = \begin{bmatrix} cos\theta & sin\theta \\ -sin\theta & cos\theta \end{bmatrix}$ is a rotation matrix, which is orthogonal, namely $PP^T = P^T P = I$. (Note the difference between coordinate rotation and the rotation of a point. In coordinate rotation, the point is fixed in space, while the axis of the coordinate system is rotated. The rotation of a point means that the point is rotated, while the coordinate system remains fixed in space.)

2.2.2 Velocity gradient tensor in the rotated coordinate system

If a rotation matrix P is used to rotate the xy-frame to the XY-frame, the velocity gradient tensor in the XY-frame $\nabla \vec{V}$ is related to the velocity gradient tensor in the xy-frame $\nabla \vec{v}$ through the following expression:

$$\nabla \vec{V} = P \nabla \vec{v} P^{-1} = P \nabla \vec{v} P^T. \tag{2.5}$$

The last equation uses the orthogonality of the rotation matrix P, namely $P^T P = PP^T = I$. Eq. (2.5) also works in three-dimensional case. In the XY-frame, $\nabla \vec{V}$ can be written as

$$\nabla \vec{V} = \begin{bmatrix} \dfrac{\partial U}{\partial X} & \dfrac{\partial U}{\partial Y} \\ \dfrac{\partial V}{\partial X} & \dfrac{\partial V}{\partial Y} \end{bmatrix}. \tag{2.6}$$

For example, $\nabla \vec{V}$ in the rotated coordinate system as shown in Fig. 2.1 is

$$\nabla \vec{V} = \begin{bmatrix} \dfrac{\partial U}{\partial X} & \dfrac{\partial U}{\partial Y} \\ \dfrac{\partial V}{\partial X} & \dfrac{\partial V}{\partial Y} \end{bmatrix} = \boldsymbol{P}\nabla\vec{v}\boldsymbol{P}^T = \begin{bmatrix} cos\theta & sin\theta \\ -sin\theta & cos\theta \end{bmatrix} \begin{bmatrix} \dfrac{\partial u}{\partial x} & \dfrac{\partial u}{\partial y} \\ \dfrac{\partial v}{\partial x} & \dfrac{\partial v}{\partial y} \end{bmatrix} \begin{bmatrix} cos\theta & -sin\theta \\ sin\theta & cos\theta \end{bmatrix}$$

$$\frac{\partial U}{\partial X} = \frac{\partial u}{\partial x}cos^2\theta + \frac{\partial v}{\partial x}sin\theta cos\theta + \frac{\partial u}{\partial y}sin\theta cos\theta + \frac{\partial v}{\partial y}sin^2\theta$$

$$\frac{\partial U}{\partial Y} = -\frac{\partial u}{\partial x}sin\theta cos\theta - \frac{\partial v}{\partial x}sin^2\theta + \frac{\partial u}{\partial y}cos^2\theta + \frac{\partial v}{\partial y}sin\theta cos\theta$$

$$\frac{\partial V}{\partial X} = -\frac{\partial u}{\partial x}sin\theta cos\theta + \frac{\partial v}{\partial x}cos^2\theta - \frac{\partial u}{\partial y}sin^2\theta + \frac{\partial v}{\partial y}sin\theta cos\theta$$

$$\frac{\partial V}{\partial Y} = \frac{\partial u}{\partial x}sin^2\theta - \frac{\partial v}{\partial x}sin\theta cos\theta - \frac{\partial u}{\partial y}sin\theta cos\theta + \frac{\partial v}{\partial y}cos^2\theta$$

2.2.3 Eigenvalues of the velocity gradient tensor

For a 2D velocity gradient tensor

$$\nabla \vec{v} = \begin{bmatrix} \dfrac{\partial u}{\partial x} & \dfrac{\partial u}{\partial y} \\ \dfrac{\partial v}{\partial x} & \dfrac{\partial v}{\partial y} \end{bmatrix}, \tag{2.7}$$

the characteristic equation is

$$\left| \nabla \vec{v} - \lambda \boldsymbol{I} \right| = \begin{vmatrix} \dfrac{\partial u}{\partial x} - \lambda & \dfrac{\partial u}{\partial y} \\ \dfrac{\partial v}{\partial x} & \dfrac{\partial v}{\partial y} - \lambda \end{vmatrix} = 0 \tag{2.8}$$

where $\boldsymbol{I}$ is a 2×2 identity matrix.

This characteristic equation can be written as a general form:

$$\lambda^2 + b\lambda + c = (\lambda - \lambda_1)(\lambda - \lambda_2) = 0, \tag{2.9}$$

where $b = -\left(\frac{\partial u}{\partial x} + \frac{\partial v}{\partial y}\right)$, $c = \frac{\partial u}{\partial x}\frac{\partial v}{\partial y} - \frac{\partial u}{\partial y}\frac{\partial v}{\partial x}$, $-(\lambda_1 + \lambda_2) = b$ and $\lambda_1\lambda_2 = c$.

According to the discriminant of Eq. (2.9), there are three cases:

(1) If $\Delta = b^2 - 4c > 0$, the second-order polynomial has two different real roots

$$\lambda_1 = \frac{-b + \sqrt{b^2 - 4c}}{2} = \frac{-b + \sqrt{\Delta}}{2}$$

$$\lambda_2 = \frac{-b - \sqrt{b^2 - 4c}}{2} = \frac{-b - \sqrt{\Delta}}{2}.$$

(2.10)

In order to avoid subtraction of two similar numbers in practice, the following formula is used:

$$q = -\frac{1}{2}\left[b + sign(b)\sqrt{b^2 - 4c}\right].$$

So,

$$\lambda_1 = q$$

$$\lambda_2 = \frac{c}{q}.$$

(2.11)

(2) If $\Delta = b^2 - 4c = 0$, there are two repeated roots

$$\lambda_1 = \lambda_2 = -\frac{b}{2}.$$

(2.12)

(3) $\Delta = b^2 - 4c < 0$, the second-order polynomial has two conjugated complex roots:

$$\lambda_1 = \frac{-b + i\sqrt{4c - b^2}}{2} = \frac{-b + i\sqrt{-\Delta}}{2}$$

$$\lambda_2 = \frac{-b - i\sqrt{4c - b^2}}{2} = \frac{-b - i\sqrt{-\Delta}}{2}$$

(2.13)

2.2.4 Rigid rotation

Before diving into the general fluid motion involving stretching/compression, shearing, and rotation, the simplest rotational motion, namely the rigid rotation, is first considered. The definition is given by the following

Definition. A rigid rotation is defined as fluid motion like a rigid body.

For two-dimensional case, one possible rigid rotation is shown in Fig. 2.2 and the velocity is

$$\begin{cases} U = -\omega Y \\ V = \omega X \end{cases}$$

(2.14)

Here, ω is a constant representing the angular velocity. The corresponding velocity gradient tensor is

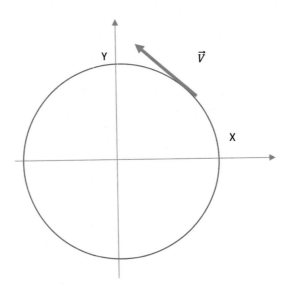

FIGURE 2.2

Rigid rotation.

$$\nabla \vec{V} = \begin{bmatrix} 0 & \dfrac{\partial U}{\partial Y} \\ \dfrac{\partial V}{\partial X} & 0 \end{bmatrix} = \begin{bmatrix} 0 & -\omega \\ \omega & 0 \end{bmatrix} \tag{2.15}$$

In this simplest rotation motion, the diagonal terms of the velocity gradient tensor are both zero, which implies no stretching/compression effect. And the off-diagonal terms have the same absolute value but different signs, which implies no shearing effect. Accordingly, the velocity gradient tensor presents a very simple form in the case of rigid rotation. Moreover, even if a coordinate rotation is applied, Eq. (2.15) remains the same, which means the rigid rotation has an identical form in all (rotated) coordinate systems.

2.2.5 Principal coordinates

However, for a general fluid motion, there is no direct way to distinguish the rotational motion from stretching, compression, and shearing effects. Furthermore, the components of the velocity gradient tensor rely on the specific coordinate system. For coordinate rotation clockwise through an angle θ, the velocity gradient tensor in the rotated coordinate will become $\nabla \vec{V} = P \nabla \vec{v} P^T = \begin{bmatrix} \dfrac{\partial U}{\partial X} & \dfrac{\partial U}{\partial Y} \\ \dfrac{\partial V}{\partial X} & \dfrac{\partial V}{\partial Y} \end{bmatrix}$ and

$P = \begin{bmatrix} cos\theta & -sin\theta \\ sin\theta & cos\theta \end{bmatrix}$. It means that the elements of $\nabla \vec{V}$ could be a function of θ

or, in other words, the elements of $\nabla \vec{V}$ is dependent on θ. So, to analyze the velocity gradient tensor of the general fluid motion, a particular matrix $\nabla \vec{V_\theta}$, which has the same diagonal elements, i.e., $\frac{\partial U}{\partial X_\theta} = \frac{\partial V}{\partial Y_\theta}$ is pursued.

Definition. A principal coordinate of a 2×2 velocity gradient tensor is defined as a coordinate system in which $\nabla \vec{V} = P \nabla \vec{v} P^T$, $P = \begin{bmatrix} \cos\theta & -\sin\theta \\ \sin\theta & \cos\theta \end{bmatrix}$, and $\frac{\partial U}{\partial X_\theta} = \frac{\partial V}{\partial Y_\theta}$.

Every 2×2 velocity gradient tensor should have a principal coordinate as θ can be easily determined by $\frac{\partial U}{\partial X_\theta} = \frac{\partial V}{\partial Y_\theta}$ although it may not be unique. For simplicity, the subscript θ is dropped, and assume that the analysis is performed in the principal coordinate system in the following. In the principal coordinate system,

$$\nabla \vec{V} = P \nabla \vec{v} P^T = \begin{bmatrix} \frac{1}{2}(\lambda_1 + \lambda_2) & \frac{\partial U}{\partial Y} \\ \frac{\partial V}{\partial X} & \frac{1}{2}(\lambda_1 + \lambda_2) \end{bmatrix} = \begin{bmatrix} \lambda_{cr} & \frac{\partial U}{\partial Y} \\ \frac{\partial V}{\partial X} & \lambda_{cr} \end{bmatrix}, \qquad (2.16)$$

where the diagonal elements are the same as $\frac{1}{2}(\lambda_1 + \lambda_2) = \lambda_{cr}$.

2.2.6 Rotation in principal coordinates

Apparently, the diagonal element $\frac{\partial U}{\partial X}$ represents the fluid stretching $\left(\frac{\partial U}{\partial X} > 0 \right)$ or compression $\left(\frac{\partial U}{\partial X} < 0 \right)$ in the X-direction. It is similar for the diagonal element $\frac{\partial V}{\partial Y}$. Therefore, the diagonal elements in the principal coordinate can be zeroed out for the analysis of the rotation. There exist three cases:

(1) $\frac{\partial U}{\partial Y} \cdot \frac{\partial V}{\partial X} = 0$ No rotation exists as shown in Fig. 2.3A. In this case, $\frac{\partial V}{\partial X} = 0$ or $\frac{\partial U}{\partial Y} = 0$, which means it is a simple shear motion.

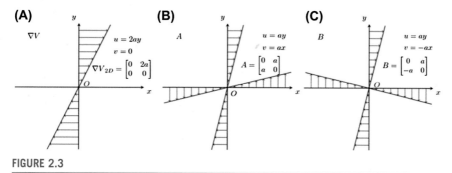

FIGURE 2.3

Fluid rotation in the 2D principal coordinates.

Reproduced from Wang, Y. and Gui, N., A review of the third-generation vortex identification method and its applications, Chinese Journal of Hydrodynamics 34(4), 2019, with the permission of Chinese Journal of Hydrodynamics.

(2) $\frac{\partial U}{\partial Y} \cdot \frac{\partial V}{\partial X} > 0$ No rotation exists as shown in Fig. 2.3B.

(3) $\frac{\partial U}{\partial Y} \cdot \frac{\partial V}{\partial X} < 0$ There exists rotational motion as shown in Fig. 2.3C.

Hence, fluid rotation is determined by the off-diagonal elements in the principal coordinate and the following theorem provides a criterion for fluid rotation:

Theorem 2.1. The necessary and sufficient condition for fluid rotation is that the 2D velocity gradient tensor has two conjugate complex eigenvalues.

Proof. The characteristic equation in the principal coordinates is $(\lambda - \lambda_{cr})^2 - \frac{\partial U}{\partial Y}\frac{\partial V}{\partial X} = 0$. If $\frac{\partial U}{\partial Y}\frac{\partial V}{\partial X} < 0$, the fluid has rotational motion, and then $\lambda = \lambda_{cr} \pm i\sqrt{-\frac{\partial U}{\partial Y}\frac{\partial V}{\partial X}}$, the velocity gradient tensor has two conjugate complex eigenvalues. On the other hand, if there are two complex eigenvalues, $\frac{\partial U}{\partial Y}\frac{\partial V}{\partial X} < 0$, the fluid has rotational motion according to the aforementioned physical analysis.

2.2.7 Fluid rotation strength

Definition. Fluid rotation strength is defined as the angular speed of the rigid rotation part.

Justification of this definition of fluid rotation strength: Consider the laminar boundary layer as shown in Fig. 2.4.

Approximately, the x-component of the velocity u only depends on the y coordinate, i.e., $u = u(y)$. Near the wall $\frac{\partial u}{\partial y} \gg 0$, $\frac{\partial v}{\partial x} \approx 0$, the velocity gradient tensor is

$$\nabla \vec{v} = \begin{bmatrix} \lambda_{cr} & \frac{\partial u}{\partial y} \\ \frac{\partial v}{\partial x} & \lambda_{cr} \end{bmatrix} = \begin{bmatrix} \lambda_{cr} & \frac{\partial u}{\partial y} \\ 0 & \lambda_{cr} \end{bmatrix}. \tag{2.17}$$

What is an appropriate measure of the rotation strength? Neither $\max\left\{\left|\frac{\partial u}{\partial y}\right|, \left|\frac{\partial v}{\partial x}\right|\right\}$ nor $\frac{1}{2}\left(\left|\frac{\partial u}{\partial y}\right| + \left|\frac{\partial u}{\partial y}\right|\right)$ is appropriate since the flow has no rotation. The only choice is $\min\left\{\left|\frac{\partial u}{\partial y}\right|, \left|\frac{\partial v}{\partial x}\right|\right\} = 0$. This example clearly shows the above definition is

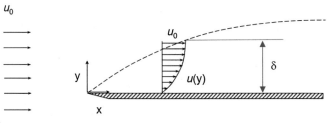

FIGURE 2.4

Laminar boundary layer.

appropriate to define the fluid rotation strength, which is minimum of $\left\{ \left| \frac{\partial u}{\partial y} \right|, \; \left| \frac{\partial v}{\partial x} \right| \right\}$ in a principal coordinates, rather than maximum or average of the off-diagonal elements.

Assume that the original velocity gradient tensor is $\nabla \vec{v} = \begin{bmatrix} \frac{\partial u}{\partial x} & \frac{\partial u}{\partial y} \\ \frac{\partial v}{\partial x} & \frac{\partial v}{\partial y} \end{bmatrix}$, if $\nabla \vec{v}$ has two complex eigenvalues, the eigenvalues and vorticity

$$\lambda_1 = \frac{-b + i\sqrt{4c - b^2}}{2} \quad \text{and} \quad \lambda_2 = \frac{-b - i\sqrt{4c - b^2}}{2}$$

where $b = -\left(\frac{\partial u}{\partial x} + \frac{\partial v}{\partial y} \right)$, $c = \frac{\partial u}{\partial x} \frac{\partial v}{\partial y} - \frac{\partial u}{\partial y} \frac{\partial v}{\partial x}$,

$$\lambda_{cr} = -\frac{b}{2} = \frac{1}{2}\left(\frac{\partial u}{\partial x} + \frac{\partial v}{\partial y} \right), \quad \lambda_{ci} = \frac{\sqrt{4c - b^2}}{2}$$

$$= \frac{1}{2}\sqrt{-\left(\frac{\partial u}{\partial x} - \frac{\partial v}{\partial y} \right)^2 - 4\frac{\partial u}{\partial y} \frac{\partial v}{\partial x}} \quad \text{and} \quad \omega_z = \frac{\partial v}{\partial x} - \frac{\partial u}{\partial y} \qquad (2.18)$$

are all uniquely determined by $\nabla \vec{v}$.

Now, look at the velocity gradient tensor in a principal coordinate after rotating the original coordinates:

$$\nabla \vec{V} = \begin{bmatrix} \lambda_{cr} & \frac{\partial U}{\partial Y} \\ \frac{\partial V}{\partial X} & \lambda_{cr} \end{bmatrix} = \begin{bmatrix} \lambda_{cr} & -R/2 \\ R/2 + \varepsilon & \lambda_{cr} \end{bmatrix} = \begin{bmatrix} 0 & -R/2 \\ R/2 & 0 \end{bmatrix} + \begin{bmatrix} \lambda_{cr} & 0 \\ \varepsilon & \lambda_{cr} \end{bmatrix} = R + NR$$

$$(2.19)$$

where $R/2 = -\frac{\partial U}{\partial Y} > 0$, and $\frac{\partial V}{\partial X} = \frac{R}{2} + \varepsilon$, if $\frac{\partial V}{\partial X} > 0$ and $\left| \frac{\partial V}{\partial X} \right| > \left| \frac{\partial U}{\partial Y} \right|$ is assumed.

Due to the rotational invariance of vorticity and eigenvalues, vorticity and eigenvalues keep invariant during the orthogonal transformation $\nabla \vec{V} = P \nabla \vec{v} P^T$, so

$$(R/2)(R/2 + \varepsilon) = \lambda_{ci}^2$$

$$(R/2 + \varepsilon) - (-R/2) = \omega_z \qquad (2.20)$$

Solve the aforementioned two equations to find R and ε

$$R = \omega_z - \sqrt{\omega_z^2 - 4\lambda_{ci}^2}$$

$$\varepsilon = \frac{2\lambda_{ci}^2}{R} - R/2 \qquad (2.21)$$

Apparently, $R/2$ is the angular speed of the rigid rotation.

2.2.8 Summary

In this section, a 2D rotation matrix P is given and the principal coordinates are defined when the two diagonal elements are equal in that coordinates. All 2×2 matrices can be rotated to obtain the principal coordinates. If the velocity gradient tensor has two real eigenvalues, the fluid will have no rotation since $\frac{\partial U}{\partial Y} \cdot \frac{\partial V}{\partial X} \geq 0$; if it has two conjugate complex eigenvalues, it can be written as

$$
\nabla \vec{V} = \begin{bmatrix} \frac{1}{2}(\lambda_1 + \lambda_2) & \frac{\partial U}{\partial Y} \\ \frac{\partial V}{\partial X} & \frac{1}{2}(\lambda_1 + \lambda_2) \end{bmatrix} = \begin{bmatrix} \lambda_{cr} & -\frac{1}{2}R \\ \frac{1}{2}R + \varepsilon & \lambda_{cr} \end{bmatrix} \quad \text{assuming } \left| \frac{\partial U}{\partial Y} \right| < \left| \frac{\partial V}{\partial X} \right|
$$

and the fluid will have rotation since $\frac{\partial U}{\partial Y} \cdot \frac{\partial V}{\partial X} < 0$. The rotation strength is defined as $R = 2 \min \left\{ \left| \frac{\partial U}{\partial Y} \right|, \left| \frac{\partial V}{\partial X} \right| \right\} = \omega - \sqrt{\omega^2 - 4\lambda_{ci}^2}$ where ω is the vorticity and λ_{ci} is the imaginary part of the eigenvalue. The rest tensor excluding the rigid rotation,

$$
\nabla \vec{V} - \begin{bmatrix} 0 & -\frac{1}{2}R \\ \frac{1}{2}R & 0 \end{bmatrix} = \begin{bmatrix} \lambda_{cr} & 0 \\ 0 & \lambda_{cr} \end{bmatrix} + \begin{bmatrix} 0 & 0 \\ \varepsilon & 0 \end{bmatrix} = CS + S, \quad \text{can be decom-}
$$

posed as a stretching/compression part and a shearing part (no rotational vorticity). For 2D incompressible flows, $\frac{\partial U}{\partial X} + \frac{\partial V}{\partial Y} = 2\lambda_{cr} = 0$ or $\lambda_{cr} = 0$, there is only rotation and shearing without stretching or compression.

2.3 3D principal coordinate system

2.3.1 3D coordinate rotation

If a rotation matrix Q is used to rotate the xyz-frame to the XYZ-frame, the velocity gradient tensor in the XYZ-frame $\nabla \vec{V}$ is related to the velocity gradient tensor in the xyz-frame $\nabla \vec{v}$ through the following expression:

$$
\nabla \vec{V} = Q \nabla \vec{v} Q^{-1} = Q \nabla \vec{v} Q^T.
$$

which is similar to the 2D case. However, the rotation matrix Q is a 3×3 matrix and is much more complicated than the 2D rotation matrix P. Actually, any 3×3 rotation matrix can be decomposed into three successive basic rotations around one axis. But this property is not used here.

2.3.2 Eigenvalues of a 3×3 matrix

For a 3D velocity gradient tensor, which is a 3×3 matrix:

$$\nabla \vec{v} = \begin{bmatrix} \dfrac{\partial u}{\partial x} & \dfrac{\partial u}{\partial y} & \dfrac{\partial u}{\partial z} \\[2mm] \dfrac{\partial v}{\partial x} & \dfrac{\partial v}{\partial y} & \dfrac{\partial v}{\partial z} \\[2mm] \dfrac{\partial w}{\partial x} & \dfrac{\partial w}{\partial y} & \dfrac{\partial w}{\partial z} \end{bmatrix},$$

the eigenvalue equation can be written as:

$$\left| \nabla \vec{v} - \lambda \mathbf{I} \right| = 0$$

or

$$\begin{vmatrix} \dfrac{\partial u}{\partial x} - \lambda & \dfrac{\partial u}{\partial y} & \dfrac{\partial u}{\partial z} \\[2mm] \dfrac{\partial v}{\partial x} & \dfrac{\partial v}{\partial y} - \lambda & \dfrac{\partial v}{\partial z} \\[2mm] \dfrac{\partial w}{\partial x} & \dfrac{\partial w}{\partial y} & \dfrac{\partial w}{\partial z} - \lambda \end{vmatrix} = 0 \tag{2.22}$$

where $|\mathbf{A}|$ stands for the determinant of $\mathbf{A}$.

$$\left(\frac{\partial u}{\partial x} - \lambda \right) \left[\left(\frac{\partial v}{\partial y} - \lambda \right) \left(\frac{\partial w}{\partial z} - \lambda \right) - \frac{\partial v}{\partial z} \frac{\partial w}{\partial y} \right] - \frac{\partial u}{\partial y} \left[\frac{\partial v}{\partial x} \left(\frac{\partial w}{\partial z} - \lambda \right) - \frac{\partial v}{\partial z} \frac{\partial w}{\partial x} \right]$$
$$+ \frac{\partial u}{\partial z} \left[\frac{\partial v}{\partial x} \frac{\partial w}{\partial y} - \left(\frac{\partial v}{\partial y} - \lambda \right) \frac{\partial w}{\partial x} \right] = 0 \tag{2.23}$$

$$\left(\frac{\partial u}{\partial x} - \lambda \right) \left[\lambda^2 - \left(\frac{\partial v}{\partial y} + \frac{\partial w}{\partial z} \right) \lambda + \frac{\partial v}{\partial y} \frac{\partial w}{\partial z} - \frac{\partial v}{\partial z} \frac{\partial w}{\partial y} \right] - \frac{\partial u}{\partial y} \left[- \frac{\partial v}{\partial x} \lambda + \frac{\partial v}{\partial x} \frac{\partial w}{\partial z} - \frac{\partial v}{\partial z} \frac{\partial w}{\partial x} \right]$$
$$+ \frac{\partial u}{\partial z} \left[\frac{\partial v}{\partial x} \frac{\partial w}{\partial y} - \frac{\partial v}{\partial y} \frac{\partial w}{\partial x} + \frac{\partial w}{\partial x} \lambda \right] = 0 \tag{2.24}$$

$$\lambda^3 - \left(\frac{\partial u}{\partial x} + \frac{\partial v}{\partial y} + \frac{\partial w}{\partial z} \right) \lambda^2 + \left(\frac{\partial u}{\partial x} \frac{\partial v}{\partial y} + \frac{\partial u}{\partial x} \frac{\partial w}{\partial z} + \frac{\partial v}{\partial y} \frac{\partial w}{\partial z} - \frac{\partial v}{\partial z} \frac{\partial w}{\partial y} - \frac{\partial u}{\partial z} \frac{\partial w}{\partial x} - \frac{\partial v}{\partial x} \frac{\partial u}{\partial y} \right) \lambda$$
$$+ \left(\frac{\partial u}{\partial x} \frac{\partial v}{\partial z} \frac{\partial w}{\partial y} + \frac{\partial u}{\partial z} \frac{\partial v}{\partial y} \frac{\partial w}{\partial x} + \frac{\partial u}{\partial y} \frac{\partial v}{\partial x} \frac{\partial w}{\partial z} - \frac{\partial u}{\partial x} \frac{\partial v}{\partial y} \frac{\partial w}{\partial z} - \frac{\partial u}{\partial y} \frac{\partial v}{\partial z} \frac{\partial w}{\partial x} - \frac{\partial u}{\partial z} \frac{\partial v}{\partial x} \frac{\partial w}{\partial y} \right) = 0 \tag{2.25}$$

The aforementioned equation is a cubic equation, which can be written is a general form:

$$\lambda^3 + a\lambda^2 + b\lambda + c = 0, \tag{2.26}$$

where

$$a = -\left(\frac{\partial u}{\partial x} + \frac{\partial v}{\partial y} + \frac{\partial w}{\partial z}\right)$$

$$b = \frac{\partial u}{\partial x}\frac{\partial v}{\partial y} + \frac{\partial u}{\partial x}\frac{\partial w}{\partial z} + \frac{\partial v}{\partial y}\frac{\partial w}{\partial z} - \frac{\partial v}{\partial z}\frac{\partial w}{\partial y} - \frac{\partial u}{\partial z}\frac{\partial w}{\partial x} - \frac{\partial v}{\partial x}\frac{\partial u}{\partial y}$$

$$c = \frac{\partial u}{\partial x}\frac{\partial v}{\partial z}\frac{\partial w}{\partial y} + \frac{\partial u}{\partial z}\frac{\partial v}{\partial y}\frac{\partial w}{\partial x} + \frac{\partial u}{\partial y}\frac{\partial v}{\partial x}\frac{\partial w}{\partial z} - \frac{\partial u}{\partial x}\frac{\partial v}{\partial y}\frac{\partial w}{\partial z} - \frac{\partial u}{\partial y}\frac{\partial v}{\partial z}\frac{\partial w}{\partial x} - \frac{\partial u}{\partial z}\frac{\partial v}{\partial x}\frac{\partial w}{\partial y}$$

A robust algorithm to solve the cubic equation can be described as follows:

(1) For a velocity gradient tensor, a, b, c are real, first compute

$$Q = \frac{a^2 - 3b}{9} \text{ and } R = \frac{2a^3 - 9ab + 27c}{54} \tag{2.27}$$

(2) If $R^2 < Q^3$, then the equation has three real roots. Compute the intermediate variable

$$\vartheta = \arccos\left(R / \sqrt{Q^3}\right)$$

Three roots are

$$\lambda_1 = -2\sqrt{Q}\cos\left(\frac{\vartheta}{3}\right) - \frac{a}{3}$$

$$\lambda_2 = -2\sqrt{Q}\cos\left(\frac{\vartheta + 2\pi}{3}\right) - \frac{a}{3} \tag{2.28}$$

$$\lambda_3 = -2\sqrt{Q}\cos\left(\frac{\vartheta - 2\pi}{3}\right) - \frac{a}{3}$$

(3) If $R^2 \geq Q^3$, compute $A = -sign(R)\left[|R| + \sqrt{R^2 - Q^3}\right]^{1/3}$ where the square root is positive. Then compute

$$B = \begin{cases} \dfrac{Q}{A} & (A \neq 0) \\ 0 & (A = 0) \end{cases}$$

One root is real and two are conjugate complex

$$\lambda_1 = (A + B) - \frac{a}{3} = \lambda_r$$

$$\lambda_2 = -\frac{1}{2}(A + B) - \frac{a}{3} + i\,\frac{\sqrt{3}}{2}(A - B) = \lambda_{cr} + i\lambda_{ci} \tag{2.29}$$

$$\lambda_3 = -\frac{1}{2}(A + B) - \frac{a}{3} - i\,\frac{\sqrt{3}}{2}(A - B) = \lambda_{cr} - i\lambda_{ci}$$

Because A and B are both real, λ_1 is the real eigenvalue and λ_2 and λ_3 are the complex eigenvalues.

2.3.3 Eigenvector corresponding to the real eigenvalue

The analytical expression of the normalized real eigenvector $\vec{r}$ corresponding to the real eigenvalue λ_r is derived here. As will be shown in the following, the fluid rotation in three-dimensional case only occurs when the velocity gradient tensor has two complex eigenvalues and one real eigenvalue. In this case, the normalized real eigenvector is unique (up to the $\pm$ sign). If A is used to denote a matrix representation of the velocity gradient tensor and $\vec{r}^* = \left[r_x^*, r_y^*, r_z^*\right]^T$ is used to denote an unnormalized eigenvector corresponding to λ_r, according to the definition of eigenvalue and eigenvector,

$$A\,\vec{r}^* = \lambda_r\,\vec{r}^* \tag{2.30}$$

Eq. (2.30) can be rewritten as

$$\begin{bmatrix} \dfrac{\partial u}{\partial x} - \lambda_r & \dfrac{\partial u}{\partial y} & \dfrac{\partial u}{\partial z} \\[2mm] \dfrac{\partial v}{\partial x} & \dfrac{\partial v}{\partial y} - \lambda_r & \dfrac{\partial v}{\partial z} \\[2mm] \dfrac{\partial w}{\partial x} & \dfrac{\partial w}{\partial y} & \dfrac{\partial w}{\partial z} - \lambda_r \end{bmatrix} \begin{bmatrix} r_x^* \\[1mm] r_y^* \\[1mm] r_z^* \end{bmatrix} = 0 \tag{2.31}$$

By checking three first minors

$$\Delta_x = \begin{vmatrix} \dfrac{\partial v}{\partial y} - \lambda_r & \dfrac{\partial v}{\partial z} \\[2mm] \dfrac{\partial w}{\partial y} & \dfrac{\partial w}{\partial z} - \lambda_r \end{vmatrix}, \Delta_y = \begin{vmatrix} \dfrac{\partial u}{\partial x} - \lambda_r & \dfrac{\partial u}{\partial z} \\[2mm] \dfrac{\partial w}{\partial x} & \dfrac{\partial w}{\partial z} - \lambda_r \end{vmatrix}, \Delta_z = \begin{vmatrix} \dfrac{\partial u}{\partial x} - \lambda_r & \dfrac{\partial u}{\partial y} \\[2mm] \dfrac{\partial v}{\partial x} & \dfrac{\partial v}{\partial y} - \lambda_r \end{vmatrix}$$

$$\tag{2.32}$$

the maximum absolute value is

$$\Delta_{max} = \max(|\Delta_x|, |\Delta_y|, |\Delta_z|) \tag{2.33}$$

(Note: not all the minors will be equal to zero, thus $\Delta_{max} > 0$. Otherwise, there exists a contradiction that the normalized real eigenvector is nonunique, or the real eigenvector is a zero vector.)

If $\Delta_{max} = |\Delta_x|$, set

$$r_x^* = 1. \tag{2.34}$$

By solving

$$
\begin{bmatrix}
\dfrac{\partial v}{\partial y} - \lambda_r & \dfrac{\partial v}{\partial z} \\[2ex]
\dfrac{\partial w}{\partial y} & \dfrac{\partial w}{\partial z} - \lambda_r
\end{bmatrix}
\begin{bmatrix}
r_y^* \\[2ex]
r_z^*
\end{bmatrix}
=
\begin{bmatrix}
-\dfrac{\partial v}{\partial x} \\[2ex]
-\dfrac{\partial w}{\partial x}
\end{bmatrix}
\tag{2.35}
$$

the other two components of $\overrightarrow{r}^*$ are

$$r_y^* = \frac{-\left(\dfrac{\partial w}{\partial z} - \lambda_r\right)\dfrac{\partial v}{\partial x} + \dfrac{\partial v}{\partial z}\dfrac{\partial w}{\partial x}}{\left(\dfrac{\partial v}{\partial y} - \lambda_r\right)\left(\dfrac{\partial w}{\partial z} - \lambda_r\right) - \dfrac{\partial v}{\partial z}\dfrac{\partial w}{\partial y}} \tag{2.36}$$

$$r_z^* = \frac{\dfrac{\partial w}{\partial y}\dfrac{\partial v}{\partial x} - \left(\dfrac{\partial v}{\partial y} - \lambda_r\right)\dfrac{\partial w}{\partial x}}{\left(\dfrac{\partial v}{\partial y} - \lambda_r\right)\left(\dfrac{\partial w}{\partial z} - \lambda_r\right) - \dfrac{\partial v}{\partial z}\dfrac{\partial w}{\partial y}} \tag{2.37}$$

Similarly, if $\Delta_{max} = |\Delta_y|$, let

$$r_y^* = 1 \tag{2.38}$$

the other two components of $\overrightarrow{r}^*$ are

$$r_x^* = \frac{-\left(\dfrac{\partial w}{\partial z} - \lambda_r\right)\dfrac{\partial u}{\partial y} + \dfrac{\partial u}{\partial z}\dfrac{\partial w}{\partial y}}{\left(\dfrac{\partial u}{\partial x} - \lambda_r\right)\left(\dfrac{\partial w}{\partial z} - \lambda_r\right) - \dfrac{\partial u}{\partial z}\dfrac{\partial w}{\partial x}} \tag{2.39}$$

$$r_z^* = \frac{\dfrac{\partial w}{\partial x}\dfrac{\partial u}{\partial y} - \left(\dfrac{\partial u}{\partial x} - \lambda_r\right)\dfrac{\partial w}{\partial y}}{\left(\dfrac{\partial u}{\partial x} - \lambda_r\right)\left(\dfrac{\partial w}{\partial z} - \lambda_r\right) - \dfrac{\partial u}{\partial z}\dfrac{\partial w}{\partial x}} \tag{2.40}$$

In the case of $\Delta_{max} = |\Delta_z|$, if

$$r_z^* = 1 \tag{2.41}$$

$$r_x^* = \frac{-\left(\dfrac{\partial v}{\partial y} - \lambda_r\right)\dfrac{\partial u}{\partial z} + \dfrac{\partial u}{\partial y}\dfrac{\partial v}{\partial z}}{\left(\dfrac{\partial u}{\partial x} - \lambda_r\right)\left(\dfrac{\partial v}{\partial y} - \lambda_r\right) - \dfrac{\partial u}{\partial y}\dfrac{\partial v}{\partial x}} \tag{2.42}$$

$$r_y^* = \frac{\dfrac{\partial v}{\partial x}\dfrac{\partial u}{\partial z} - \left(\dfrac{\partial u}{\partial x} - \lambda_r\right)\dfrac{\partial v}{\partial z}}{\left(\dfrac{\partial u}{\partial x} - \lambda_r\right)\left(\dfrac{\partial v}{\partial y} - \lambda_r\right) - \dfrac{\partial u}{\partial y}\dfrac{\partial v}{\partial x}} \tag{2.43}$$

And the normalized real eigenvector $\vec{r}$ will be

$$\vec{\mathbf{r}} = \vec{r}^* / \left|\vec{r}^*\right| \tag{2.44}$$

2.3.4 Real eigenvector as the Z-axis

A 3×3 matrix can have either three real eigenvalues or one real and two conjugate complex eigenvalues, depending on the sign of Δ. In the 2D case, for points with fluid rotation, the velocity gradient matrix has $\Delta < 0$ and the matrix will have two conjugate complex eigenvalues. In other words, any 3×3 real matrix must have at least one real eigenvalue and one corresponding real eigenvector. The general form of three eigenvalues in the rotation points can be written as:
λ_r, $\lambda_{cr} + i\lambda_{ci}$, $\lambda_{cr} - i\lambda_{ci}$, where i disappears and $\lambda_{ci} = \frac{\sqrt{\Delta}}{2}$ when all three eigenvalues are real

$$\nabla\vec{v}\cdot\vec{r} = \lambda_r\vec{r}\,(\text{Since both }\vec{r}\text{ and } -\vec{r}\text{ are the real eigenvectors,}$$
$$\text{we require }\vec{\omega}\cdot\vec{r}\,0\text{ to keep uniqueness.}) \tag{2.45}$$

Only the case when the 3×3 matrix has one real and two conjugate complex eigenvalues, or $\Delta < 0$ is considered in this section as the rotational points are studied.

For a 3D velocity gradient tensor, which is a 3×3 matrix:

$$\nabla\vec{v} = \begin{bmatrix} \dfrac{\partial u}{\partial x} & \dfrac{\partial u}{\partial y} & \dfrac{\partial u}{\partial z} \\[2mm] \dfrac{\partial v}{\partial x} & \dfrac{\partial v}{\partial y} & \dfrac{\partial v}{\partial z} \\[2mm] \dfrac{\partial w}{\partial x} & \dfrac{\partial w}{\partial y} & \dfrac{\partial w}{\partial z} \end{bmatrix},$$

there exists a 3D rotation matrix Q, which is orthogonal, to rotate the original coordinates to coincide the Z-coordinate with the real eigenvector $\vec{r}$ (Fig. 2.5), or in other words, the new matrix $\nabla\vec{V}$ in the rotated coordinates (X, Y, Z) has the Z-axis aligned with its eigenvector $\vec{r}$.

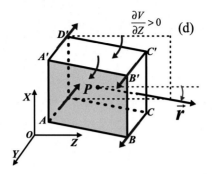

FIGURE 2.5

3D rotation, which makes the Z-axis aligned with the real eigenvector $\vec{r}$.

In general, $\nabla \vec{V} = \begin{bmatrix} \dfrac{\partial U}{\partial X} & \dfrac{\partial U}{\partial Y} & \dfrac{\partial U}{\partial Z} \\[2mm] \dfrac{\partial V}{\partial X} & \dfrac{\partial V}{\partial Y} & \dfrac{\partial V}{\partial Z} \\[2mm] \dfrac{\partial W}{\partial X} & \dfrac{\partial W}{\partial Y} & \dfrac{\partial W}{\partial Z} \end{bmatrix}$ and $\nabla \vec{V} = Q \nabla \vec{v} Q^T$.

where Q is the rotation matrix from the xyz-frame to the XYZ-frame. So, the coordinate of a vector in the rotated coordinate system is

$$\begin{bmatrix} X \\ Y \\ Z \end{bmatrix} = Q \begin{bmatrix} x \\ y \\ z \end{bmatrix}$$

The new matrix $\nabla \vec{V}\, X, Y, Z$ will satisfy

$$\nabla \vec{V} \begin{bmatrix} 0 \\ 0 \\ 1 \end{bmatrix} = \lambda_r \begin{bmatrix} 0 \\ 0 \\ 1 \end{bmatrix}, \tag{2.46}$$

since the Z-axis is the real eigenvector of $\nabla \vec{V}$.

In order to satisfy the aforementioned condition

$$\begin{bmatrix} \dfrac{\partial U}{\partial X} & \dfrac{\partial U}{\partial Y} & \dfrac{\partial U}{\partial Z} \\[2mm] \dfrac{\partial V}{\partial X} & \dfrac{\partial V}{\partial Y} & \dfrac{\partial V}{\partial Z} \\[2mm] \dfrac{\partial W}{\partial X} & \dfrac{\partial W}{\partial Y} & \dfrac{\partial W}{\partial Z} \end{bmatrix} \cdot \begin{bmatrix} 0 \\ 0 \\ 1 \end{bmatrix} = \lambda_r \begin{bmatrix} 0 \\ 0 \\ 1 \end{bmatrix},$$

it must be

$$\frac{\partial U}{\partial Z} = 0 \tag{2.47}$$

$$\frac{\partial V}{\partial Z} = 0 \tag{2.48}$$

which means the velocity gradient tensor in the new *XYZ*-frame has the following form

$$\nabla \vec{V} = \begin{bmatrix} \dfrac{\partial U}{\partial X} & \dfrac{\partial U}{\partial Y} & 0 \\[2mm] \dfrac{\partial V}{\partial X} & \dfrac{\partial V}{\partial Y} & 0 \\[2mm] \dfrac{\partial W}{\partial X} & \dfrac{\partial W}{\partial Y} & \dfrac{\partial W}{\partial Z} \end{bmatrix} = \begin{bmatrix} \dfrac{\partial U}{\partial X} & \dfrac{\partial U}{\partial Y} & 0 \\[2mm] \dfrac{\partial V}{\partial X} & \dfrac{\partial V}{\partial Y} & 0 \\[2mm] \dfrac{\partial W}{\partial X} & \dfrac{\partial W}{\partial Y} & \lambda_r \end{bmatrix} \tag{2.49}$$

2.3.5 3D principal coordinates

Similar to the 2D case, the components of the velocity gradient tensor are related to the coordinate system, so a 3D principal coordinate should be determined when a velocity gradient tensor has one real and two conjugate complex eigenvalues.

Definition. A principal coordinate associated with the velocity gradient tensor, which has rotation at one point, is defined as a coordinate system with its Z-axis aligned with the real eigenvector $\vec{r}$ and with a 2D principal coordinate in the 2D plane perpendicular to the Z-axis.

In order to find the 3D principal coordinates of $\nabla \vec{v}$ at one point with rotation, the real eigenvector is obtained first and then a rotation matrix Q is used to make the Z-axis aligned with $\vec{r}$. After that, a second rotation in the plane perpendicular to the Z-axis is performed to get the 2D principal coordinates:

$$\nabla \vec{V}_\theta = P \nabla \vec{V} P^{-1}$$

where P is a rotation matrix around the Z-axis

$$P = \begin{bmatrix} cos\theta & sin\theta & 0 \\ -sin\theta & cos\theta & 0 \\ 0 & 0 & 1 \end{bmatrix}, \quad P^{-1} = \begin{bmatrix} cos\theta & -sin\theta & 0 \\ sin\theta & cos\theta & 0 \\ 0 & 0 & 1 \end{bmatrix} \tag{2.50}$$

As addressed in the previous sections, this will lead to 3D principal coordinates (X, Y, Z) in which the original velocity gradient tensor becomes

$$\nabla \vec{V}_\theta = P \nabla \vec{V} P^{-1} = P\left(Q \nabla \vec{v} Q^T\right) P^{-1} = \begin{bmatrix} \lambda_{cr} & -\dfrac{1}{2}R & 0 \\[2mm] \dfrac{1}{2}R + \varepsilon & \lambda_{cr} & 0 \\[2mm] \dfrac{\partial W}{\partial X} & \dfrac{\partial W}{\partial Y} & \lambda_r \end{bmatrix} \tag{2.51}$$

As discussed earlier, λ_r is the real eigenvalue, λ_{cr} is the real part of the conjugated complex eigenvalues.

In the 3D case,

$$\frac{\partial V}{\partial X_\theta} = \frac{1}{2}\left(\sqrt{\left(\vec{\omega}\cdot\vec{r}\right)^2 - 4\lambda_{ci}^2} + \left(\vec{\omega}\cdot\vec{r}\right)\right)$$

$$\frac{\partial U}{\partial Y_\theta} = \frac{1}{2}\left(\sqrt{\left(\vec{\omega}\cdot\vec{r}\right)^2 - 4\lambda_{ci}^2} - \left(\vec{\omega}\cdot\vec{r}\right)\right)$$

The rotation strength is

$$R = 2\cdot min\left\{\left|\frac{\partial U}{\partial Y_\theta}\right|, \left|\frac{\partial V}{\partial X_\theta}\right|\right\}$$

$$= \left(\vec{\omega}\cdot\vec{r}\right) - \sqrt{\left(\vec{\omega}\cdot\vec{r}\right)^2 - 4\lambda_{ci}^2}\ \left(\text{assuming } \left|\frac{\partial U}{\partial Y_\theta}\right| < \left|\frac{\partial V}{\partial X_\theta}\right|\right)$$

$$\vec{\omega} = \nabla \times \vec{v} \qquad (2.52)$$

Here, Q is called 3D rotation matrix and P is called 2D rotation matrix. P and Q can be determined by $\nabla\vec{v}$. However, the eigenvalues, λ_r, $\lambda_{cr} \pm i\lambda_{ci}$, and $\vec{\omega}$ are independent of coordinate rotation. Actually, the explicit form of the rotation matrix P and Q is not necessary except for $\frac{\partial W}{\partial X}$ and $\frac{\partial W}{\partial Y}$ in the principal coordinates.

So, a unique principal coordinate is obtained and a unique 3×3 velocity gradient tensor (principal tensor) in the principal coordinates is

$$\nabla\vec{V}_\theta = \begin{bmatrix} \lambda_{cr} & -\frac{1}{2}R & 0 \\ \frac{1}{2}R + \varepsilon & \lambda_{cr} & 0 \\ \xi & \eta & \lambda_r \end{bmatrix} \qquad (2.53)$$

2.4 Summary

The elements of any velocity gradient tensor is dependent on coordinate systems. However, there is a coordinate system determined by the velocity gradient tensor with one axis as the real eigenvector and two other diagonal elements are equal to each other. In such a coordinate system, a unique velocity tensor, which is called principal velocity gradient tensor, can be obtained.

Liutex—a new mathematical definition of fluid rotation

3.1 Short review of vortex definition

It is no doubt of the ubiquitousness of vortices in nature. For example, Table 3.1 presents the size range of a variety of vortices, which clearly demonstrates that the vortex can occur from quantum scale to galaxy scale. Intuitively, one can easily perceive vortices as the rotational/swirling motion of the fluids (Fig. 3.1). However, a precise and rational definition of vortex is deceptively complicated and remains an open issue for a long time. The lack of a consensus on the vortex definition has caused considerable confusions in visualizing, analyzing, and understanding the vortical structures, their evolution, and the interaction in complex vortical flows, especially in turbulence. The previous efforts to find an accurate definition for vortex have been briefly described in Chapter 1. However, a proper definition had not been achieved until the concept of Liutex was proposed by Liu et al. in 2018.

The main idea of Liutex is to extract the rigid rotation part from the whole fluid motion. This idea originated from Liu et al. (2014, 2016) and the idea of vorticity decomposition to rigid rotation and antisymmetric shear has been conveyed by Liu et al. in 2016. Note that Kolář (2007) has similar idea to take shear out from vorticity (triple decomposition). These ideas eventually become a reality only after Liu et al. gave the definition of Rortex, which is later changed to "Liutex".

From the very beginning, Liutex is a vector representing rigid rotation extracted from the fluid motion. Liutex has its corresponding scalar and tenser forms as well. It is a new defined physical quantity to describe fluid rotation or vortex. Although Liutex is not exactly equivalent to vortex, they are conjugated, or in other words they coexist. No vortex implies no Liutex and vice versa. However, the presence of vorticity does not necessarily mean the presence of vortex, such as the laminar boundary layer on a flat plate. Therefore, Liutex is extremely important to define vortex and is extremely important to turbulence research as it is a mathematical definition for fluid rotation. This is the first time for scientists to find a unique physical quantity to describe fluid rotation, or vortex, after around two-century-long efforts to find an appropriate definition for vortex.

Liutex and Its Applications in Turbulence Research. https://doi.org/10.1016/B978-0-12-819023-4.00008-2
Copyright © 2021 Elsevier Inc. All rights reserved.

Table 3.1 The size range of a variety of vortices.

Vortex type	Scale
Quantum vortex in liquid helium	1–8 μm
Small vortex in turbulence	~0.1 cm
Vortex induced by insect	0.1~10 cm
Vortex in the tide	1–10 m
Vortex ring of volcanic eruption	100~1000 m
Hurricane	100~2000 km
Ocean circulation	2000~5000 km
Saturn's ring	282,000 km
Galaxy vortex	Light years

3.2 Definition of Liutex

Since the rigid rotation must have local rotation axis and local angular speed, Liutex must be defined as a vector with both direction and magnitude. Of course, it should have its corresponding scalar (magnitude) and tensor forms.

3.2.1 Four principles

To reasonably define Liutex, four principles are followed:

(1) Local. Although a vortex is regarded as rotation of a group of fluid particles, the presence of viscosity in real flows leads to the continuity of the kinematic

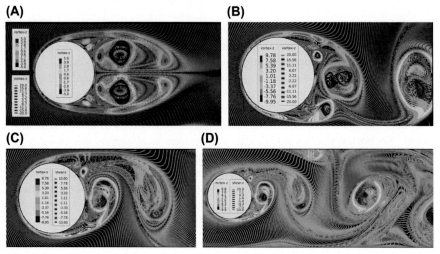

(A) **(B)**

(C) **(D)**

FIGURE 3.1

Direct numerical simulation for Karman vortex street.

features of the flow field, and numerous studies have suggested that the cores of vortical structures in turbulent flows are well localized in space. Nonlocality usually implies much more complexity in computation.

(2) Galilean Invariant. It means that the definition is the same in all inertial co-ordinate frames. This principle is followed by almost all Eulerian vortex identification criteria.

(3) Unique. The description of the local rigidly rotation must be accurate and unique. It is inappropriate to use two or more different quantities to represent fluid rotation. It requires the exclusion of the contamination by shearing.

(4) Systematical. The definition will provide a scalar giving the strength of the rigid rotation, a vector indicating both the rotation axis and rotation strength, and a tensor representing the rigid rotation part of the velocity gradient tensor.

3.2.2 Local rotation axis

Definition. A local rotation axis is defined as the direction of $\vec{r}$ where $d\vec{v} = \alpha d\vec{r}$.

This means the fluid can rotate around the axis, but the rotation axis cannot rotate by itself or $d\vec{v_n} = 0$, which leads the fluid motion to become stretching (compression) only along the local rotation axis (Fig. 3.2).

Theorem 3.1. The local rotation axis is the real eigenvector of $\nabla \vec{v}$.
Proof. According to the definition of the velocity gradient tensor,

$$d\vec{v} = \nabla \vec{v} \cdot d\vec{r}.$$

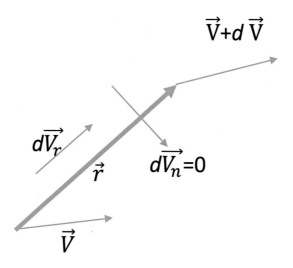

FIGURE 3.2

Local rotation axis.

Combined with the definition of local rotation axis, it must be

$$d\vec{v} = \nabla\vec{v}\cdot d\vec{r} = \alpha d\vec{r} \tag{3.1}$$

which means α is the real eigenvalue of $\nabla\vec{v}$ and $\vec{r}$ is the corresponding real eigenvector.

3.2.3 Definition of Liutex

Definition. Liutex is defined as a rigid rotation part of fluid motion

$$\vec{R} = R\vec{r} \text{ and } \vec{\omega}\cdot\vec{r} > 0 \tag{3.2}$$

$$R = (\vec{\omega} \cdot \vec{r}) - \sqrt{(\vec{\omega}\cdot\vec{r})^2 - 4\lambda_{ci}^2} \tag{3.3}$$

Here, $\vec{r}$ is the real eigenvalue of $\nabla\vec{v}$ and the condition $\vec{\omega}\cdot\vec{r} > 0$ is used to keep the definition unique and consistent when the fluid motion is pure rotation. Here, $\vec{\omega} = \nabla \times \vec{v}$ is the vorticity and λ_{ci} is the imaginary part of the conjugate complex eigenvalues of $\nabla\vec{v}$. According to Chapter 2, in the principal coordinate,

$$\nabla\vec{V}_\theta = \begin{bmatrix} 0 & -\frac{1}{2}R & 0 \\ \frac{1}{2}R & 0 & 0 \\ 0 & 0 & 0 \end{bmatrix} + \begin{bmatrix} \lambda_{cr} & 0 & 0 \\ \varepsilon & \lambda_{cr} & 0 \\ \xi & \eta & \lambda_r \end{bmatrix} = \boldsymbol{R} + \boldsymbol{NR}.$$

The physical meaning of Liutex is clear that Liutex is a vector, the direction of Liutex is the local rotation axis, and the magnitude of Liutex is twice the angular speed of local rigid rotation as shown by R.

3.3 Revisit of Helmholtz velocity decomposition

3.3.1 Helmholtz velocity decomposition and Cauchy—Stokes tensor decomposition

Helmholtz's theorem is known as the fundamental theorem of vector calculus, which states that any sufficiently smooth, rapidly decaying vector field in three dimensions can be resolved into the sum of an irrotational (curl-free) vector field and a solenoidal (divergence-free) vector field:

For a smooth vector function $\vec{F}(\vec{r})$, which vanishes faster than $\frac{1}{r}$ as $r \to \infty$, $\vec{F}(\vec{r})$ can be decomposed to two parts:

$$\vec{F}(\vec{r}) = -\nabla\Phi + \nabla \times \vec{A} \tag{3.4}$$

where $\Phi = \frac{1}{4\pi}\int_\tau \frac{D\left(\vec{r}\right)}{\delta r}d\tau'$ and $\vec{A} = \frac{1}{4\pi}\int_\tau \frac{\vec{C}\left(\vec{r}\right)}{\delta r}d\tau'$, $\delta r = \vec{r} - \vec{r}'$, $D = \nabla\cdot\vec{F}, \vec{C} = \nabla \times \vec{F}$.

As an irrotational vector field has a scalar potential and a solenoidal vector field has a vector potential, the Helmholtz decomposition states that a vector field (satisfying appropriate smoothness and decay conditions) can be decomposed as the sum of the form $\vec{F} = -\nabla\Phi + \nabla \times \vec{A}$, where Φ is a scalar field, called scalar potential, and $\vec{A}$ is a vector field, called a vector potential.

According to Helmholtz, a fluid particle motion can be decomposed to three parts: translation, deformation, and rotation (Fig. 3.3).

However, from the Cauchy–Stokes decomposition,

$$d\vec{v} = \nabla\vec{v}\cdot d\vec{r} = (A + B)\cdot d\vec{r} = A\cdot d\vec{r} + (\nabla \times \vec{v}) \times d\vec{r}. \tag{3.5}$$

It is equivalent to Helmholtz decomposition. The second part is traditionally called rotational part and the first part is called deformation part. People recognize

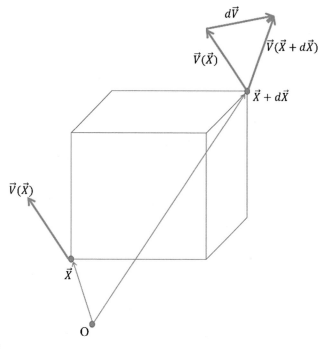

FIGURE 3.3

Velocity increment.

that the nonrotational (actually no-curl) vector field has potential and the rotational vector has no source. Therefore, people just use the Cauchy—Stokes decomposition to replace Helmholtz decomposition.

According to Liu's opinion (2014, 2016), the first part is potential, which is mistakenly called nonrotational, and the second part is mistakenly called rotational or solenoidal (divergence free). However, the real physical meaning of $\nabla \times \vec{v}$ is the curl of velocity field, which does not mean fluid rotation since fluids is different from solid. Fluid rotation cannot be measured by vorticity. Fluid can rotate only when vorticity overtakes fluid deformation. A new concept for fluid rotation must be introduced to measure the fluid rotation, instead of vorticity. As discussed in Chapter 2, $\nabla \times \vec{v}$ is extremely large in the laminar boundary layer near the wall and the channel flow near the wall, but no fluid rotation is found. These clearly show that $\nabla \times \vec{v} \neq 0$ does not mean the flow is really rotational. This is a misunderstanding for centuries since Helmholtz (1858).

3.3.2 Identification of fluid rotation

As addressed earlier, vorticity cannot be used to judge if the fluid is rotational or nonrotational and the magnitude of vorticity cannot be used to measure the strength of fluid rotation. Liutex is a rigid rotational part extracted out from the complete fluid motion. Liutex is a new physical quantity, which can be used to judge if the fluid is rotational or is nonrotational and can be used to measure the strength of fluid rotation as the Liutex magnitude is twice the fluid angular speed.

(1) From the Liutex definition:

$$\vec{R} = R\vec{r} \text{ and } \vec{\omega} \cdot \vec{r} > 0$$

$$R = (\vec{\omega} \cdot \vec{r}) - \sqrt{(\vec{\omega} \cdot \vec{r})^2 - 4\lambda_{ci}^2}$$

$R(x,y,z) > 0$: fluid is rotational at that point.
$R(x,y,z) = 0$: fluid is nonrotational even if $|\vec{\omega}| > 0$.

(2) From critical point theory:

According to Chong and Perry (1990), a point is rotational if and only if the velocity gradient tensor has two conjugate complex eigenvalues: $\lambda = \lambda_r \pm i\lambda_{ci}$ and $\lambda_{ci} \neq 0$.

Qualitatively, these two criteria are essentially the same. $\lambda_{ci} \neq 0$ implies $R > 0$. On the other hand, if $R > 0$, it can be easily obtained that $\lambda_{ci} \neq 0$ from Eq. (3.3). And if $\lambda_{ci} = 0$, $R = 0$.

(3) From the physical perspective:

Consider a 2D plane. In the principal coordinates:

$$\nabla \vec{V} = \begin{bmatrix} \lambda_{cr} & \dfrac{\partial U}{\partial Y} \\ \dfrac{\partial V}{\partial X} & \lambda_{cr} \end{bmatrix} \tag{3.6}$$

If there exists fluid rotation, it must be $\frac{\partial U}{\partial Y}\cdot\frac{\partial V}{\partial X} < 0$. If $\frac{\partial U}{\partial Y}\cdot\frac{\partial V}{\partial X} = 0$, the fluid has shear without rotation. If $\frac{\partial U}{\partial Y}\cdot\frac{\partial V}{\partial X} > 0$, fluid has deformation without rotation (Fig. 3.4). Comparing with the eigenvalue equation:

$(\lambda_{cr} - \lambda)^2 - \frac{\partial U}{\partial Y}\cdot\frac{\partial V}{\partial X} = 0$, the fluid rotation requires $\frac{\partial U}{\partial Y}\cdot\frac{\partial V}{\partial X} < 0$, which leads to two complex eigenvalues.

From the earlier analyses, the Liutex definition is consistent with the physical perspective and critical point theory and provides the same conclusion on fluid rotation. Therefore, the introduction of Liutex is appropriate to represent fluid rotation. Actually, Liutex is zero at all nonrotational points and therefore Liutex is a new physical quantity defined for all points in the whole flow field.

3.3.3 Vorticity cannot represent fluid rotation

ω_Z can be larger than zero even for points where $\frac{\partial V}{\partial X} = 0$ or $\frac{\partial U}{\partial Y}\cdot\frac{\partial V}{\partial X} > 0$ and no fluid rotation exists. Therefore, vorticity cannot be used to judge fluid rotation or

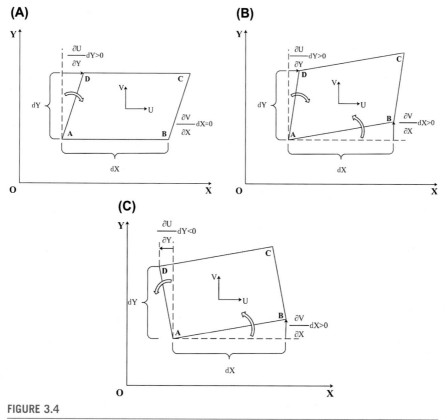

FIGURE 3.4

Fluid rotation (A) no rotation $\frac{\partial U}{\partial Y}\cdot\frac{\partial V}{\partial X} = 0$, (B) no rotation $\frac{\partial U}{\partial Y}\cdot\frac{\partial V}{\partial X} > 0$, (C) rotation $\frac{\partial U}{\partial Y}\cdot\frac{\partial V}{\partial X} < 0$.

nonrotation. Both Helmholtz fluid motion decomposition and the Cauchy–Stokes decomposition indicate vorticity as fluid rotation, which is fundamentally incorrect and cannot be used for fluid motion decomposition as the foundation of fluid kinetics. The vorticity decomposition will be elaborated in Chapter 5. In Section 3.6, the correlation coefficient between Liutex and vorticity will provide another evidence that vorticity cannot represent fluid rotation.

3.4 Fluid rotation in 3D coordinates

3.4.1 Determination of the direction of the possible rotational axis

As discussed before, if the point is located inside a vortex, the velocity gradient tensor must have two complex eigenvalues and one real eigenvalue. If the original coordinate system is rotated so that the Z-axis of the XYZ-frame is aligned with the eigenvector $\vec{r}$ corresponding to the real eigenvalue, there is a relation between these two coordinate systems:

$$\nabla \vec{V} = Q \nabla \vec{v} Q^{-1} \tag{3.7}$$

$$\nabla \vec{v} = \begin{bmatrix} \dfrac{\partial u}{\partial x} & \dfrac{\partial u}{\partial y} & \dfrac{\partial u}{\partial z} \\[2ex] \dfrac{\partial v}{\partial x} & \dfrac{\partial v}{\partial y} & \dfrac{\partial v}{\partial z} \\[2ex] \dfrac{\partial w}{\partial x} & \dfrac{\partial w}{\partial y} & \dfrac{\partial w}{\partial z} \end{bmatrix}, \quad \nabla \vec{V} = \begin{bmatrix} \dfrac{\partial U}{\partial X} & \dfrac{\partial U}{\partial Y} & \dfrac{\partial U}{\partial Z} \\[2ex] \dfrac{\partial V}{\partial X} & \dfrac{\partial V}{\partial Y} & \dfrac{\partial V}{\partial Z} \\[2ex] \dfrac{\partial W}{\partial X} & \dfrac{\partial W}{\partial Y} & \dfrac{\partial W}{\partial Z} \end{bmatrix}$$

where Q is the rotation matrix from the xyz-frame to the XYZ-frame. The necessary condition for a velocity gradient tensor to have rotation around the Z-axis only is

$$\frac{\partial U}{\partial Z} = 0 \tag{3.8}$$

$$\frac{\partial V}{\partial Z} = 0 \tag{3.9}$$

This means the XYZ-frame has the following form

$$\nabla \vec{V} = \begin{bmatrix} \dfrac{\partial U}{\partial X} & \dfrac{\partial U}{\partial Y} & 0 \\[2ex] \dfrac{\partial V}{\partial X} & \dfrac{\partial V}{\partial Y} & 0 \\[2ex] \dfrac{\partial W}{\partial X} & \dfrac{\partial W}{\partial Y} & \dfrac{\partial W}{\partial Z} \end{bmatrix} \tag{3.10}$$

Theorem 3.2. If a tensor has $\frac{\partial U}{\partial Z} = 0$, there is no rotation around the Y-axis.
Proof.

$$
\begin{bmatrix}
\dfrac{\partial U}{\partial X} & \dfrac{\partial U}{\partial Y} & 0 \\[2mm]
\dfrac{\partial V}{\partial X} & \dfrac{\partial V}{\partial Y} & \dfrac{\partial V}{\partial Z} \\[2mm]
\dfrac{\partial W}{\partial X} & \dfrac{\partial W}{\partial Y} & \dfrac{\partial W}{\partial Z}
\end{bmatrix}
=
\begin{bmatrix}
\dfrac{\partial U}{\partial X} & \dfrac{1}{2}\left(\dfrac{\partial U}{\partial Y}+\dfrac{\partial V}{\partial X}\right) & \dfrac{1}{2}\dfrac{\partial W}{\partial X} \\[3mm]
\dfrac{1}{2}\left(\dfrac{\partial V}{\partial X}+\dfrac{\partial U}{\partial Y}\right) & \dfrac{\partial V}{\partial Y} & \dfrac{1}{2}\left(\dfrac{\partial W}{\partial Y}+\dfrac{\partial V}{\partial Z}\right) \\[3mm]
\dfrac{1}{2}\dfrac{\partial W}{\partial X} & \dfrac{1}{2}\left(\dfrac{\partial W}{\partial Y}+\dfrac{\partial V}{\partial Z}\right) & \dfrac{\partial W}{\partial Z}
\end{bmatrix}
$$

$$
+
\begin{bmatrix}
0 & -\dfrac{1}{2}\left(\dfrac{\partial V}{\partial X}-\dfrac{\partial U}{\partial Y}\right) & -\dfrac{1}{2}\dfrac{\partial W}{\partial X} \\[3mm]
\dfrac{1}{2}\left(\dfrac{\partial V}{\partial X}-\dfrac{\partial U}{\partial Y}\right) & 0 & -\dfrac{1}{2}\left(\dfrac{\partial W}{\partial Y}-\dfrac{\partial V}{\partial Z}\right) \\[3mm]
\dfrac{1}{2}\dfrac{\partial W}{\partial X} & \dfrac{1}{2}\left(\dfrac{\partial W}{\partial Y}-\dfrac{\partial V}{\partial Z}\right) & 0
\end{bmatrix}
\tag{3.11}
$$

$$
B =
\begin{bmatrix}
0 & -\dfrac{1}{2}\left(\dfrac{\partial V}{\partial X}-\dfrac{\partial U}{\partial Y}\right) & -\dfrac{1}{2}\dfrac{\partial W}{\partial X} \\[3mm]
\dfrac{1}{2}\left(\dfrac{\partial V}{\partial X}-\dfrac{\partial U}{\partial Y}\right) & 0 & -\dfrac{1}{2}\left(\dfrac{\partial W}{\partial Y}-\dfrac{\partial V}{\partial Z}\right) \\[3mm]
\dfrac{1}{2}\dfrac{\partial W}{\partial X} & \dfrac{1}{2}\left(\dfrac{\partial W}{\partial Y}-\dfrac{\partial V}{\partial Z}\right) & 0
\end{bmatrix}
$$

$$
=
\begin{bmatrix}
0 & 0 & -\dfrac{1}{2}\dfrac{\partial W}{\partial X} \\[3mm]
0 & 0 & 0 \\[3mm]
\dfrac{1}{2}\dfrac{\partial W}{\partial X} & 0 & 0
\end{bmatrix}
+
\begin{bmatrix}
0 & -\dfrac{1}{2}\left(\dfrac{\partial V}{\partial X}-\dfrac{\partial U}{\partial Y}\right) & 0 \\[3mm]
\dfrac{1}{2}\left(\dfrac{\partial V}{\partial X}-\dfrac{\partial U}{\partial Y}\right) & 0 & 0 \\[3mm]
0 & 0 & 0
\end{bmatrix}
$$

$$
+
\begin{bmatrix}
0 & 0 & 0 \\[3mm]
0 & 0 & -\dfrac{1}{2}\left(\dfrac{\partial W}{\partial Y}-\dfrac{\partial V}{\partial Z}\right) \\[3mm]
0 & \dfrac{1}{2}\left(\dfrac{\partial W}{\partial Y}-\dfrac{\partial V}{\partial Z}\right) & 0
\end{bmatrix}
$$

$$
= C + D + E
\tag{3.12}
$$

First consider tensor C. For the Y-axis, $\frac{\partial U}{\partial Z}\cdot\frac{\partial W}{\partial X} = 0\cdot\frac{\partial W}{\partial X} = 0$. According to the fluid rotation criterion, there is no rotation around the Y-axis for tensor C. Applying the rotation criterion to tensor D and tensor E, there is no rotation around the Y-axis for tensor $\boldsymbol{D}$ and tensor $\boldsymbol{E}$. Therefore, for tensor $\boldsymbol{B}$, there is no rotation around the Y-axis. Similarly, if a tensor has $\frac{\partial V}{\partial Z} = 0$, there is no rotation around the X-axis. Also, if a tensor has $\frac{\partial U}{\partial Y} = 0$, there is no rotation around the Z-axis.

Corollary 3.3. If a tensor has $\frac{\partial U}{\partial Z} = 0$ and $\frac{\partial V}{\partial Z} = 0$, there is no rotation around the Y-axis and the X-axis. The only possible rotational axis is the Z-axis.

Proof.

$$
\begin{bmatrix}
\dfrac{\partial U}{\partial X} & \dfrac{\partial U}{\partial Y} & 0 \\[2mm]
\dfrac{\partial V}{\partial X} & \dfrac{\partial V}{\partial Y} & 0 \\[2mm]
\dfrac{\partial W}{\partial X} & \dfrac{\partial W}{\partial Y} & \dfrac{\partial W}{\partial Z}
\end{bmatrix}
=
\begin{bmatrix}
\dfrac{\partial U}{\partial X} & \dfrac{1}{2}\left(\dfrac{\partial V}{\partial X}+\dfrac{\partial U}{\partial Y}\right) & \dfrac{1}{2}\dfrac{\partial W}{\partial X} \\[3mm]
\dfrac{1}{2}\left(\dfrac{\partial V}{\partial X}+\dfrac{\partial U}{\partial Y}\right) & \dfrac{\partial V}{\partial Y} & \dfrac{1}{2}\dfrac{\partial W}{\partial Y} \\[3mm]
\dfrac{1}{2}\dfrac{\partial W}{\partial X} & \dfrac{1}{2}\dfrac{\partial W}{\partial Y} & \dfrac{\partial W}{\partial Z}
\end{bmatrix}
$$

$$
+
\begin{bmatrix}
0 & -\dfrac{1}{2}\left(\dfrac{\partial V}{\partial X}-\dfrac{\partial U}{\partial Y}\right) & -\dfrac{1}{2}\dfrac{\partial W}{\partial X} \\[3mm]
\dfrac{1}{2}\left(\dfrac{\partial V}{\partial X}-\dfrac{\partial U}{\partial Y}\right) & 0 & -\dfrac{1}{2}\dfrac{\partial W}{\partial Y} \\[3mm]
\dfrac{1}{2}\dfrac{\partial W}{\partial X} & \dfrac{1}{2}\dfrac{\partial W}{\partial Y} & 0
\end{bmatrix}
$$

$$
= \boldsymbol{A} + \boldsymbol{B} \tag{3.13}
$$

$$
\boldsymbol{B} =
\begin{bmatrix}
0 & -\dfrac{1}{2}\left(\dfrac{\partial V}{\partial X}-\dfrac{\partial U}{\partial Y}\right) & -\dfrac{1}{2}\dfrac{\partial W}{\partial X} \\[3mm]
\dfrac{1}{2}\left(\dfrac{\partial V}{\partial X}-\dfrac{\partial U}{\partial Y}\right) & 0 & -\dfrac{1}{2}\dfrac{\partial W}{\partial Y} \\[3mm]
\dfrac{1}{2}\dfrac{\partial W}{\partial X} & \dfrac{1}{2}\dfrac{\partial W}{\partial Y} & 0
\end{bmatrix}
$$

$$
=
\begin{bmatrix}
0 & -\dfrac{1}{2}\left(\dfrac{\partial V}{\partial X}-\dfrac{\partial U}{\partial Y}\right) & 0 \\[3mm]
\dfrac{1}{2}\left(\dfrac{\partial V}{\partial X}-\dfrac{\partial U}{\partial Y}\right) & 0 & 0 \\[3mm]
0 & 0 & 0
\end{bmatrix}
+
\begin{bmatrix}
0 & 0 & -\dfrac{1}{2}\dfrac{\partial W}{\partial X} \\[3mm]
0 & 0 & 0 \\[3mm]
\dfrac{1}{2}\dfrac{\partial W}{\partial X} & 0 & 0
\end{bmatrix}
$$

$$
+
\begin{bmatrix}
0 & 0 & 0 \\[3mm]
0 & 0 & -\dfrac{1}{2}\dfrac{\partial W}{\partial Y} \\[3mm]
0 & \dfrac{1}{2}\dfrac{\partial W}{\partial Y} & 0
\end{bmatrix}
$$

$$
= \boldsymbol{C} + \boldsymbol{D} + \boldsymbol{E} \tag{3.14}
$$

There is only tensor C, which possibly has rotation while tensors D and E have no rotation.

3.4.2 Rotation strength in the 2D plane

Through the coordinate rotation Q, the possible rotational direction $\vec{r} = Z$ is obtained. But the rotation strength, which represents the angular speed of local rigid rotation, remains to be determined. This can be achieved by a second coordinate rotation in the XY-plane, which is called the coordinate rotation P. As known, vorticity, which can be written as $\frac{\partial V}{\partial X} - \frac{\partial U}{\partial Y}$ in a 2D flow, is a Galilean-invariant quantity, but $\frac{\partial V}{\partial X}$ or $\frac{\partial U}{\partial Y}$ itself is not invariant and will change with the rotation of the reference frame. When the XYZ-frame is rotated around the Z-axis through an angle θ, the new velocity gradient tensor is

$$\nabla \vec{V}_\theta = P \nabla \vec{V} P^{-1} \tag{3.15}$$

where P is the ratation matrix around the Z-axis

$$P = \begin{bmatrix} \cos\theta & \sin\theta & 0 \\ -\sin\theta & \cos\theta & 0 \\ 0 & 0 & 1 \end{bmatrix}, \quad P^{-1} = \begin{bmatrix} \cos\theta & -\sin\theta & 0 \\ \sin\theta & \cos\theta & 0 \\ 0 & 0 & 1 \end{bmatrix}$$

So,

$$\nabla \vec{V}_\theta = P \nabla \vec{V} P^{-1} = P \left(Q \nabla \vec{v} Q^{-1} \right) P^{-1} = \begin{bmatrix} \left.\frac{\partial U}{\partial X}\right|_\theta & \left.\frac{\partial U}{\partial Y}\right|_\theta & 0 \\ \left.\frac{\partial V}{\partial X}\right|_\theta & \left.\frac{\partial V}{\partial Y}\right|_\theta & 0 \\ \frac{\partial W}{\partial X} & \frac{\partial W}{\partial Y} & \frac{\partial W}{\partial Z} \end{bmatrix}$$

$$\left.\frac{\partial V}{\partial X}\right|_\theta = \alpha \sin(2\theta + \varphi) + \beta \tag{3.16}$$

$$\left.\frac{\partial U}{\partial Y}\right|_\theta = \alpha \sin(2\theta + \varphi) - \beta \tag{3.17}$$

where

$$\alpha = \frac{1}{2} \sqrt{ \left(\frac{\partial V}{\partial Y} - \frac{\partial U}{\partial X} \right)^2 + \left(\frac{\partial V}{\partial X} + \frac{\partial U}{\partial Y} \right)^2 } \tag{3.18}$$

$$\beta = \frac{1}{2} \left(\frac{\partial V}{\partial X} - \frac{\partial U}{\partial Y} \right) \tag{3.19}$$

and

$$\varphi = \begin{cases} a\cos\left(\dfrac{\frac{1}{2}\left(\frac{\partial V}{\partial Y}-\frac{\partial U}{\partial X}\right)}{\alpha}\right), & \dfrac{\partial V}{\partial X}+\dfrac{\partial U}{\partial Y}>0 \\[3ex] a\sin\left(\dfrac{\frac{1}{2}\left(\frac{\partial V}{\partial X}+\frac{\partial U}{\partial Y}\right)}{\alpha}\right), & \dfrac{\partial V}{\partial X}+\dfrac{\partial U}{\partial Y}<0,\ \dfrac{\partial V}{\partial Y}-\dfrac{\partial U}{\partial X}\geq 0 \\[3ex] a\sin\left(\dfrac{-\frac{1}{2}\left(\frac{\partial V}{\partial X}+\frac{\partial U}{\partial Y}\right)}{\alpha}\right)+\pi, & \dfrac{\partial V}{\partial Y}-\dfrac{\partial U}{\partial X}=0,\ \dfrac{\partial V}{\partial X}+\dfrac{\partial U}{\partial Y}<0 \end{cases} \tag{3.20}$$

(NOTE: If $\alpha=0$, $\frac{\partial V}{\partial Y}-\frac{\partial U}{\partial X}=0$, $\frac{\partial V}{\partial X}+\frac{\partial U}{\partial Y}=0$, $\frac{\partial V}{\partial X}=\beta$, $\frac{\partial U}{\partial Y}=-\beta$ for any θ, so φ is not needed.) Accordingly, $2\beta=\frac{\partial V}{\partial X}-\frac{\partial U}{\partial Y}$ represents the 2D vorticity in the XY-plane. According to the criterion for fluid rotation,

$$g_{z\theta}=-\left.\frac{\partial V}{\partial X}\right|_\theta\left.\frac{\partial U}{\partial Y}\right|_\theta=\beta^2-\alpha^2\sin^2(2\theta+\varphi)>0. \tag{3.21}$$

In order to satisfy this condition for all θ, the condition $\beta^2>\alpha^2$ should be satisfied for all θ. In order to decompose the tensor C to be a rigid rotation, it must be required to get $\min\{|\frac{\partial U}{\partial Y}|_\theta,\ |\frac{\partial V}{\partial X}|_\theta\}$ for all θ. Since α is always positive, if $\beta>0$, $\min\{|\frac{\partial U}{\partial Y}|_\theta,\ |\frac{\partial V}{\partial X}|_\theta\}=|\frac{\partial V}{\partial X}|_{\theta\min}=\beta-\alpha$ and the rigid rotation strength is $R=2\frac{\partial V}{\partial X}|_{\theta\min}=2(\beta-\alpha)$. For simplicity, $R=2\frac{\partial V}{\partial X}$ is used, which should be understood as $R=2\frac{\partial V}{\partial X}|_{\theta\min}$ after the P rotation. If $\beta<0$, $\min\{|\frac{\partial U}{\partial Y}|_\theta,\ |\frac{\partial V}{\partial X}|_\theta\}=\frac{\partial U}{\partial Y}|_{\theta\min}=-(\beta+\alpha)$ and then $R=2\frac{\partial U}{\partial Y}|_{\theta\min}$.

Note that $\nabla\vec{V}_\theta=P\nabla\vec{V}P^{-1}=P(Q\nabla\vec{v}Q^{-1})P^{-1}$

$$=\begin{bmatrix} \left.\frac{\partial U}{\partial X}\right|_\theta & \left.\frac{\partial U}{\partial Y}\right|_\theta & 0 \\[2ex] \left.\frac{\partial V}{\partial X}\right|_\theta & \left.\frac{\partial V}{\partial Y}\right|_\theta & 0 \\[2ex] \frac{\partial W}{\partial X} & \frac{\partial W}{\partial Y} & \frac{\partial W}{\partial Z} \end{bmatrix}=\begin{bmatrix} \lambda_{cr} & -\frac{1}{2}R & 0 \\[2ex] \frac{1}{2}R+\varepsilon & \lambda_{cr} & 0 \\[2ex] \frac{\partial W}{\partial X} & \frac{\partial W}{\partial Y} & \lambda_r \end{bmatrix}$$

$$R = 2\frac{\partial U}{\partial Y}\Big|_{\theta\min} = (\vec{\omega}\cdot\vec{r}) - \sqrt{(\vec{\omega}\cdot\vec{r})^2 - 4\lambda_{ci}^2}$$

They are consistent with the definition given by Eqs. (3.2) and (3.3).

Definition 3.6.   A Vortex is a connected area where $R > 0$.

3.5 Calculation procedures of Liutex

(1) Compute the velocity gradient tensor $\nabla\vec{v}$ in the original coordinate system;

(2) Calculate the real eigenvalue λ_r and the conjugate complex eigenvalues of the velocity gradient tensor $\nabla\vec{v}$ when the complex eigenvalues exist (the analytical expression has been presented in Chapter 2);

(3) Calculate the (normalized) real eigenvector $\vec{r} = [r_x, r_y, r_z]^T$ corresponding to the real eigenvalue λ_r (the analytical expression has been presented in Chapter 2);

(4) Compute Liutex $\vec{R}$ via Eqs. (3.2) and (3.3).

3.6 Comparison of Liutex and eigenvalue-based vortex identification criteria (second-generation methods)

3.6.1 Analytical relation and comparison between Liutex, Q criterion, and λ_{ci} criterion

The analytical relation between R and λ_{ci} in the principal coordinate is given by

$$\lambda_{ci} = \sqrt{\frac{R}{2}\left(\frac{R}{2} + \varepsilon\right)} \tag{3.22}$$

Since ϵ represents shearing, Eq. (3.22) implies the λ_{ci} criterion will consider the shearing ϵ as part of rotation. In the principal coordinate,

$$\nabla\vec{V} = \begin{bmatrix} \lambda_{cr} & -\frac{1}{2}R & 0 \\ \frac{1}{2}R + \varepsilon_A & \lambda_{cr} & 0 \\ \xi_A & \eta_A & \lambda_r \end{bmatrix}$$

The Q criterion has been discussed in Chapter 1. The analytical relation between R and Q can be obtained as

$$Q = \lambda_1\lambda_2 + \lambda_2\lambda_3 + \lambda_3\lambda_1$$
$$= (\lambda_{cr} + i\lambda_{ci})(\lambda_{cr} - i\lambda_{ci}) + (\lambda_{cr} - i\lambda_{ci})\lambda_r + \lambda_r(\lambda_{cr} + i\lambda_{ci})$$

$$= \lambda_{cr}^2 + \lambda_{ci}^2 + 2\lambda_{cr}\lambda_r$$

$$= \lambda_{cr}^2 + \frac{R}{2}\left(\frac{R}{2} + \varepsilon\right) + 2\lambda_{cr}\lambda_r \tag{3.23}$$

From the UTA R-NR decomposition (Chapter 5), another relation between R and Q can be written as

$$Q = 2(R/2 + \varepsilon/2)^2 - \left[2\lambda_{cr}^2 + \lambda_r^2\right]$$

The aforementioned expressions mean that Q considers shearing ϵ and the stretching/compression λ_{cr}, λ_r as part of rotation or contaminated by shear and stretching/compression.

Since λ_{ci} and Q are eigenvalue-based, the same eigenvalues always yield the same values of λ_{ci} and Q. In contrast, Liutex is not exclusively determined by eigenvalues. Assume that two velocity gradient tensors $\nabla\vec{v}\big|_A$ and $\nabla\vec{v}\big|_B$ have the same eigenvalues $\lambda_{cr} + i\lambda_{ci}$, $\lambda_{cr} - i\lambda_{ci}$, and λ_r but different real eigenvectors. Through two appropriate rotation matrices $Q\big|_A$ and $P\big|_A$, $\nabla\vec{V}_{\theta\min}\big|_A$ can be written as

$$\nabla\vec{V}_{\theta\min}\big|_A = P\big|_A Q\big|_A \nabla\vec{v}\big|_A (Q\big|_A)^{\mathrm{T}} (P\big|_A)^{\mathrm{T}} = \begin{bmatrix} \lambda_{cr} & -\frac{1}{2}R_A & 0 \\ \frac{1}{2}R_A + \varepsilon_A & \lambda_{cr} & 0 \\ \xi_A & \eta_A & \lambda_r \end{bmatrix}$$

$$= \begin{bmatrix} 0 & -\frac{1}{2}R_A & 0 \\ \frac{1}{2}R_A & 0 & 0 \\ 0 & 0 & 0 \end{bmatrix} + \begin{bmatrix} \lambda_{cr} & 0 & 0 \\ \varepsilon_A & \lambda_{cr} & 0 \\ \xi_A & \eta_A & \lambda_r \end{bmatrix} \tag{3.24}$$

Similarly, through two appropriate rotation matrices $Q\big|_B$ and $P\big|_B$, $\nabla\vec{V}_{\theta\min}\big|_B$ can be written as

$$\nabla\vec{V}_{\theta\min}\big|_B = P\big|_B Q\big|_B \nabla\vec{v}\big|_B (Q\big|_B)^{\mathrm{T}} (P\big|_B)^{\mathrm{T}} = \begin{bmatrix} \lambda_{cr} & -\frac{1}{2}R_B & 0 \\ \frac{1}{2}R_B + \varepsilon_B & \lambda_{cr} & 0 \\ \xi_B & \eta_B & \lambda_r \end{bmatrix}$$

$$= \begin{bmatrix} 0 & -\frac{1}{2}R_B & 0 \\ \frac{1}{2}R_B & 0 & 0 \\ 0 & 0 & 0 \end{bmatrix} + \begin{bmatrix} \lambda_{cr} & 0 & 0 \\ \varepsilon_B & \lambda_{cr} & 0 \\ \xi_B & \eta_B & \lambda_r \end{bmatrix} \tag{3.25}$$

Since the eigenvalues are identical,

$$Q|_A = Q|_B \tag{3.26}$$

$$\lambda_{ci}|_A = \lambda_{ci}|_B \tag{3.27}$$

and the following conditions

$$R_A(R_A + \varepsilon_A) = 4\lambda_{ci}^2 \tag{3.28}$$

$$R_B(R_B + \varepsilon_B) = 4\lambda_{ci}^2 \tag{3.29}$$

However, there is no further relation of R_A and R_B, since four unknowns, i.e., R_A, R_B, ε_A, ε_B cannot be uniquely determined by two Eqs. (3.28) and (3.29). Therefore, in general, the rotational strength $R_A \neq R_B$ unless $\omega_A = \omega_B$. According to the Liutex definition, Liutex magnitude is uniquely determined by λ_{ci} and $\vec{\omega} \cdot \vec{r}$, which is clearly related to vorticity and eigenvector. Consider a specific case. Two matrices

$$\nabla \vec{V}_{\theta min}|_A = \begin{bmatrix} 1 & -2 & 0 \\ 2 & 1 & 0 \\ \xi_A & \eta_A & 2 \end{bmatrix} = \begin{bmatrix} 0 & -2 & 0 \\ 2 & 0 & 0 \\ 0 & 0 & 0 \end{bmatrix} + \begin{bmatrix} 1 & 0 & 0 \\ 0 & 1 & 0 \\ \xi_A & \eta_A & 2 \end{bmatrix} \tag{3.30}$$

$$\nabla \vec{V}_{\theta min}|_B = \begin{bmatrix} 1 & -1 & 0 \\ 4 & 1 & 0 \\ \xi_B & \eta_B & 2 \end{bmatrix} = \begin{bmatrix} 0 & -1 & 0 \\ 1 & 0 & 0 \\ 0 & 0 & 0 \end{bmatrix} + \begin{bmatrix} 1 & 0 & 0 \\ 3 & 1 & 0 \\ \xi_B & \eta_B & 2 \end{bmatrix} \tag{3.31}$$

have the same eigenvalues $1 + 2i$, $1 - 2i$ and 2. Certainly, $Q|_A = Q|_B = 9$ and $\lambda_{ci}|_A = \lambda_{ci}|_B = 2$. But the rotational strengths are quite different: $R_A = 4$ and $R_B = 2$.

From Eq. (3.21), it can be found that the shearing effect ϵ always exists in the imaginary part of the complex eigenvalues. Therefore, as long as eigenvalue-based criteria are dependent on the complex eigenvalues, they will be inevitably contaminated by shearing. The earlier analyses indicate the shearing effect on λ_{ci} and Q, respectively. The investigation of this contamination in some simple examples and realistic flows will be given in the following.

3.6.2 Comparison for simple examples

(1) Rigid rotation

First, consider a 2D rigid rotation. The velocity in the polar coordinate system can be expressed as

$$\begin{cases} v_r = \omega r \\ v_\theta = 0 \end{cases} \tag{3.32}$$

Here, ω is a constant and represents the angular velocity. Assume $\omega > 0$, which means the flow field is rotating in counter-clockwise order. Then, the velocity in the Cartesian coordinate system will be written as

$$\begin{cases} u = -\omega y \\ v = \omega x \end{cases} \tag{3.33}$$

In this simple case, Liutex, Q, and λ_{ci} can be analytically expressed as

$$R = 2\omega \tag{3.34}$$

$$Q = \omega^2 \tag{3.35}$$

$$\lambda_{ci} = \omega \tag{3.36}$$

It can be found that Liutex is exactly equal to vorticity.

Now consider the superposition of a prograde shearing motion, which is given by

$$\begin{cases} u = 0 \\ v = \sigma x, \quad \sigma > 0 \end{cases} \tag{3.37}$$

$\sigma > 0$ implies that the shearing motion is consistent with the clockwise rigid rotation. The velocity becomes

$$\begin{cases} u = -\omega y \\ v = (\omega + \sigma)x \end{cases} \tag{3.38}$$

It can be easily verified that Eq. (3.37) fulfills the 2D vorticity equations. The velocity gradient tensor is decomposed to

$$\begin{bmatrix} 0 & -\omega & 0 \\ \omega+\sigma & 0 & 0 \\ 0 & 0 & 0 \end{bmatrix} = \begin{bmatrix} 0 & -\omega & 0 \\ \omega & 0 & 0 \\ 0 & 0 & 0 \end{bmatrix} + \begin{bmatrix} 0 & 0 & 0 \\ \sigma & 0 & 0 \\ 0 & 0 & 0 \end{bmatrix} \tag{3.39}$$

which exactly presents the rigidly rotational part and the shearing part. The explicit expressions of Liutex, Q, and λ_{ci} are given by

$$R = 2\omega \tag{3.40}$$

$$Q = \omega(\omega + \sigma) \tag{3.41}$$

$$\lambda_{ci} = \sqrt{\omega(\omega + \sigma)} \tag{3.42}$$

It is expected that in this case the rigidly rotational part of fluids should not be affected by the shear motion. Only Liutex remains the same as no-shearing case and provides the precise rigidly rotational strength as expected, whereas Q and λ_{ci} are altered by the shearing effect σ. Obviously, the stronger shearing will result in the larger alteration of Q and λ_{ci}, as shown in Fig. 3.5. In Fig. 3.5, the shearing effect σ is normalized by the angular velocity ω. With the increase of the shearing effect σ, Q and λ_{ci} both indicate significant deviations from the values in the no-shearing case, which implies these criteria are prone to contamination by shearing and cannot reasonably represent the local rotation. In contrast, Liutex excludes the shearing effect and remains exactly twice the angular velocity ω as the no-shearing case. Note that here, ω is not vorticity but represents the fluid angular speed here.

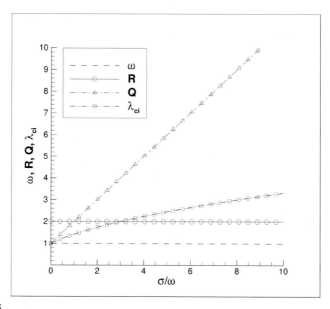

FIGURE 3.5

R (Liutex), Q, and λ_{ci} as functions of $\boldsymbol{\sigma/\omega}$ for 2D rigid rotation superposed by a prograde shearing motion.

Reproduced from Y. Gao and C. Liu, Rortex and comparison with eigenvalue-based vortex identification criteria, Phys. Fluids 30 (085107), 2018, with the permission of AIP Publishing.

(2) Burgers vortex

Here the Burgers vortex is examined. This vortex has been widely used for modeling fine scales of turbulence. The Burgers vortex is an exact steady solution of the Navier–Stokes equation, where the radial viscous diffusion of vorticity is dynamically balanced by vortex stretching due to an axisymmetric strain. The velocity components in the cylindrical coordinates for a Burgers vortex can be written as

$$v_r = -\xi r$$

$$v_\theta = \frac{\Gamma}{2\pi r}\left(1 - e^{\frac{-r^2\xi}{2\nu}}\right)$$

$$v_z = 2\xi z \tag{3.43}$$

where Γ is the circulation, ξ the axisymmetric strain rate, and ν the kinematic viscosity. The Reynolds number for the vortex is defined as $\mathrm{Re} = \Gamma/(2\pi\nu)$. The velocity in the Cartesian coordinate system will be written as

$$u = -\xi x - \frac{\Gamma}{2\pi r^2}\left(1 - e^{\frac{-r^2\xi}{2\nu}}\right)y$$

$$v = -\xi y + \frac{\Gamma}{2\pi r^2}\left(1 - e^{\frac{-r^2\xi}{2\nu}}\right)x$$

$$w = 2\xi z \qquad\qquad (3.44)$$

The analytical expressions of Liutex, Q, and λ_{ci} are given by

$$R = 2\mathrm{Re}\xi\zeta$$

$$Q = \xi^2\left[\mathrm{Re}^2\zeta(\zeta + \varepsilon) - 3\right]$$

$$\lambda_{ci} = \mathrm{Re}\xi\sqrt{\zeta(\zeta + \varepsilon)} \qquad\qquad (3.45)$$

where $\tilde{r} = r\sqrt{\xi/\nu}$ and

$$\zeta = \frac{1}{\tilde{r}^2}\left[\left(1 + \tilde{r}^2\right)e^{-\frac{\tilde{r}^2}{2}} - 1\right]$$

$$\varepsilon = \frac{2}{\tilde{r}^2}\left[1 - \left(1 + \frac{\tilde{r}^2}{2}\right)e^{-\frac{\tilde{r}^2}{2}}\right]$$

Since Liutex and λ_{ci} are equivalent to the Δ criterion with a zero threshold, the existence conditions of Liutex and λ_{ci} are identical, namely $\zeta > 0$, which yields a nondimensional vortex size of. $\tilde{r}_0 = 1.5852$.

Eq. (3.44) indicates that the shearing part ϵ will affect Q and λ_{ci}. To investigate this shearing effect, the superposition of a shearing motion (with an appropriate external force term to fulfill the Navier–Stokes equations) is considered, which is given by

$$\begin{cases} u = -C\dfrac{\mathrm{Re}\xi}{\tilde{r}_0^2}y \\[2mm] v = 0 \\[2mm] w = 0 \end{cases} \qquad\qquad (3.46)$$

where C is a user-specified constant. Here, $\mathrm{Re} = 10$ and $\xi = 1$ are selected. Figs. 3.6 and 3.7 demonstrate the iso-contours of Liutex, λ_{ci}, and Q in the xy-plane for the Burgers vortex superposed with the shearing motion when C is set to 1 and 10, respectively. It can be obviously seen that the increase of the shearing motion slightly modifies the distribution of Liutex, but the rotational strength near the central part nearly remains constant. On the other hand, the distribution of λ_{ci} and Q is significantly disturbed. The value of λ_{ci} near the central part is increased from 6 to 14 and the value of Q near the center is increased from 40 to 180. This significant deviation demonstrates that these two criteria are prone to the contamination by shearing and the λ_{ci} criterion is not a reliable measure of the local swirling strength at least when high shear strain exists.

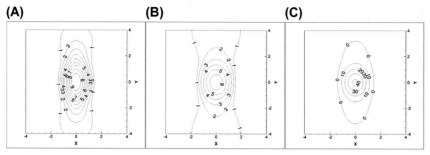

FIGURE 3.6

Iso-contours of Liutex (A), λ_{ci} (B), and Q (C) in the xy plane for the Burgers vortex superposed with the shearing motion ($C = 1$).

Reproduced from Y. Gao and C. Liu, Rortex and comparison with eigenvalue-based vortex identification criteria, Phys. Fluids 30 (085107), 2018, with the permission of AIP Publishing.

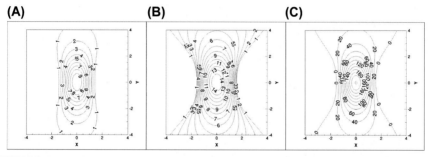

FIGURE 3.7

Iso-contours of Liutex (A), λ_{ci} (B), and Q (C) in the xy plane for the Burgers vortex superposed with the shearing motion ($C = 10$).

Reproduced from Y. Gao and C. Liu, Rortex and comparison with eigenvalue-based vortex identification criteria, Phys. Fluids 30 (085107), 2018, with the permission of AIP Publishing.

(3) Sullivan vortex

The Sullivan vortex is an exact solution to the Navier–Stokes equations for a three-dimensional axisymmetric two-celled vortex. The two-celled vortex has an inner cell in which air flow descends from above and flows outward to meet a separate airflow that is converging radially. The mathematical form of the Sullivan vortex is

$$v_r = -ar + \frac{6\nu}{r}\left(1 - e^{-\frac{ar^2}{2\nu}}\right)$$

$$v_\theta = \frac{\Gamma}{2\pi r}\left(\frac{H\left(\frac{ar^2}{2\nu}\right)}{H(\infty)}\right)$$

$$v_z = 2az\left(1 - 3e^{-\frac{ar^2}{2\nu}}\right) \tag{3.47}$$

where

$$H(x) = \int_0^x e^{-t+3\int_0^t [(1-e^{-\tau})/\tau]d\tau} dt \qquad (3.48)$$

In this case, $a = 1$, $\Gamma = 10$, and $\nu = 0.001$ are selected. Fig. 3.8 shows the Liutex vector lines on the iso-surface, which represent the local rotational axis. It can be seen that the local axis given by Liutex is consistent with the global rotation axis, i.e., the z axis, which means the direction of Liutex is physically reasonable.

3.6.3 Comparison for flow transition

Here the DNS data of late boundary layer transition on a flat plate is used to compare Liutex with Q and λ_{ci}. The DNS data are generated by an inhouse DNS code called DNSUTA. A sixth-order compact scheme is applied in the streamwise and normal directions. In the spanwise direction, where periodic conditions are applied, the pseudo-spectral method is used. In order to eliminate the spurious numerical oscillations caused by central difference schemes, an implicit sixth-order compact filter is applied to the primitive variables after a specified number of time steps. The simulation was performed with near 60 million grid points and over 400,000 time steps at a free stream Mach number of 0.5. For the detailed case setup, please refer to Liu et al. (2014).

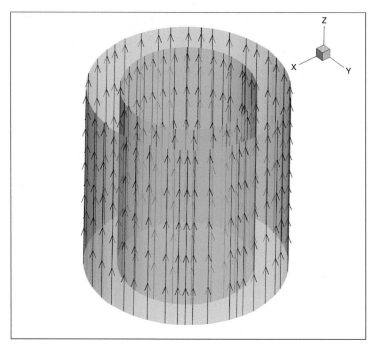

FIGURE 3.8

Liutex vector lines for Sullivan vortex.

Reproduced from Y. Gao and C. Liu, Rortex and comparison with eigenvalue-based vortex identification criteria, Phys. Fluids 30 (085107), 2018, with the permission of AIP Publishing.

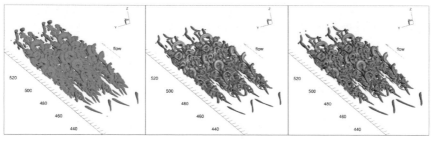

FIGURE 3.9

Isosufaces of Liutex, λ_{ci}, and Q for late boundary layer transition.

*Reproduced from Y. Gao and C. Liu, Rortex and comparison with eigenvalue-based vortex identification criteria,
Phys. Fluids 30 (085107), 2018, with the permission of AIP Publishing.*

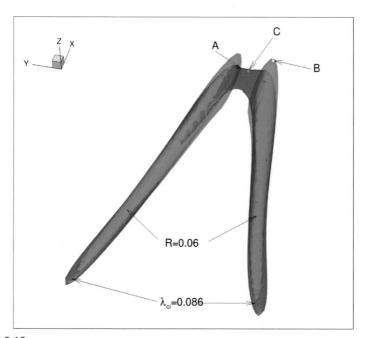

FIGURE 3.10

Iso-surfaces of Liutex and λ_{ci}.

*Reproduced from Y. Gao and C. Liu, Rortex and comparison with eigenvalue-based vortex identification criteria,
Phys. Fluids 30 (085107), 2018, with the permission of AIP Publishing.*

Although all methods illustrate the similar iso-surfaces of vortical structures as shown in Fig. 3.9 the values of λ_{ci} and Q can be found contaminated by shearing. Examine three points A, B, and C on the Liutex and λ_{ci} iso-surfaces of the quasi-streamwise vortical structure as shown in Fig. 3.10 Point A is located on both the

Table 3.2 Eigenvalues (λ_r, λ_{cr}, and λ_{ci}), Liutex strengths (R), and shearing components (ε) of point A, B, and C.

	A	B	C
λ_r	0.0197	0.0162	0.0281
λ_{cr}	−0.00104	−0.00843	−0.0151
λ_{ci}	0.086	0.059	0.086
R	0.06	0.06	0.018
ε	0.218	0.0866	0.81

Reproduced from Y. Gao and C. Liu, Rortex and comparison with eigenvalue-based vortex identification criteria, Phys. Fluids 30 (085107), 2018, with the permission of AIP Publishing.

Liutex and λ_{ci} iso-surfaces, B on the Liutex iso-surface (green) only, and C on the λ_{ci} iso-surface (blue) only. The corresponding velocity gradient tensors of A, B, and C are given by Eq. (3.48). The eigenvalues, the magnitudes of Liutex and shearing components are provided in Table 3.2. From Table 3.2, it can be found that A and B possess the same local rotational strength with different eigenvalues, while A and C have the same imaginary value of the complex eigenvalues but different local rotational strength. The shearing parts are so strong that the λ_{ci} criterion will be seriously contaminated. Especially for point C, the shearing component $\epsilon = 0.81$ is significantly larger than the actual local rotation strength $R = 0.018$, making point C being mistaken for a point with large swirling strength by the λ_{ci} criterion. From Fig. 3.10, it can be found that point B, which has a strong rotation ($R = 0.06$), is missed by the λ_{ci} criterion, but point C, which contains a weak rotation ($R = 0.018$), is misidentified by the λ_{ci} criterion. The Q criterion as shown in Fig. 3.11 will indicate a similar result of contamination, so the detailed analysis is omitted here.

$$\nabla \vec{v}|_A = \begin{bmatrix} 0.0622 & 0.215 & 0.560 \\ -0.00733 & 0.00312 & 0.0562 \\ -0.0121 & -0.0514 & -0.0664 \end{bmatrix}$$

$$\nabla \vec{v}|_B = \begin{bmatrix} 0.0498 & 0.211 & 0.361 \\ -0.00109 & -0.00986 & 0.0697 \\ -0.00625 & -0.0417 & -0.0406 \end{bmatrix}$$

$$\nabla \vec{v}|_C = \begin{bmatrix} 0.0372 & -0.000897 & 0.818 \\ 0.000064 & 0.0281 & 0.00012 \\ -0.0124 & 0.000250 & -0.0674 \end{bmatrix} \tag{3.49}$$

Since Liutex is a vector quantity, the local rotation axis of the vortex structures can be visualized by Liutex vectors. Figs. 3.12 and 3.13 demonstrate that the Liutex vector is actually tangent to the iso-surface of Liutex. Assume that a point P is located on the iso-surface and a point P^* is on the direction of Liutex vector at P,

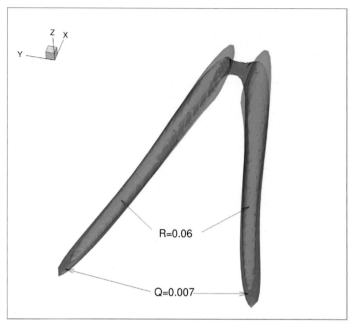

FIGURE 3.11

Iso-surfaces of Liutex (green) and *Q (blue)*.

Reproduced from Y. Gao and C. Liu, Rortex and comparison with eigenvalue-based vortex identification criteria,
Phys. Fluids 30 (085107), 2018, with the permission of AIP Publishing.

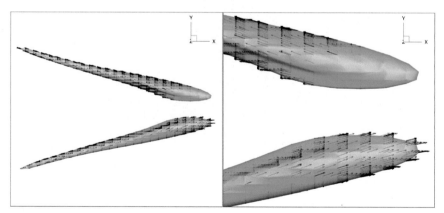

FIGURE 3.12

Liutex vector on the leg part of the vorical structure.

Reproduced from Y. Gao and C. Liu, Rortex and comparison with eigenvalue-based vortex identification criteria,
Phys. Fluids 30 (085107), 2018, with the permission of AIP Publishing.

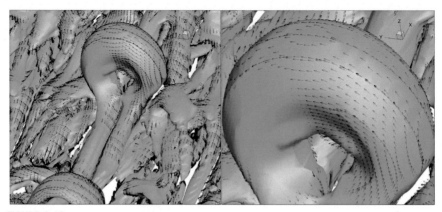

FIGURE 3.13

Liutex vector on the vortex ring.

Reproduced from Y. Gao and C. Liu, Rortex and comparison with eigenvalue-based vortex identification criteria, Phys. Fluids 30 (085107), 2018, with the permission of AIP Publishing.

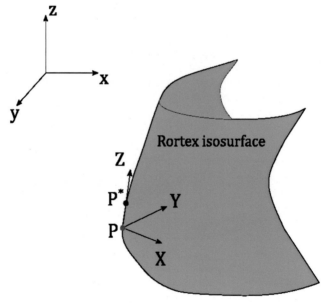

FIGURE 3.14

Illustration of Liutex vector at point *P*.

Reproduced from Y. Gao and C. Liu, Rortex and comparison with eigenvalue-based vortex identification criteria, Phys. Fluids 30 (085107), 2018, with the permission of AIP Publishing.

as shown in Fig. 3.14. According to the definition of Liutex, when P^* limits toward P, only the velocity along the local rotation axis Z can change. Correspondingly, only the component along the local rotation axis Z of the velocity gradient tensor can change. So, the component of the velocity gradient tensor in the XY-plane will not change, which means P^* will be located on the same iso-surface in the limit and the Liutex vector is tangent to the iso-surface of Liutex at point P.

Fig. 3.15 shows the structures of Liutex lines and demonstrates vorticity lines. Both pass the same seed points. As can be seen in Fig. 3.16, vorticity lines can only represent the ring part of the hairpin vortex. In contrast, Liutex lines can provide a skeleton of the whole hairpin vortex. It is expected that Liutex lines will offer a new perspective to analyze the vortical structures.

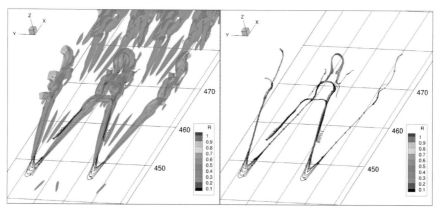

FIGURE 3.15

Liutex lines for hairpin vortex.

Reproduced from Y. Gao and C. Liu, Rortex and comparison with eigenvalue-based vortex identification criteria, Phys. Fluids 30 (085107), 2018, with the permission of AIP Publishing.

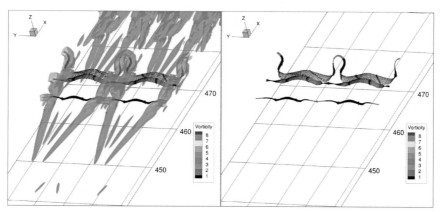

FIGURE 3.16

Vorticity lines for hairpin vortex.

Reproduced from Y. Gao and C. Liu, Rortex and comparison with eigenvalue-based vortex identification criteria, Phys. Fluids 30 (085107), 2018, with the permission of AIP Publishing.

From the aforementioned computational examples and analysis, it can be found that Q, λ_{ci} and other second generation of vortex identification methods are seriously contaminated by shears and, actually, by stretching and compression as well except for λ_{ci}. In addition, the vortex field must be defined as a vector field. Any scalar field cannot exactly represent vortex field since vortex has the axis with directions and, therefore, the second generation in principal cannot be used to represent vortex field as they are all scalar and seriously contaminated by stretching/compression and shear. Of course, the first generation cannot as well as discussed in Chapter 1.

3.7 Correlation coefficient between Liutex and vorticity

Correlation coefficient is a statistical concept revealing how much two groups' data are related. If two quantities are totally irrelevant issues, their correlation coefficient will approach to 0. On the contrary, if two quantities are 100% relevant, their correlation coefficient is 1. Therefore, it is a quantized tool to judge the relation between Liutex, the correct quantity to detect vortex, and vorticity, which has been regarded as the indicator of vortex for a long time.

3.7.1 Basic concepts of correlation

Expectation: $\vec{X}$ is an $n \times 1$ vector, $E(\vec{X})$ is the expectation of $\vec{X}$ if:

$$E(\vec{X}) = \frac{1}{n} \sum_{i=1}^{n} X_i \tag{3.50}$$

where i is the ith component of $\vec{X}$.

Variance: $\vec{X}$ is an $n \times 1$ vector, $V(\vec{X})$ is the variance of $\vec{X}$ if:

$$V(\vec{X}) = E\left[(\vec{X} - E(\vec{X}))^2\right] \tag{3.51}$$

Standard deviation: $\vec{X}$ is an $n \times 1$ vector, $\sigma(\vec{X})$ is the standard deviation of $\vec{X}$ if:

$$\sigma(\vec{X}) = \sqrt{\mathrm{Var}(\vec{X})} \tag{3.52}$$

Covariance: $\vec{X}$, $\vec{Y}$ are $n \times 1$ vectors, $\mathrm{cov}(\vec{X}, \vec{Y})$ is the covariance of $\vec{X}$ and $\vec{Y}$ if:

$$\mathrm{cov}(\vec{X}, \vec{Y}) = E\left[(\vec{X} - E(\vec{X}))(\vec{Y} - E(\vec{Y}))\right] \tag{3.53}$$

Correlation coefficient: $\vec{X}$, $\vec{Y}$ are $n \times 1$ vectors, $\rho(\vec{X}, \vec{Y})$ is the correlation coefficient of $\vec{X}$ and $\vec{Y}$ if:

$$\rho(\vec{X}, \vec{Y}) = \frac{\mathrm{cov}(\vec{X}, \vec{Y})}{\sigma(\vec{X})\sigma(\vec{Y})} \tag{3.54}$$

3.7.2 Correlation between Liutex and vorticity

Based on a DNS data of boundary layer transition, the value of Liutex and vorticity of one point shown in Fig. 3.17 is recorded for 200 time steps. The value of Liutex forms the variable X and the value of vorticity forms the variable Y, and the correlation coefficient between X and Y is calculated.

The coordinate of the point selected is:

$x = 454.83544404915006 \quad y = 8.4218553060388146 \quad z = 1.0781576896284206$

The analysis time is taken from $6.005T$ to $8T$ where T is the period of Tollmien−Schlichting wave.

The correlation efficient is:

$$\rho(X, Y) = 0.1017575, \tag{3.55}$$

which shows Liutex, which is the rigid rotation of fluids, and vorticity, is almost irrelevant. Therefore, it is a solid evidence that vorticity is not related to vortex. They are in general irrelevant.

In addition, the correlation analysis between Liutex and Q method as well as vorticity and Q method is conducted for the same point and same time period.

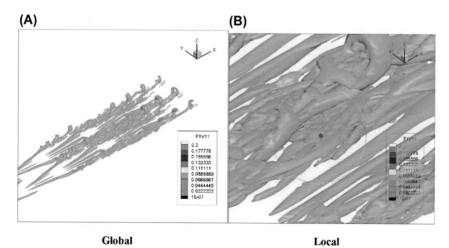

(A)   **(B)**

Global   Local

FIGURE 3.17

Selected point in a DNS case of boundary layer transition. (A) Global (B) Local.

The correlation between vorticity and Q is:

$$\rho = 0.1405711 \qquad (3.56)$$

The correlation between Liutex and Q method is:

$$\rho = 0.5985837 \qquad (3.57)$$

It should be pointed out that the correlation is conducted at a point close to the wall, where the shear is large. In general, $\nabla \times \vec{v} = \vec{R} + \vec{S}$ or vorticity equals to Liutex plus shear. In the place near the wall, $\nabla \times \vec{v} \sim \vec{S}$ and vorticity is not correlated to rotation or Liutex. However, at the point that is located far away from the wall, shear is small which means $\nabla \times \vec{v} \sim \vec{R}$ and then vorticity may be correlated to rotation or Liutex.

These analyses clearly show that vorticity is in general irrelevant with vortex or fluid rotation (Liutex or Q) unless shear is very small. On the other hand, the outcome also shows the correlation between Liutex and Q is not very high as Q is contaminated by stretching and shearing while Liutex is exactly twice the fluid angular speed of the fluid rigid rotation without any contamination by either stretching or shearing.

3.8 Summary

A consistent and systematic interpretation of scalar, vector, and tensor versions of Liutex is presented to provide a unified characterization of the local fluid rotation. Several conclusions are summarized as follows:

(1) The real eigenvector of the velocity gradient tensor is used to determine the direction of Liutex, which represents the axis of the local fluid rotation, and the rotational strength obtained in the plane perpendicular to the local axis is defined as the magnitude of Liutex.

(2) Eigenvalue-based criteria (second generation of vortex identification methods) are exclusively determined by the eigenvalues of the velocity gradient tensor. If two points have the same eigenvalues, they are located on the same iso-surface. But Liutex cannot be exclusively determined by the eigenvalues. Even if two points have the same eigenvalues, the magnitudes of Liutex are generally different.

(3) The existing eigenvalue-based methods can be seriously contaminated by shearing. Since shearing always manifests its effect on the imaginary part of the complex eigenvalues, any criterion associated with the complex eigenvalues will be prone to contamination by shear, while Liutex eliminates the contamination and thus can accurately quantify the local rotational strength.

(4) Liutex can identify the local rotational axis and provide the precise local rotational strength, thereby can reasonably represent the local rigidly rotation of fluids or vortex.

(5) Since both local rotation axis and magnitude of Liutex are uniquely deter-mined by the velocity gradient tensor without any dynamics involved, Liutex is a mathematical definition of fluid kinematics.

(6) The calculation of Liutex is quite simple, and the computational efficiency is comparable to that of the λ_{ci} criterion.

(7) In contrast to eigenvalue-based criteria, as a vector quantity, not only the iso-surface of Liutex but also the Liutex vector can be used to show the vortex structure, including Liutex vector field, Liutex lines, Liutex tubes, and Liutex surfaces.

(8) Liutex lines are parallel to vortex structure unlike vorticity lines, which can penetrate the vortex structure.

(9) Not only Liutex iso-surface, which has the same magnitude, can be applied to describe the vortex structures, but also different strength along the Liutex lines can be applied to do analysis on physics of vortex and/or turbulence generation.

(10) There is in general no correlation between vorticity and vortex (fluid rotation). Vorticity and vortex are two totally different concepts. Since Q is contami-nated by stretching and shearing, the correlation between Q and Liutex is not high and it may be inappropriate to use Q to visualize the vortex structure near the wall when shearing is dominated but Liutex is the only one that can be applied to visualize vortex structure correctly, uniquely, and accurately.

Existence, uniqueness, continuity, stability, Galilean invariance, applicability to compressible flows

In this chapter, several important properties of Liutex are elaborated. Although seemly redundant, the existence and uniqueness are the fundamental requirements from mathematical perspective. Actually, the first definition of Liutex is based on coordinate rotation (Liu et al., 2018a), which is an implicit definition, so the existence and uniqueness are not at all obvious. The introduction of the explicit definition (Wang et al., 2019a; Xu et al., 2019a) significantly simplifies the proof. Continuity is also important and is derived from the continuity of eigenvalues and eigenvectors of the velocity gradient tensor. Stability is also a mathematical property but reveals an essential and unique feature of Liutex in physics: the fluid rotation is stable and hard to change, while the nonrotational state is sensitive to the perturbation and quite easy to transfer to the fluid-rotational state. Galilean invariance is the fundamental requirement from physical perspective for any vortex identification method since it is natural to require the captured vortices independent of the observer in an inertial system. Applicability to compressible flows means Liutex is valid for compressible flows without any modification, while several second-generation vortex identification methods are questionable for compressible flows.

4.1 Existence

The first proof of the existence of Liutex is provided in Liu et al. (2018a). The proof relies on the real Schur decomposition (Golub and Van Loan, 2012) and coordinate rotation. The main idea is to prove that for any velocity gradient tensor $\nabla \vec{v}$, there is a proper rotation matrix Q, such that

$$\nabla \vec{V} = Q \nabla \vec{v} Q^{-1} = \begin{bmatrix} \dfrac{\partial U}{\partial X} & \dfrac{\partial U}{\partial Y} & 0 \\[2mm] \dfrac{\partial V}{\partial X} & \dfrac{\partial V}{\partial Y} & 0 \\[2mm] \dfrac{\partial W}{\partial X} & \dfrac{\partial W}{\partial Y} & \dfrac{\partial W}{\partial Z} \end{bmatrix}, \quad \dfrac{\partial U}{\partial Z} = 0, \quad \dfrac{\partial V}{\partial Z} = 0 \qquad (4.1)$$

Liutex and Its Applications in Turbulence Research. https://doi.org/10.1016/B978-0-12-819023-4.00013-6
Copyright © 2021 Elsevier Inc. All rights reserved.

So, the Z-axis, which is the (possible) rotation axis, serves as the direction of Liutex. And the magnitude of Liutex is obtained by the constructive definition of the fluid rotation strength (see Chapter 2) in the XY-plane perpendicular to the Z-axis.

A simpler proof of the existence can be directly obtained from the explicit definition of Liutex given by Eqs. (3.2) and (3.3):

Theorem 4.1. If a velocity gradient tensor has one real and two conjugate complex eigenvalues, there exists a Liutex vector, given by Eqs. (3.2) and (3.3), which represents the local rigid rotation.

Proof. According to Eqs. (3.2) and (3.3), Liutex relies on the imaginary part of the complex eigenvalue, the eigenvector corresponding to the real eigenvalue, and vorticity. Obviously, vorticity always exists. If the velocity gradient tensor has two complex eigenvalues and one real eigenvalue, the complex eigenvalue and the eigenvector corresponding to the real eigenvalue exist.

The remaining step is to prove the existence of Eq. (3.3), namely $\left(\vec{\omega}\cdot\vec{r}\right)^2 \geq 4\lambda_{ci}^2$. In 3D principal coordinate system, the velocity gradient tensor has the form

$$
\nabla \vec{V}_\theta = \begin{bmatrix} \lambda_{cr} & \dfrac{\partial U}{\partial Y}\Big|_\theta & 0 \\[2ex] \dfrac{\partial V}{\partial X}\Big|_\theta & \lambda_{cr} & 0 \\[2ex] \xi & \eta & \lambda_r \end{bmatrix}. \tag{4.2}
$$

Therefore,

$$
\vec{\omega}\cdot\vec{r} = \frac{\partial V}{\partial X}\Big|_\theta - \frac{\partial U}{\partial Y}\Big|_\theta \tag{4.3}
$$

The characteristic equation of the 2×2 upper left submatrix in Eq. (4.2) can be written as

$$
\left| \begin{bmatrix} \lambda_{cr} & \dfrac{\partial U}{\partial Y}\Big|_\theta \\[2ex] \dfrac{\partial V}{\partial X}\Big|_\theta & \lambda_{cr} \end{bmatrix} - \lambda \begin{bmatrix} 1 & 0 \\ 0 & 1 \end{bmatrix} \right| = 0 \tag{4.4}
$$

or

$$
\lambda^2 - 2\lambda_{cr}\lambda + \lambda_{cr}^2 - \frac{\partial U}{\partial Y}\Big|_\theta \frac{\partial V}{\partial X}\Big|_\theta = 0 \tag{4.5}
$$

Since the roots of Eq. (4.5) are $\lambda_1 = \lambda_{cr} + i\lambda_{ci}$ and $\lambda_2 = \lambda_{cr} - i\lambda_{ci}$, $\frac{\partial U}{\partial Y}\Big|_\theta \frac{\partial V}{\partial X}\Big|_\theta = -\lambda_{ci}^2$.

$$\left(\overrightarrow{\omega} \cdot \overrightarrow{r}\right)^2 - 4\lambda_{ci}^2 = \left(\left.\frac{\partial V}{\partial X}\right|_\theta - \left.\frac{\partial U}{\partial Y}\right|_\theta\right)^2 + 4\left.\frac{\partial U}{\partial Y}\right|_\theta \left.\frac{\partial V}{\partial X}\right|_\theta = \left(\left.\frac{\partial V}{\partial X}\right|_\theta + \left.\frac{\partial U}{\partial Y}\right|_\theta\right)^2 \geq 0$$

(4.6)

It completes the proof.

On the other hand, if the velocity gradient tensor has three real eigenvalues, Liutex is equal to zero. This implies Liutex is well defined for any velocity gradient tensor. The existence of the tensor form is obvious since the definition of the tensor form is consistent with the vector form.

4.2 Uniqueness

Theorem 4.2. The Liutex vector is unique.

Proof. First, consider the velocity gradient tensor has two complex eigenvalues. Similar to the proof of the existence, since the Liutex vector is defined by the imaginary part of the complex eigenvalue λ_{ci}, the eigenvector corresponding to the real eigenvalue $\overrightarrow{r}$, and the vorticity $\overrightarrow{\omega}$, it needs to prove the uniqueness of these three terms. Assume two Liutex vectors $\overrightarrow{R}_1$ and $\overrightarrow{R}_2$ have the identical velocity gradient tensor. Apparently, the vorticity $\overrightarrow{\omega}_1$ is equal to $\overrightarrow{\omega}_2$, which means the uniqueness of the vorticity. When the velocity gradient tensor has two complex eigenvalues, three eigenvalues are distinct, the imaginary part of the complex eigenvalue must be the same, namely

$$\lambda_{ci1} = \lambda_{ci2}.$$

Also, in this case, the algebraic multiplicity and the geometric multiplicity for any eigenvalue are equal to 1 (Stewart and Sun, 1990). According to the definition of the geometric multiplicity, the maximal number of linearly independent eigenvectors associated with any eigenvalue is one, including the real eigenvalue. Hence, the (normalized) eigenvector corresponding to the real eigenvalue is unique up to the $\pm$ sign. And the condition $\overrightarrow{\omega} \cdot \overrightarrow{r} > 0$ can uniquely determine the direction. So, it must be

$$\overrightarrow{r}_1 = \overrightarrow{r}_2,$$

which implies

$$\overrightarrow{R}_1 = \overrightarrow{R}_2.$$

On the other hand, if the velocity gradient tensor has three real eigenvalues, Liutex is equal to zero. So, for any velocity gradient tensor in the flow field, the Liutex vector is uniquely defined.

4.3 Continuity

Continuity is important to a new defined physical quantity, i.e., Liutex. The meaning of continuity is given by the following definition:

Definition. For a point in the flow field, a nonzero quantity Y, which can be a scalar, vector, or tensor, is a function of the velocity gradient tensor $\nabla \overrightarrow{v}$, namely $Y = f\left(\nabla \overrightarrow{v}\right)$.

For any $\varepsilon > 0$, if there exists a $\delta > 0$ so that a small perturbation E with $\|E\| \langle \delta$ ($\| \cdot \|$ denotes a matrix norm) will result in

$$\|\Delta Y\| = \left\| f\left(\nabla \vec{v} + E \right) - f\left(\nabla \vec{v} \right) \right\| < \varepsilon,$$

the quantity Y is continuous at that point.

Since the Liutex vector is uniquely determined by the vorticity vector $\vec{\omega}$, the imaginary part of the conjugate complex eigenvalue λ_{ci}, and the eigenvector corresponding to the real eigenvalue $\vec{r}$ when Liutex exists or $R > 0$, it requires to prove the continuity of these three terms.

Theorem 4.3. The vorticity is continuous when the velocity gradient tensor has one real and two conjugate complex eigenvalues.

Proof. This is obvious since the vorticity vector $\vec{\omega}$ exists everywhere and is continuously dependent on the velocity gradient tensor.

The continuity of eigenvalues has been extensively discussed in matrix perturbation theory. The following theorem (Stewart and Sun, 1990) indicates the simple eigenvalue is a differentiable function of the perturbation of the velocity gradient tensor:

Theorem 4.4. (Stewart and Sun, 1990, Chapter 4, Theorem 2.3) Let λ be a simple eigenvalue of the matrix A, with right and left eigenvectors $\vec{x}$ and $\vec{y}$, and let $\tilde{A} = A + E$ be a perturbation of A. Then there is a unique eigenvalue $\tilde{\lambda}$ of $\tilde{A}$ such that

$$\tilde{\lambda} = \lambda + \frac{\vec{y}^H E \vec{x}}{\vec{y}^H \vec{x}} + O\left(\|E\|^2 \right) \tag{4.7}$$

From Theorem 4.4, one can obtain the following theorem:

Theorem 4.5. (Stewart and Sun, 1990, Chapter 4, Corollary 2.4) Under the hypotheses of Theorem 4.4, the eigenvalue λ is a differentiable function of A. Moreover,

$$\frac{\partial \lambda}{\partial \alpha_{ij}} = \frac{\overline{\eta_i} \xi_j}{\vec{y}^H \vec{x}} \tag{4.8}$$

Here, α_{ij} represents the (i, j)-element of the matrix A, η_i the i-element of the left eigenvector $\vec{y}$, and ξ_j the j-element of the right eigenvector $\vec{x}$.

Theorem 4.6. The imaginary part of the conjugate complex eigenvalue is continuous when the velocity gradient tensor has one real and two conjugate complex eigenvalues.

Proof. Apparently, all three eigenvalues are simple eigenvalues when the velocity gradient tensor has two complex eigenvalues and one real eigenvalue. According to Eq. (4.8), the complex eigenvalue is continuously dependent on the element of the velocity gradient tensor, so is the imaginary part of the complex eigenvalue. It must satisfy the definition of continuity. Definition 4.3.

According to the theory of invariant subspaces, the simple eigenvector (the eigenvector corresponding to the simple eigenvalue) is a differentiable function of the perturbation, which can be written as (Stewart and Sun, 1990, Page 240)

$$\tilde{\vec{x}} = \vec{x} + X_2(\lambda I - L_2)^{-1} Y_2^H E \vec{x} + O\left(\|E\|^2\right) \tag{4.9}$$

Here, $\tilde{\vec{x}}$ represents the simple eigenvector of the matrix $\tilde{A} = A + E$, $\vec{x}$ the simple eigenvector of the matrix A, and λ the simple eigenvalue of the matrix A. The detailed meanings of the matrices X_2, L_2, and Y_2 should be found in Chapter 5 of Stewart and Sun (1990). Since the real eigenvector $\vec{r}$ is a simple eigenvector when Liutex exists, Eq. (4.9) implies the real eigenvector $\vec{r}$ is continuously dependent on the perturbation matrix E. Therefore, the following theorem is obtained:

Theorem 4.7. The real eigenvector $\vec{r}$ is continuous when the velocity gradient tensor has one real and two conjugate complex eigenvalues.

From the earlier discussion, $\vec{\omega}$, λ_{ci}, and $\vec{r}$ are all continuously dependent on the elements of the velocity gradient tensor, the following theorem is obtained:

Theorem 4.8. The Liutex vector $\vec{R}$ given by Eqs. (3.2) and (3.3) is continuous when the velocity gradient tensor has one real and two conjugate complex eigenvalues.

On the other hand, if the velocity gradient tensor has three real eigenvalues, Liutex remains zero. Therefore, Liutex is continuous in the whole flow field. The above proof shows that Liutex is continuous and differentiable in the vortex area where the velocity gradient tensor has one real and two conjugate complex eigenvalues, which is stronger than the continuity. However, there must be boundaries between vortex area and non-rotational area, where Theorem 4.8 only shows Liutex is continuous but does not prove it is differentiable.

4.4 Stability of Liutex in physics

Liutex is defined as a rigid rotation part of fluid motion, and in general any velocity gradient tensor can be decomposed to Liutex part and a nonrotation part or UTA R-NR decomposition which will be thoroughly discussed in Chapter 5.

$$\nabla \vec{V}_\theta = P \nabla \vec{V} P^{-1} = P\left(Q \nabla \vec{v} Q^T\right) P^{-1} = \begin{bmatrix} \lambda_{cr} & -\dfrac{1}{2}R & 0 \\ \dfrac{1}{2}R + \varepsilon & \lambda_{cr} & 0 \\ \dfrac{\partial W}{\partial X} & \dfrac{\partial W}{\partial Y} & \lambda_r \end{bmatrix} = R + NR \tag{4.10}$$

$$R = \begin{bmatrix} 0 & -R/2 & 0 \\ R/2 & 0 & 0 \\ 0 & 0 & 0 \end{bmatrix}$$

$$NR = \begin{bmatrix} \lambda_{cr} & 0 & 0 \\ s & \lambda_{cr} & 0 \\ \xi & \eta & \lambda_r \end{bmatrix}$$

Definition. A state is stable when it is kept unchanged with small disturbance. Otherwise, the state is unstable.

A stone is stable if it sits in a valley and unstable if it sits on the top of a hill. This is because the stone will still keep stationary when it is placed in the valley if you kick it, but will fall down when it is placed on the top of a hill if you kick it. This example shows some state is stable and some state is unstable. In a fluid flow field, some points are identified as rotational ($R > 0$) and some are nonrotational ($R = 0$).

Theorem 4.9. The nonrotational state ($R = 0$) is unstable.

Proof. Assume Point A is nonrotational, $R = 0$, in a principal coordinate, the velocity gradient tensor can be written as

$$\nabla \vec{V}_\theta = NR = \begin{bmatrix} \lambda_1 & 0 & 0 \\ s & \lambda_1 & 0 \\ \xi & \eta & \lambda_2 \end{bmatrix} \tag{4.11}$$

Let us put a very small perturbation in $\frac{\partial U}{\partial Y}$:

$$E = \begin{bmatrix} 0 & \varepsilon & 0 \\ 0 & 0 & 0 \\ 0 & 0 & 0 \end{bmatrix} \quad (\text{assume } \varepsilon \ll s \text{ and } s, \varepsilon > 0) \tag{4.12}$$

$$\nabla \vec{V}_\theta + E = NR + E = \begin{bmatrix} \lambda_1 & \varepsilon & 0 \\ s & \lambda_1 & 0 \\ \xi & \eta & \lambda_2 \end{bmatrix} = \begin{bmatrix} 0 & \varepsilon & 0 \\ -\varepsilon & 0 & 0 \\ 0 & 0 & 0 \end{bmatrix} + \begin{bmatrix} \lambda_1 & 0 & 0 \\ s+\varepsilon & \lambda_1 & 0 \\ \xi & \eta & \lambda_2 \end{bmatrix} = \widetilde{R} + \widetilde{NR} \tag{4.13}$$

So, Point A becomes rotational as $\widetilde{R} \neq 0$.

In contrast, if Point A is a rotational point,

$$\nabla \vec{V}_\theta = R + NR = \begin{bmatrix} 0 & -R/2 & 0 \\ R/2 & 0 & 0 \\ 0 & 0 & 0 \end{bmatrix} + \begin{bmatrix} \lambda_{cr} & 0 & 0 \\ s & \lambda_{cr} & 0 \\ \xi & \eta & \lambda_r \end{bmatrix}. \tag{4.14}$$

Using the same perturbation

$$E = \begin{bmatrix} 0 & \varepsilon & 0 \\ 0 & 0 & 0 \\ 0 & 0 & 0 \end{bmatrix} \quad (\text{assume } \varepsilon \ll s \text{ and } s, \varepsilon > 0) \tag{4.15}$$

$$\nabla \vec{V}_\theta + E = \begin{bmatrix} 0 & -\dfrac{R}{2} + \varepsilon & 0 \\ \dfrac{R}{2} - \varepsilon & 0 & 0 \\ 0 & 0 & 0 \end{bmatrix} + \begin{bmatrix} \lambda_{cr} & 0 & 0 \\ s + \varepsilon & \lambda_{cr} & 0 \\ \xi & \eta & \lambda_r \end{bmatrix} = \widetilde{R} + \widetilde{NR} \qquad (4.16)$$

Point A remains in its state as a rotational point since $\varepsilon \ll s$ and $\varepsilon \ll R$ (the perturbation is assumed much smaller than the elements of the velocity gradient tensor).

Now consider a general small perturbation, which means a small perturbation could be added to any elements except for the upper right two elements, which will be zero after Q rotation of the frame

$$E = \begin{bmatrix} \varepsilon_{11} & \varepsilon_{12} & 0 \\ \varepsilon_{21} & \varepsilon_{22} & 0 \\ \varepsilon_{31} & \varepsilon_{32} & \varepsilon_{33} \end{bmatrix} \qquad (4.17)$$

$$\nabla \vec{V}_\theta + E = \begin{bmatrix} 0 & -\dfrac{R}{2} + \varepsilon_{12} & 0 \\ \dfrac{R}{2} - \varepsilon_{12} & 0 & 0 \\ 0 & 0 & 0 \end{bmatrix} + \begin{bmatrix} \lambda_{cr} + \varepsilon_{11} & 0 & 0 \\ s + \varepsilon_{21} + \varepsilon_{12} & \lambda_{cr} + \varepsilon_{22} & 0 \\ \xi + \varepsilon_{31} & \eta + \varepsilon_{32} & \lambda_r + \varepsilon_{33} \end{bmatrix}$$

$$(4.18)$$

Since $\varepsilon \ll |a_{ij}|$, Point A keeps its state as a rotational point with a minor frame rotation:

$$\nabla \vec{V}_\theta + E = \begin{bmatrix} 0 & -\dfrac{\widetilde{R}}{2} & 0 \\ \dfrac{\widetilde{R}}{2} & 0 & 0 \\ 0 & 0 & 0 \end{bmatrix} + \begin{bmatrix} \widetilde{\lambda_{cr}} & 0 & 0 \\ \widetilde{s} & \widetilde{\lambda_{cr}} & 0 \\ \widetilde{\xi} & \widetilde{\eta} & \widetilde{\lambda_r} \end{bmatrix} = \widetilde{R} + \widetilde{NR} \qquad (4.19)$$

The state can only be changed if and only if $O(\varepsilon) \geq O(R)$. If so, ε is not small perturbation, which is a contradiction.

The proof is completed.

In the universe, rotation or vortex is very common everywhere, which is dedicated by the Liutex stability. Liutex is easy to be produced by a small perturbation, but is not easy to be removed unless the perturbation has the same order as the rotation strength. This bias property of Liutex will be considered as a possible origin of the turbulence generation presented in Chapter 8.

4.5 Galilean invariance

For vortex identification, Galilean invariance implies the vortical region captured is identical under the Galilean transformation. Lugt (1983) has pointed out closed or spiraling streamlines cannot be applied to identify vortices since the topology of streamlines can change dramatically under the Galilean transformation. Without any doubt, a reasonable definition should be at least Galilean invariant since vortices are independent of the observer. In this section, the Galilean invariance of Liutex is presented. Wang, Y., et al. (2018) provide the first proof of the Galilean invariance of Liutex through a coordinate rotation approach. Using the explicit expressions given by Eqs. (3.2) and (3.3), a simpler proof is presented here.

Theorem 4.10.   The direction of Liutex is invariant under the Galilean transformation.

Proof. The Galilean transformation is a composition of a uniform relative motion, a spatial rotation, and a translation in space and time of the reference frame. The Galilean transformation between two reference frames can be represented by

$$
\begin{bmatrix} x' \\ y' \\ z' \end{bmatrix} = \boldsymbol{Q}_c \begin{bmatrix} x \\ y \\ z \end{bmatrix} + \vec{c_1} t + \vec{c_2}
\tag{4.20}
$$

where x, y, z represent the coordinates of the original reference frame and x', y', z' represent the coordinates of the new reference frame. To simplify the derivation, here assume that the time variable is identical for these two reference frames, i.e., $t' = t$ in that it is trivial to show that Liutex is invariant to a constant time shift. $\boldsymbol{Q}_c$ is a constant 3×3 orthogonal matrix, representing the spatial rotation of the reference frame and possessing the property of $\boldsymbol{Q}_c^{-1} = \boldsymbol{Q}_c^{\mathrm{T}}$. $\vec{c_1}$ is the constant relative motion speed and $\vec{c}_2$ represents the constant translation in space. Therefore, the velocity in the new reference frame is written as

$$
\begin{bmatrix} u' \\ v' \\ w' \end{bmatrix} = \boldsymbol{Q}_c \begin{bmatrix} u \\ v \\ w \end{bmatrix} + \vec{c_1}
\tag{4.21}
$$

where u, v, w represent the velocity components in the original reference frame and u', v', w' the velocity components in the new reference frame. Thus, there exists a relationship between the velocity gradient tensors in two reference frames

$$
\nabla \vec{v}' = \begin{bmatrix} \dfrac{\partial u'}{\partial x'} & \dfrac{\partial u'}{\partial y'} & \dfrac{\partial u'}{\partial z'} \\[2mm] \dfrac{\partial v'}{\partial x'} & \dfrac{\partial v'}{\partial y'} & \dfrac{\partial v'}{\partial z'} \\[2mm] \dfrac{\partial w'}{\partial x'} & \dfrac{\partial w'}{\partial y'} & \dfrac{\partial w'}{\partial z'} \end{bmatrix} = \boldsymbol{Q}_c \nabla \vec{v} \boldsymbol{Q}_c^{-1} = \boldsymbol{Q}_c \begin{bmatrix} \dfrac{\partial u}{\partial x} & \dfrac{\partial u}{\partial y} & \dfrac{\partial u}{\partial z} \\[2mm] \dfrac{\partial v}{\partial x} & \dfrac{\partial v}{\partial y} & \dfrac{\partial v}{\partial z} \\[2mm] \dfrac{\partial w}{\partial x} & \dfrac{\partial w}{\partial y} & \dfrac{\partial w}{\partial z} \end{bmatrix} \boldsymbol{Q}_c^{\mathrm{T}}
\tag{4.22}
$$

where $\nabla \vec{v}'$ is the velocity gradient tensor in the new reference frame and $\nabla \vec{v}$ is the velocity gradient tensor in the original reference frame.

Let $\vec{r}$ denote the unit vector pointing to the direction of Liutex in the original reference frame. According to the definition of Liutex, $\vec{r}$ is the eigenvector of the velocity gradient tensor corresponding to the real eigenvalue given that the other two eigenvalues are complex conjugates,

$$\nabla \vec{v} \cdot \vec{r} = \lambda_r \vec{r} \tag{4.23}$$

where λ_r represents the real eigenvalue of velocity gradient tensor in the original reference frame. Based on Eq. (4.22),

$$\boldsymbol{Q}_c \nabla \vec{v} \cdot \vec{r} = \boldsymbol{Q}_c \nabla \vec{v} \left(\boldsymbol{Q}_c^{-1} \boldsymbol{Q}_c \right) \cdot \vec{r} = \left(\boldsymbol{Q}_c \nabla \vec{v} \boldsymbol{Q}_c^{-1} \right) \boldsymbol{Q}_c \cdot \vec{r} = \nabla \vec{v}' \cdot \left(\boldsymbol{Q}_c \vec{r} \right) \tag{4.24}$$

According to Eq. (4.23),

$$\boldsymbol{Q}_c \nabla \vec{v} \cdot \vec{r} = \lambda_r \boldsymbol{Q}_c \vec{r} \tag{4.25}$$

Thus,

$$\nabla \vec{v}' \cdot \left(\boldsymbol{Q}_c \vec{r} \right) = \lambda_r \left(\boldsymbol{Q}_c \vec{r} \right) \tag{4.26}$$

which indicates that if λ_r and $\vec{r}$ are an eigenvalue and eigenvector pair of $\nabla \vec{v}$, then λ_r and $\boldsymbol{Q}_c \vec{r}$ would be the corresponding eigenvalue and eigenvector pair of $\nabla \vec{v}'$. To illustrate that $\boldsymbol{Q}_c \vec{r}$ is the transformed vector in the new reference frame corresponding to $\vec{r}$ in the original reference frame, we denote $(x_0, y_0, z_0)^T$ and $(x_1, y_1, z_1)^T$ to be the starting and ending points of $\vec{r}$. Thus, according to Eq. (4.20), the corresponding starting and ending points in the new reference frame are

$$\begin{bmatrix} x_0' \\ y_0' \\ z_0' \end{bmatrix} = \boldsymbol{Q}_c \begin{bmatrix} x_0 \\ y_0 \\ z_0 \end{bmatrix} + \vec{c}_1 t + \vec{c}_2 \tag{4.27}$$

$$\begin{bmatrix} x_1' \\ y_1' \\ z_1' \end{bmatrix} = \boldsymbol{Q}_c \begin{bmatrix} x_1 \\ y_1 \\ z_1 \end{bmatrix} + \vec{c}_1 t + \vec{c}_2 \tag{4.28}$$

Therefore, the corresponding vector $\vec{r}' = (x_1', y_1', z_1')^T - (x_0', y_0', z_0')^T = \boldsymbol{Q}_c \left[(x_1, y_1, z_1)^T - (x_0, y_0, z_0)^T \right] = \boldsymbol{Q}_c \vec{r}$. This confirms that the real eigenvector or the direction of Liutex is invariant under the Galilean transformation.

Eigenvalues are invariant under similarity transformation, thus are invariant under Galilean transformation. Also, it can be easily proved that vorticity is Galilean invariant. Thus, the following theorem is proved:

Theorem 4.11. The magnitude of Liutex R is invariant under the Galilean transformation.

From Theorem 4.10 and Theorem 4.11, it is obvious that Liutex is invariant under the Galilean transformation.

In the following, three examples are provided to illustrate the Galilean invariance. As the first example, the Burgers vortex is used to test our method. The velocity components in the cylindrical coordinate system can be written as

$$\begin{cases} V_r = -ar \\[2mm] V_\theta = \dfrac{\Gamma_0}{2\pi r}\left[1 - e^{-\frac{ar^2}{2\nu}}\right] \\[2mm] V_z = 2az \end{cases}$$

where Γ_0 is the circulation, a the axisymmetric strain rate, and ν the kinematic viscosity. Consider a Burgers vortex with $a = 1s^{-1}$, $\nu = 0.02m^2/s$, and $\Gamma_0 = 5m^2/s$. Several typical streamlines are shown in Fig. 4.1A to illustrate the spiral motion of the fluids.

To study a general Galilean transformation, a rotation matrix $\boldsymbol{Q}_c$, which is a combination of three basic rotations of the x-, y-, and z-axes, is employed

$$\boldsymbol{Q}_c = \boldsymbol{Q}_{cx}(\alpha)\boldsymbol{Q}_{cy}(\beta)\boldsymbol{Q}_{cz}(\gamma)$$

$$= \begin{bmatrix} 1 & 0 & 0 \\ 0 & \cos\gamma_1 & \sin\gamma_1 \\ 0 & -\sin\gamma_1 & \cos\gamma_1 \end{bmatrix}\begin{bmatrix} \cos\gamma_2 & 0 & -\sin\gamma_2 \\ 0 & 1 & 0 \\ \sin\gamma_2 & 0 & \cos\gamma_2 \end{bmatrix}\begin{bmatrix} \cos\gamma_3 & \sin\gamma_3 & 0 \\ -\sin\gamma_3 & \cos\gamma_3 & 0 \\ 0 & 0 & 1 \end{bmatrix}$$

where γ_1, γ_2, and γ_3 are the corresponding rotational angles along the axes. It is randomly selected that $\gamma_1 = 37°$, $\gamma_2 = 128°$, and $\gamma_3 = 173°$, so one has

$$\boldsymbol{Q}_c = \begin{bmatrix} 0.6111 & -0.0750 & -0.7880 \\ -0.5680 & -0.5513 & -0.7349 \\ -0.3705 & 0.6740 & -0.4917 \end{bmatrix}$$

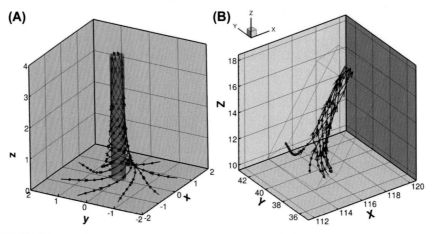

FIGURE 4.1

The Burgers vortex in the original reference frame (A) and the new reference frame (B).

Reproduced from Y. Wang, Y. Gao and C. Liu, Letter: Galilean invariance of Rortex, Phys. Fluids 30 (111701), 2018, with the permission of AIP Publishing.

In addition, the relative velocity is set between the new reference frame and the original reference frame $\vec{c_1} = (5, 3, 7)^T$, the spatial translation $\vec{c_2} = (101, 30, 2)^T$, and t $= 3$. Eq. (4.20) and Eq. (4.21) are used to calculate the coordinates and velocity in the new reference frame, which is shown in Fig. 4.1B. The blue box in Fig. 4.1B is the flow domain of interest, which is corresponding to $-2 \leq x \leq 2, -2 \leq y \leq 2$ and $-4 \leq z \leq 4$ in the original reference frame. It should be noted that only half of the domain of interest is shown in Fig. 4.1A because of the symmetry about the plane $z = 0$. The seeding points of stream traces shown in Fig. 4.1B are corresponding to the seeding points in Fig. 4.1A after the Galilean transformation. It is shown that the geometry of streamlines can be greatly changed after the Galilean transformation, which indicates that the streamlines are inappropriate for vortex identification.

Fig. 4.2 shows the directions and the magnitudes of Liutex in the two reference frames. It is clear that despite the rotation, relative motion and spatial translation together form a Galilean transformation, and the Liutex can always capture the vortical motion correctly. It should be noted that the direction of Liutex vector in Fig. 4.2A is pointing to the top while that in Fig. 4.2B is pointing to the bottom. This is essentially due to the rotation of reference frame, which we introduced in the Galilean transformation.

Also as an exact solution to Navier–Stokes equation, the Sullivan vortex is a two-celled vortex aimed to describe the flow in an intense tornado with a central downdraft. The fluid in the inner cell descends from above and flows outward to meet a separate flow (the outer cell) that is converging radially. Both flows rise at the point of meeting. The mathematical form of the Sullivan vortex is

$$\begin{cases} V_r = -ar + \dfrac{6v}{r}\left[1 - \exp\left(-\dfrac{ar^2}{2v}\right)\right] \\[2mm] V_\theta = \dfrac{\Gamma_0}{2\pi r}\dfrac{H(ar^2/2v)}{H(\infty)} \\[2mm] V_z = 2az\left[1 - 3\exp\left(-\dfrac{ar^2}{2v}\right)\right] \end{cases}$$

where $H(\eta) = \int_0^\eta \exp(-s + 3\int_0^s \frac{1-e^{-\tau}}{\tau}d\tau)ds$ and thus $H(\infty) = 37.905$. Consider a Sullivan vortex with $a = 1s^{-1}$, $v = 0.02m^2/s$, and $\Gamma_0 = 5m^2/s$. The typical streamlines in the outer cell and inner cell of the considered Sullivan vortex are shown in Fig. 4.3A and B, respectively.

To employ a different Galilean transformation, we select $\gamma_1 = 97°$, $\gamma_2 = 45°$, and $\gamma_3 = 65°$, thus the rotational matrix becomes

$$Q_c = \begin{bmatrix} 0.2988 & 0.6409 & -0.7071 \\ 0.4071 & 0.5846 & 0.7018 \\ 0.8631 & -0.4976 & -0.0862 \end{bmatrix}$$

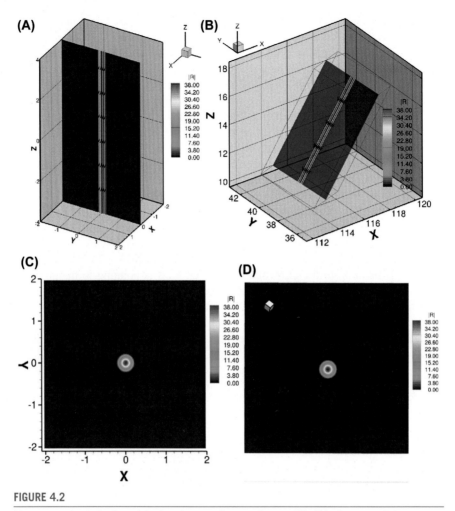

FIGURE 4.2

Liutex magnitude in the original and new reference frame. (A), (B) slices parallel to the rotational axis; (C), (D) slices perpendicular to the rotational axis.

Reproduced from Y. Wang, Y. Gao and C. Liu, Letter: Galilean invariance of Rortex, Phys. Fluids 30 (111701), 2018, with the permission of AIP Publishing.

   In addition, let the relative velocity between the new frame and reference frame $\vec{c_1} = (1, 0, 0)^T$, the spatial translation $\vec{c_2} = (0, 10, 0)^T$, and t $= 1$. Similarly, Eq. (4.20) and Eq. (4.21) are used to calculate the coordinates and velocity in the new reference frame. The streamlines are shown in Fig. 4.4 with the corresponding seeding points of that in the original reference frame shown in Fig. 4.3. It can be seen that

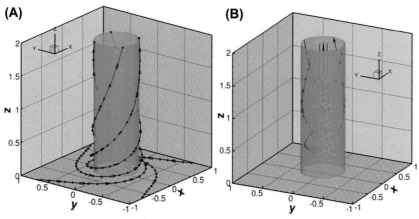

FIGURE 4.3

The typical streamlines in the Sullivan vortex with iso-surface of $V_r = 0$. (A) Streamlines in the outer cell; (B) streamlines in the inner cell.

Reproduced from Y. Wang, Y. Gao and C. Liu, Letter: Galilean invariance of Rortex, Phys. Fluids 30 (111701), 2018, with the permission of AIP Publishing.

the geometry of streamlines is severely distorted, which again states that quantities that are not Galilean invariant cannot be used to identify vortices. However, it is shown in Fig. 4.5 that the direction and magnitude of Liutex are invariant under the Galilean transformation.

In the third example, data from direct numerical simulation of boundary layer transition on a flat plate are used to demonstrate the Galilean invariance of Liutex

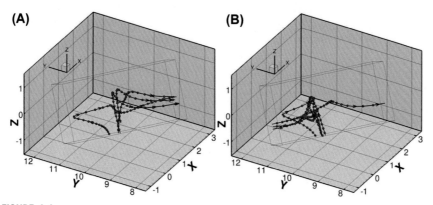

FIGURE 4.4

The streamlines in the new reference frame corresponding to the typical streamlines in the original reference frame from the outer cell (A) and inner cell (B).

Reproduced from Y. Wang, Y. Gao and C. Liu, Letter: Galilean invariance of Rortex, Phys. Fluids 30 (111701), 2018, with the permission of AIP Publishing.

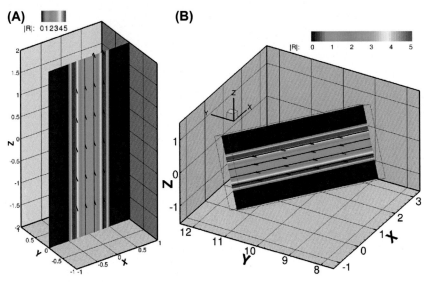

FIGURE 4.5

Liutex lines on the central plane contoured by magnitude of Liutex in the original reference frame (A) and the new reference frame (B).

Reproduced from Y. Wang, Y. Gao and C. Liu, Letter: Galilean invariance of Rortex, Phys. Fluids 30 (111701), 2018, with the permission of AIP Publishing.

in the case of compressible flows. The simulation was performed with near 60 million grid points and over 400,000 time steps at a free stream Mach number of 0.5. It is known that hairpin vortex is a typical coherent structure in boundary layer transition, which can be clearly captured by Liutex lines as shown in Fig. 4.6A. In addition, the Liutex lines are contoured by the magnitude of Liutex. The Galilean transformation employed here is the same as the one for the considered Burgers vortex, and the corresponding Liutex lines contoured by the magnitude in the new reference frame are shown in Fig. 4.6B for comparison. To further quantitatively illustrate the invariance, the Liutex vectors at point $A(462.05, 10.31, 1.49)$ in the original reference frame and the corresponding point $A^*(396.40, -231.59, -234.51)$ in the transformed frame as shown in Fig. 4.6 are given for comparison. The two vectors are $\vec{R} = R\vec{r} = 0.368 \times (0.9534, -0.1519, 0.2607)^T$ and $\vec{R}^* = R^*\vec{r}^* = 0.368 \times (0.3885, -0.5266, -0.7562)^T$. The magnitude R remains the same and the direction satisfies the relation $\vec{r}^* = Q_c\vec{r}$.

In conclusion, Liutex is invariant under the Galilean transformation, i.e., a transformation between two reference frames, which differ by a composition of a spatial rotation, a translation in space and time, and a uniform relative motion.

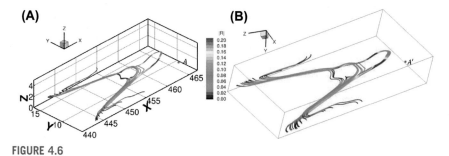

FIGURE 4.6

Liutex lines representing the hairpin vortex in the original reference frame (A) and new reference frame (B).

Reproduced from Y. Wang, Y. Gao and C. Liu, Letter: Galilean invariance of Rortex, Phys. Fluids 30 (111701), 2018, with the permission of AIP Publishing

4.6 Applicability to compressible flows

Most of the currently popular Eulerian vortex identification methods are exclusively determined by the eigenvalues or invariants of the velocity gradient tensor (Gao et al., 2018). If λ_1, λ_2, and λ_3 are used to denote three eigenvalues of the velocity gradient tensor, the characteristic equation can be written as

$$\lambda^3 + \widetilde{P}\lambda^2 + \widetilde{Q}\lambda + \widetilde{R} = 0 \tag{4.29}$$

where

$$\widetilde{P} = -(\lambda_1 + \lambda_2 + \lambda_3) = -tr\left(\nabla \overrightarrow{v}\right) \tag{4.30}$$

$$\widetilde{Q} = \lambda_1\lambda_2 + \lambda_2\lambda_3 + \lambda_3\lambda_1 = -\frac{1}{2}\left(tr\left(\nabla \overrightarrow{v}^2\right) - tr\left(\nabla \overrightarrow{v}\right)^2\right) \tag{4.31}$$

$$\widetilde{R} = -\lambda_1\lambda_2\lambda_3 = -\det\left(\nabla \overrightarrow{v}\right) \tag{4.32}$$

$\widetilde{P}$, $\widetilde{Q}$, and $\widetilde{R}$ are three invariants. And the velocity gradient tensor is usually decomposed to a symmetric part (strain rate tensor) and an antisymmetric part (vorticity tensor) as

$$\nabla \overrightarrow{v} = \begin{bmatrix} \dfrac{\partial u}{\partial x} & \dfrac{\partial u}{\partial y} & \dfrac{\partial u}{\partial z} \\[2mm] \dfrac{\partial v}{\partial x} & \dfrac{\partial v}{\partial y} & \dfrac{\partial v}{\partial z} \\[2mm] \dfrac{\partial w}{\partial x} & \dfrac{\partial w}{\partial y} & \dfrac{\partial w}{\partial z} \end{bmatrix} = A + B$$

$$A = \frac{1}{2}\left(\nabla \vec{v} + \nabla \vec{v}^T\right) = \begin{bmatrix} \frac{\partial u}{\partial x} & \frac{1}{2}\left(\frac{\partial u}{\partial y} + \frac{\partial v}{\partial x}\right) & \frac{1}{2}\left(\frac{\partial u}{\partial z} + \frac{\partial w}{\partial x}\right) \\ \frac{1}{2}\left(\frac{\partial v}{\partial x} + \frac{\partial u}{\partial y}\right) & \frac{\partial v}{\partial y} & \frac{1}{2}\left(\frac{\partial v}{\partial z} + \frac{\partial w}{\partial x}\right) \\ \frac{1}{2}\left(\frac{\partial w}{\partial x} + \frac{\partial u}{\partial z}\right) & \frac{1}{2}\left(\frac{\partial w}{\partial y} + \frac{\partial v}{\partial z}\right) & \frac{\partial w}{\partial z} \end{bmatrix}$$

(4.33)

$$B = \frac{1}{2}\left(\nabla \vec{v} - \nabla \vec{v}^T\right) = \begin{bmatrix} 0 & \frac{1}{2}\left(\frac{\partial u}{\partial y} - \frac{\partial v}{\partial x}\right) & \frac{1}{2}\left(\frac{\partial u}{\partial z} - \frac{\partial w}{\partial x}\right) \\ \frac{1}{2}\left(\frac{\partial v}{\partial x} - \frac{\partial u}{\partial y}\right) & 0 & \frac{1}{2}\left(\frac{\partial v}{\partial z} - \frac{\partial w}{\partial x}\right) \\ \frac{1}{2}\left(\frac{\partial w}{\partial x} - \frac{\partial u}{\partial z}\right) & \frac{1}{2}\left(\frac{\partial w}{\partial y} - \frac{\partial v}{\partial z}\right) & 0 \end{bmatrix}$$

(4.34)

For incompressible flows, from the continuous equation,

$$\tilde{P} = -(\lambda_1 + \lambda_2 + \lambda_3) = -tr\left(\nabla \vec{v}\right) = -\left(\frac{\partial u}{\partial x} + \frac{\partial v}{\partial y} + \frac{\partial w}{\partial z}\right) = 0 \qquad (4.35)$$

But for compressible flows, the divergence of the velocity is no longer equal to zero. Therefore, Kolář (2009) and Epps (2017) suggest that the criterion should only depend on the deviatoric part of the velocity gradient tensor, which can be written as

$$\nabla \vec{v}_D \equiv \nabla \vec{v} - kI = \begin{bmatrix} \frac{\partial u}{\partial x} - k & \frac{\partial u}{\partial y} & \frac{\partial u}{\partial z} \\ \frac{\partial v}{\partial x} & \frac{\partial v}{\partial y} - k & \frac{\partial v}{\partial z} \\ \frac{\partial w}{\partial x} & \frac{\partial w}{\partial y} & \frac{\partial w}{\partial z} - k \end{bmatrix} \qquad (4.36)$$

$$k = \frac{1}{3}tr\left(\nabla \vec{v}\right) = -\frac{1}{3}\tilde{P} = \frac{1}{3}(\lambda_1 + \lambda_2 + \lambda_3) \qquad (4.37)$$

Here, $k > 0$ means isentropic expansion while $k < 0$ means compression. And the deviatoric part of the strain rate tensor is

$$S_D \equiv S - kI = \begin{bmatrix} \dfrac{\partial u}{\partial x} - k & \dfrac{1}{2}\left(\dfrac{\partial u}{\partial y} + \dfrac{\partial v}{\partial x}\right) & \dfrac{1}{2}\left(\dfrac{\partial u}{\partial z} + \dfrac{\partial w}{\partial x}\right) \\[3mm] \dfrac{1}{2}\left(\dfrac{\partial v}{\partial x} + \dfrac{\partial u}{\partial y}\right) & \dfrac{\partial v}{\partial y} - k & \dfrac{1}{2}\left(\dfrac{\partial v}{\partial z} + \dfrac{\partial w}{\partial y}\right) \\[3mm] \dfrac{1}{2}\left(\dfrac{\partial w}{\partial x} + \dfrac{\partial u}{\partial z}\right) & \dfrac{1}{2}\left(\dfrac{\partial w}{\partial y} + \dfrac{\partial v}{\partial z}\right) & \dfrac{\partial w}{\partial z} - k \end{bmatrix} \tag{4.38}$$

According to Eqs. (4.36)–(4.38), it could be obtained that

$$tr(\nabla \vec{v}_D) = tr(S_D) = \left(\dfrac{\partial u}{\partial x} - k\right) + \left(\dfrac{\partial v}{\partial y} - k\right) + \left(\dfrac{\partial w}{\partial z} - k\right) = 0 \tag{4.39}$$

Liutex is uniquely determined by the vorticity vector $\vec{\omega}$, the imaginary part of the conjugate complex eigenvalue λ_{ci}, and the eigenvector $\vec{r}$ corresponding to the real eigenvalue. Accordingly, in the following, the deviatoric part of the velocity gradient tensor is examined to study the effect on $\vec{\omega}$, λ_{ci}, and $\vec{r}$.

Obviously, the vorticity vector $\vec{\omega}$ only depends on the off-diagonal terms of the velocity gradient, so $\vec{\omega}$ remains the same if the deviatoric part of the velocity gradient tensor is applied.

The eigendecomposition can be written as

$$\nabla \vec{v} = \begin{bmatrix} \vec{r} & \vec{v}_{cr} & \vec{v}_{ci} \end{bmatrix} \begin{bmatrix} \lambda_r & 0 & 0 \\ 0 & \lambda_{cr} & \lambda_{ci} \\ 0 & -\lambda_{ci} & \lambda_{cr} \end{bmatrix} \begin{bmatrix} \vec{r} & \vec{v}_{cr} & \vec{v}_{ci} \end{bmatrix}^{-1} \tag{4.40}$$

According to the definition of eigenvalue and Eq. (4.25),

$$\nabla \vec{v} \cdot (\vec{v}_{cr} + i\vec{v}_{ci}) = (\lambda_{cr} + i\lambda_{ci})(\vec{v}_{cr} + i\vec{v}_{ci}) \tag{4.41}$$

Therefore,

$$\begin{aligned} \nabla \vec{v}_D \cdot (\vec{v}_{cr} + i\vec{v}_{ci}) &= \left(\nabla \vec{v} - kI\right) \cdot (\vec{v}_{cr} + i\vec{v}_{ci}) = \nabla \vec{v} \cdot (\vec{v}_{cr} + i\vec{v}_{ci}) - kI \cdot (\vec{v}_{cr} + i\vec{v}_{ci}) \\ &= (\lambda_{cr} + i\lambda_{ci})(\vec{v}_{cr} + i\vec{v}_{ci}) - k(\vec{v}_{cr} + i\vec{v}_{ci}) = (\lambda_{cr} - k + i\lambda_{ci})(\vec{v}_{cr} + i\vec{v}_{ci}) \end{aligned} \tag{4.42}$$

Eq. (4.42) indicates that the uniform dilatation given by k does not alter the imaginary part of the complex eigenvalue; λ_{ci} and λ_{ci} remain the same for the deviatoric part of the velocity gradient tensor.

For the real eigenvector,

$$\nabla \vec{v} \cdot \vec{r} = \lambda_r \vec{r} \tag{4.43}$$

$$\nabla \vec{v}_D \cdot \vec{r} = \left(\nabla \vec{v} - kI\right) \cdot \vec{r} = \nabla \vec{v} \cdot \vec{r} - kI \cdot \vec{r} = \lambda_r \vec{r} - k\vec{r} = (\lambda_r - k)\vec{r} \tag{4.44}$$

Eq. (4.44) implies that the real eigenvalue is altered but the real eigenvector of the original velocity gradient tensor retains the real eigenvector of the deviatoric part of the velocity gradient tensor.

Consequently $\vec{\omega}$, λ_{ci}, and $\vec{r}$ remain the same even if the deviatoric part of the of velocity gradient tensor is applied, independent of the compressibility effect. So, all the scalar, vector, and tensor forms of Liutex are valid for compressible flows without any modification.

Here the implicit large-eddy simulation (ILES) data of MVG at Mach 2.5 and $Re_\theta = 5760$ (Reynolds number based on momentum thickness) are examined to compare Liutex with the λ_{ci} and Q_D criteria. The ILES data are generated by a LES code called LESUTA (Li and Liu, 2011). A fifth-order bandwidth-optimized WENO scheme is applied for the convective terms and the traditional fourth-order central scheme is used twice to compute the second-order derivatives for viscous terms. Without explicitly using the subgrid scale (SGS) model, the intrinsic dissipation of the numerical method is utilized to dissipate the turbulent energy accumulated at the unresolved scales with high wave numbers. For time integration, the explicit third-order TVD-type Runge–Kutta scheme is employed. The grid number for the whole field is $n_{\text{spanwise}} \times n_{\text{normal}} \times n_{\text{streamwise}} = 137 \times 192 \times 1600$. For the detailed case setup, please refer to Yan and Liu (2014).

Fig. 4.7 illustrates iso-surfaces obtained by Q_D, λ_{ci}, and Liutex with small thresholds. It can be found that vortical structures identified by three methods are very similar (the structures identified by λ_{ci} and Liutex are not so smooth as Q_D since Q_D is a smooth function while λ_{ci} and Liutex are truncated according to their definitions). This implies that all methods present a similar (approximated) vortex boundary for compressible flows. However, λ_{ci} and Q_D would be severely contaminated by shearing. To investigate the effect of the contamination by shearing, three points A, B, and C on the iso-surface of Liutex ($R = 1.0$), as shown in Fig. 4.8, are examined. Point A is located immediately after MVG. Point B is located in the middle part and Point C is located after the weakened shock wave. Table 4.1 lists the eigenvalues, Q_D, Liutex, and shearing components of Points A, B, C obtained by Liutex-based tensor decomposition (detailed discuss in Chapter 5), which demonstrate the local rotation, expansion or compression, and shearing effects. It is obvious that the shearing components s in the plane perpendicular to the local rotational axis are relatively large and can be much larger than the magnitude of Liutex, which implies that the shearing effect is so strong that large portions of Q_D and λ_{ci} are severely contaminated by shearing. It should be also noted that the real eigenvalue and the real part of the complex eigenvalue, which indicate the expansion or compression effects, are no long small and may be important for compressible flows. The Liutex-based tensor decomposition can provide the complete rotation, expansion or compression, and shearing effects of any point, manifesting the superiority over the scalar-based criteria, which can only present iso-surfaces to detect vortical structures.

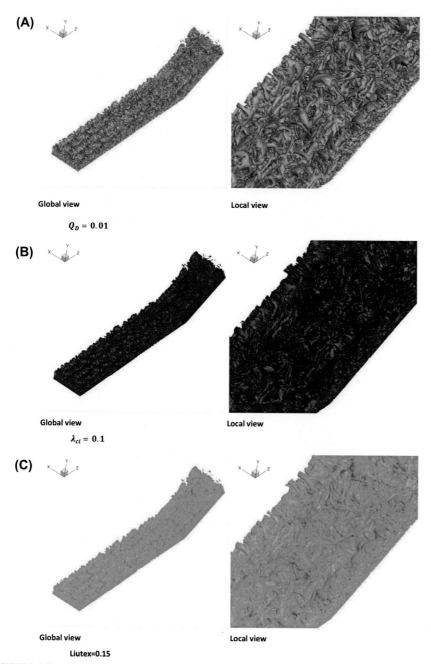

FIGURE 4.7

Iso-surfaces of Q_D, λ_{ci}, and Liutex for the MVG case with small thresholds. (A) $Q_D = 0.01$, (B) $\lambda_{ci} = 0.1$, (C) Liutex $= 0.15$.

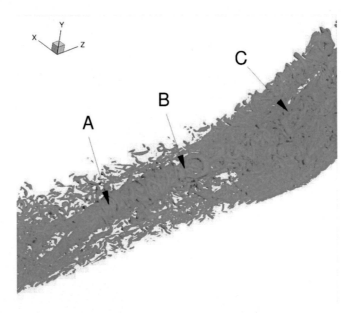

FIGURE 4.8

Iso-surfaces of Liutex with $R = 1.0$. Three points are examined to study the contamination by shearing.

Table 4.1 Eigenvalues (λ_r, λ_{cr}, λ_{ci}), Q_D, Liutex, and shearing components of Point A, B, C.

Variable	Point A	Point B	Point C
λ_r	0.671	0.258	−0.076
λ_{cr}	−0.191	−0.123	−0.523
λ_{ci}	1.716	1.046	1.391
Q_D	2.807	1.778	1.853
R	1.0	1.0	1.0
s	5.610	1.776	3.454
ξ	1.247	0.031	1.082
η	−0.372	−0.158	−0.924

4.7 Summary

In this chapter, the existence, uniqueness, continuity, stability in physics, Galilean invariance, applicability to compressible flow are all proved. The stability of Liutex in nature can be observed in anywhere. This bias feature of Liutex may cause the asymmetry of turbulence, which will be discussed in Chapter 8.

Liutex-based velocity gradient tensor decomposition and vorticity R-S decomposition

5.1 Short review on velocity and velocity gradient tensor decomposition

The Cauchy–Stokes tensor decomposition or $\nabla \vec{v} = A + B = \frac{1}{2}\left(\nabla\vec{v} + \nabla\vec{v}^T\right) + \frac{1}{2}\left(\nabla\vec{v} - \nabla\vec{v}^T\right)$, which is equivalent to Helmholtz velocity decomposition, has been widely accepted as the foundation of fluid kinematics for long time (Batchelor, 2000). However, there are several problems with these decompositions, which cannot be neglected. Firstly, the Cauchy–Stokes decomposition itself is not Galilean invariant, which means under different coordinate frames, the elements of the velocity tensor are varying with the frame rotation, showing that stretching (compression) and deformation are dependent on the angle of the frame rotation when $\nabla\vec{V} = Q\nabla\vec{v}Q^T$, where Q is an orthogonal and rotation matrix (Liu et al., 2018a; Gao et al., 2018, 2019a; Yu et al., 2020). Another problem is that the antisymmetric part of the velocity gradient tensor is not the proper quantity to represent fluid rotation as discussed many times in the previous chapters. Similarly, Helmholtz's velocity decomposition divides the fluid velocity to two parts, one has potential (deformation but no rotation) and the other has rotation (vorticity but no divergence). They are correct in mathematics. However, these decompositions induce many questions as vorticity has no correlation with flow rotation and $\nabla \times \vec{v} \neq 0$ does not mean the flow is rotational. On the other hand, vorticity has part of deformation as well, which is antisymmetric shear. Helmholtz (1858) first considered a vorticity tube with infinitesimal cross section as a vortex filament, which was followed by Lamb (1932) to simply call a vortex filament as a vortex in his classic monograph. Since vorticity is well defined, vorticity dynamics has been systematically developed for the generation and evolution of vorticity and applied in the study of vortical-flow stability and vortical structures in transitional and turbulent flows (Wu et al., 2006; Saffman, 1992; Majda and Bertozzi, 2001). However, the use of vorticity will meet severe difficulties in viscous flows, especially in turbulence: (1) vorticity is unable to distinguish between a real rotational region and a shear layer region; (2) it has been noticed by several researchers that the local vorticity vector is not always aligned with the direction of vortical structures in turbulent wall-bounded flows, especially at locations close to the wall (Zhou et al., 1999; Pirozzoli et al., 2008; Gao et al., 2011); (3) the maximum vorticity does not necessarily occur in the central region of vortical structures (Robinson, 1990a,b; Wang et al., 2017). In general, the previous "vortex dynamics" work is linked to vorticity dynamics and has nothing to do with vortex dynamics since vortex is not correlated to vorticity. On the other hand, all eigenvalue-based vortex identification methods are scalar and seriously contaminated by shears and/or stretching (compression) as all elements of the velocity gradient tensor make contributions to the vortex identification criteria like Q-critera. This issue prompts Kolář (2007) to formulate a triple decomposition from which the residual vorticity can be obtained after the extraction of an effective pure shearing motion and represents a direct and accurate measure of the pure rigid-body rotation of a fluid element. However, the triple decomposition requires a basic reference frame to be first determined. Searching for the basic reference frame in 3D cases will result in an expensive optimization problem for every point in the flow field, which limits the applicability of the method (Epps, 2017). Kolář et al. (2013) also introduced the concepts of the maximum corotation and the average corotation of line segments near a point and apply these methods for vortex identification. However, the maxima and the averaged corotation vector is evaluated by integration over a unit sphere, which makes it difficult to be used to study the transport property of vortex. Anyway, a perfect solution for the optimization has not been found yet.

The vector and tensor decomposition should be unique and have a clear physical meaning, such as rotation, stretching (compression), and shear. This must require a unique coordinate and unique decomposition, which have been introduced in previous chapters. They are called "Principal Coordinate" and "Principal Decomposition" of vorticity and velocity gradient tensor. These decompositions are not only unique but also Galilean invariant. The principal decomposition, which is based on Liutex vector and Liutex tensor, is unique for correct vorticity and tensor decomposition. They should replace the traditional Helmholtz decomposition and Cauchy–Stokes decomposition, which have caused countless confusion in fluid dynamics.

5.2 Basic form of fluid motion

First, it is necessary to make it clear which part of the velocity gradient represents rigid body rotation, stretching, shear deformation, and symmetric deformation. For simplicity, one can consider a 2-D velocity gradient tensor:

$$\nabla v = \begin{bmatrix} \dfrac{\partial u}{\partial x} & \dfrac{\partial u}{\partial y} \\[2mm] \dfrac{\partial v}{\partial x} & \dfrac{\partial v}{\partial y} \end{bmatrix} \tag{5.1}$$

Definition 5.1. In a 2D x-y frame, the velocity gradient corresponds to the rigid body rotation if:

$$(1)\frac{\partial u}{\partial y} = -\frac{\partial v}{\partial x}$$

$$(2)\frac{\partial u}{\partial x} = \frac{\partial v}{\partial y} = 0 \tag{5.2}$$

Suppose the velocity at O is 0 and $\frac{\partial u}{\partial y} = -\frac{\partial v}{\partial x} = -\omega$, the coordinates at any point on the circle could be expressed as $(\cos\theta, \sin\theta)$ and the velocity at that point is:

$$\vec{v} = \begin{bmatrix} u \\ v \end{bmatrix} = \nabla v \begin{bmatrix} \cos\theta \\ \sin\theta \end{bmatrix} = \begin{bmatrix} 0 & -\omega \\ \omega & 0 \end{bmatrix}\begin{bmatrix} \cos\theta \\ \sin\theta \end{bmatrix} = \begin{bmatrix} -\omega\sin\theta \\ \omega\cos\theta \end{bmatrix} \tag{5.3}$$

The magnitude of the velocity is:

$$\left|\vec{v}\right| = \sqrt{(-\omega\sin\theta)^2 + (\omega\cos\theta)^2} = \omega \tag{5.4}$$

For each point on the circle, the direction of the velocity is tangent to the circle and the magnitude of velocity is ω as shown in Fig. 5.1. Therefore, the movement is a rigid body rotation (see Figs. 5.1 and 5.2).

Definition 5.2. The velocity gradient corresponds to stretching if:

$$(1)\quad \frac{\partial u}{\partial y} = \frac{\partial v}{\partial x} = 0$$

$$\tag{5.5}$$

$$(2)\quad \text{At least one of } \frac{\partial u}{\partial x} \text{ and } \frac{\partial v}{\partial y} \text{ is nonzero.}$$

Suppose the velocity at O is 0, $\frac{\partial u}{\partial x} = 0$ and $\frac{\partial v}{\partial y} = c$. Then, $v_a = ac$ and $v_b = -ac$. After a small time step, the control volume will become longer as shown in Fig. 5.2. Therefore, this movement is stretching.

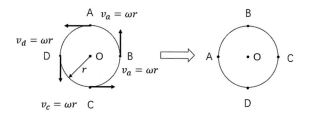

FIGURE 5.1

Movement of velocity gradient corresponding to rigid rotation.

Reproduced from Yu, Y., et al., Principal coordinates and principal velocity gradient tensor decomposition, Journal of Hydrodynamics 32, 2020, with the permission of Journal of Hydrodynamics.

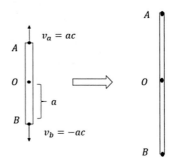

FIGURE 5.2

Movement of velocity gradient tensor corresponding to stretching.

*Reproduced from Yu, Y., et al., Principal coordinates and principal velocity gradient tensor decomposition,
Journal of Hydrodynamics 32, 2020, with the permission of Journal of Hydrodynamics.*

Definition 5.3.   The velocity gradient corresponds to shear deformation if:

$$(1)\ \text{Either}\ \frac{\partial u}{\partial y}\ \text{or}\ \frac{\partial v}{\partial x}\ \text{is 0 but the other one is nonzero.}$$

$$(5.6)$$

$$(2)\ \frac{\partial u}{\partial x} = \frac{\partial v}{\partial y} = 0$$

Suppose $\frac{\partial u}{\partial y} = c > 0$, $\frac{\partial v}{\partial x} = 0$ and velocity at O is 0, then $u_A = ac$, $u_B = -ac$ and $v = 0$ in the whole domain. After a small time step, the shape of the control volume will become the right part of Fig. 5.3. Therefore, it is a shear deformation.

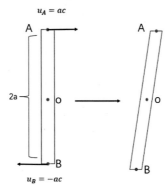

FIGURE 5.3

Movement of velocity gradient tensor corresponding to shear deformation.

*Reproduced from Yu, Y., et al., Principal coordinates and principal velocity gradient tensor decomposition,
Journal of Hydrodynamics 32, 2020, with the permission of Journal of Hydrodynamics.*

Definition 5.4. The velocity gradient corresponds to symmetric deformation if:

$$(1)\ \ \frac{\partial u}{\partial y} = \frac{\partial v}{\partial x}$$

$$(2)\ \ \frac{\partial u}{\partial x} = \frac{\partial v}{\partial y} = 0 \tag{5.7}$$

Suppose the velocity at O is 0, $\frac{\partial u}{\partial y} = \frac{\partial v}{\partial x} = c$. As shown in Fig. 5.4, the velocity distributes linearly along the two legs and the behavior is symmetric deformation. Also, symmetric deformation could be considered as a combination of two shear deformations. One is $\frac{\partial u}{\partial y} = c$ and $\frac{\partial v}{\partial x} = 0$, and the other is $\frac{\partial v}{\partial x} = c$ and $\frac{\partial u}{\partial y} = 0$.

5.3 Problems with Cauchy–Stokes decomposition
5.3.1 Vorticity cannot represent fluid rotation

To illustrate that vorticity is not a proper way to represent rotation, the following 2-D Couette flow, as shown in Fig. 5.5, is considered:

$$\begin{cases} u = 2ay \\ v = 0 \end{cases} \tag{5.8}$$

Here, a is a positive constant, u and v are respectively the x-component and y-component of the velocity.

The velocity gradient tensor is:

$$\nabla \vec{v} = \begin{bmatrix} 0 & 2a \\ 0 & 0 \end{bmatrix} \tag{5.9}$$

The Cauchy–Stokes decomposition is:

$$\nabla v = \begin{bmatrix} 0 & 2a \\ 0 & 0 \end{bmatrix} = A + B = \begin{bmatrix} 0 & a \\ a & 0 \end{bmatrix} + \begin{bmatrix} 0 & a \\ -a & 0 \end{bmatrix} \tag{5.10}$$

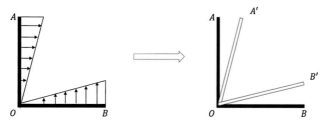

FIGURE 5.4

Movement of velocity gradient tensor corresponding to symmetric deformation.

Reproduced from Yu, Y., et al., Principal coordinates and principal velocity gradient tensor decomposition, Journal of Hydrodynamics 32, 2020, with the permission of Journal of Hydrodynamics.

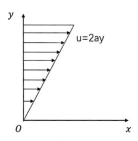

FIGURE 5.5

2-D Couette flow.

Reproduced from Yu, Y., et al., Principal coordinates and principal velocity gradient tensor decomposition, Journal of Hydrodynamics 32, 2020, with the permission of Journal of Hydrodynamics.

This decomposition is correct in mathematics; however, its physical meaning is very suspicious. From Def. 5.3, (5.9) represents a shear deformation that contains no rotation, but the Cauchy–Stokes decomposition (5.10) results in a rotational part that does not exist. Let us analyze this case in detail to see where this artificial rotation comes from. In the original matrix, $\frac{\partial u}{\partial y} = 2a$. This "$2a$" is divided into "$a$" and "$a$", then equally distributed to matrix A and matrix B. Similarly, $\frac{\partial v}{\partial x} = 0$ is split into "$-a$" and "a", and they are distributed to matrix A and matrix B, as shown in Fig. 5.6.

Let's look at "$\frac{\partial v}{\partial x} = 0$". The $\frac{\partial v}{\partial x}$ component does not have anything in the original matrix, but it is forced to give "$-a$" to matrix B to make a faked rigid body rotation, which requires $\frac{\partial u}{\partial y} = -\frac{\partial v}{\partial x}$ in Def. 5.1. Meanwhile an artificial "a" is added to matrix A for balance. This process made by the Cauchy–Stokes decomposition artificially makes something from nothing to rotation and additional deformation. This is a man-made rotation by the Cauchy–Stokes decomposition. To avoid this faked rotation, a decompose rule must be set up.

These rules of fluid motion decomposition are:

(1) If $p > 0$, then a decomposition of a $p = \sum v_i$ must satisfy $0 \leq v_i \leq p$

(2) If $p > 0$, then a decomposition of a $p = \sum v_i$ must satisfy $p \leq v_i \leq 0$

(3) If $p = 0$, then a decomposition of a $p = \sum v_i$ must satisfy $v_i = 0$

$$(5.11)$$

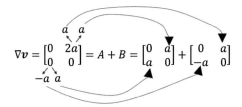

FIGURE 5.6

Distribution of Cauchy–Stokes decomposition.

Reproduced from Yu, Y., et al., Principal coordinates and principal velocity gradient tensor decomposition, Journal of Hydrodynamics 32, 2020, with the permission of Journal of Hydrodynamics.

Here, p is a sum and v_i is an element in a decomposition. Based on this decomposition criterion, the decomposition of $\frac{\partial v}{\partial x} = 0$ must be 0. From Def. 5.1, in a rotation matrix, one has $\frac{\partial u}{\partial y} = -\frac{\partial v}{\partial x}$ and $\frac{\partial u}{\partial x} = \frac{\partial v}{\partial y} = 0$. Since $\frac{\partial u}{\partial y} = 0$, the rotational part of (5.10) is $B = \begin{bmatrix} 0 & 0 \\ 0 & 0 \end{bmatrix}$, indicating there is no rotation, which is consistent with the physical fact. The Cauchy–Stokes tensor decomposition is found to violate the aforementioned criterion. This criterion is easy to understand that one can decompose 3 to be 2 and 1, but it is not appropriate to decompose 3 to be -7 and 10 although $2 + 1 = 3$ and $-7 + 10 = 3$ as well. This is the problem for Cauchy–Stokes to make a faked rotation. Apparently, the Cauchy–Stokes decomposition (Helmholtz decomposition) violates the fluid motion decomposition rule and should be strictly avoided.

So, to find the real rotational part in the decomposition of velocity gradient tensor, instead of $B = \frac{1}{2}(\nabla V - \nabla V^T)$, it seems we should choose the minimum of $\left|\frac{\partial v}{\partial x}\right|$ and $\left|\frac{\partial u}{\partial y}\right|$, otherwise the decompose criteria are not satisfied. It is reasonable to define B as:

$$ B = \begin{bmatrix} 0 & -\min\left(\left|\frac{\partial v}{\partial x}\right|, \left|\frac{\partial u}{\partial y}\right|\right) \\ \min\left(\left|\frac{\partial v}{\partial x}\right|, \left|\frac{\partial u}{\partial y}\right|\right) & 0 \end{bmatrix} \tag{5.12} $$

if $\frac{\partial u}{\partial y} < 0$

However, there is another problem, $\left|\frac{\partial v}{\partial x}\right|$ and $\left|\frac{\partial u}{\partial y}\right|$ are changing when the coordinate rotates. If the given coordinate is rotated with angle θ anticlockwise to $Ox'y'$ (Fig. 5.7), the rotation matrix is:

$$ P = \begin{bmatrix} \cos\theta & \sin\theta \\ -\sin\theta & \cos\theta \end{bmatrix} \tag{5.13} $$

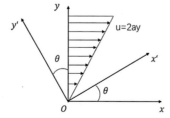

FIGURE 5.7

2-D laminar flow in different coordinates.

Reproduced from Yu, Y., et al., Principal coordinates and principal velocity gradient tensor decomposition,
Journal of Hydrodynamics 32, 2020, with the permission of Journal of Hydrodynamics.

The velocity gradient tensor under the new coordinate should be:

$$\nabla V_1 = P \nabla V P^{-1} = \begin{bmatrix} 2a \sin\theta \cos\theta & 2a \cos^2\theta \\ -2a \sin^2\theta & -2a \sin\theta \cos\theta \end{bmatrix} \tag{5.14}$$

Therefore the elements are dependent on θ, which are not Galilean invariant. The principal coordinate must be applied (see Chapter 2). The rotation strength will be uniquely defined, which requires a principal decomposition in the principal coordinate.

5.3.2 Galilean invariant

Failing to satisfy Galilean invariant is another problem with the Cauchy–Stokes decomposition. If the decomposition is different under different coordinates, then it is hard to determine which one is correct. Let's look at a 3-D example.

$$\nabla V_0 = \begin{bmatrix} 0 & 2a & 0 \\ 0 & 0 & 0 \\ 0 & 0 & 0 \end{bmatrix} \tag{5.15}$$

Performing the Cauchy–Stokes decomposition, it can be obtained that:

$$A_0 = \begin{bmatrix} 0 & a & 0 \\ a & 0 & 0 \\ 0 & 0 & 0 \end{bmatrix}$$

$$B_0 = \begin{bmatrix} 0 & a & 0 \\ -a & 0 & 0 \\ 0 & 0 & 0 \end{bmatrix} \tag{5.16}$$

$$C_0 = \begin{bmatrix} 0 & 0 & 0 \\ 0 & 0 & 0 \\ 0 & 0 & 0 \end{bmatrix}$$

where A_0 is the symmetric deformation part, B_0 is the rotation part, C_0 is the stretching part.

$$\begin{aligned} ||A_0||_1 &= a & ||A_0||_2 &= a & ||A_0||_\infty &= a \\ ||B_0||_1 &= a & ||B_0||_2 &= a & ||B_0||_\infty &= a \\ ||C_0||_1 &= 0 & ||C_0||_2 &= 0 & ||C_0||_\infty &= 0 \end{aligned} \tag{5.17}$$

where $1, 2, \infty$ represent L_1, L_2, L_∞ norms, respectively.

Look at following matrices, for example:

$$P_1 = \begin{bmatrix} \dfrac{\sqrt{3}}{2} & \dfrac{1}{2} & 0 \\[2mm] -\dfrac{1}{2} & \dfrac{\sqrt{3}}{2} & 0 \\[2mm] 0 & 0 & 1 \end{bmatrix}$$

$$P_2 = \begin{bmatrix} \dfrac{\sqrt{3}}{2} & 0 & \dfrac{1}{2} \\[2mm] 0 & 1 & 0 \\[2mm] -\dfrac{1}{2} & 0 & \dfrac{\sqrt{3}}{2} \end{bmatrix}$$

(5.18)

After rotation, the velocity gradient tensor under the new coordinates is:

$$\nabla V_1 = P_2 P_1 \nabla V_0 P_1^T P_2^T = \begin{bmatrix} \dfrac{3\sqrt{3}\,a}{8} & \dfrac{3\sqrt{3}\,a}{4} & -\dfrac{3a}{8} \\[2mm] -\dfrac{\sqrt{3}\,a}{4} & -\dfrac{\sqrt{3}\,a}{2} & \dfrac{a}{4} \\[2mm] -\dfrac{\sqrt{3}\,a}{8} & -\dfrac{3a}{4} & \dfrac{\sqrt{3}\,a}{8} \end{bmatrix}$$

(5.19)

Performing the Cauchy–Stokes decomposition,

$$A_1 = \begin{bmatrix} 0 & \dfrac{\sqrt{3}\,a}{2} & 0 \\[2mm] -\dfrac{\sqrt{3}\,a}{2} & 0 & \dfrac{a}{2} \\[2mm] 0 & -\dfrac{a}{2} & 0 \end{bmatrix}$$

$$B_1 = \begin{bmatrix} 0 & \dfrac{\sqrt{3}\,a}{4} & -\dfrac{3a}{8} \\[2mm] \dfrac{\sqrt{3}\,a}{4} & 0 & -\dfrac{a}{4} \\[2mm] -\dfrac{3a}{8} & -\dfrac{a}{4} & 0 \end{bmatrix}$$

(5.20)

$$C_1 = \begin{bmatrix} \dfrac{\sqrt{3}\,a}{8} & 0 & 0 \\[2mm] 0 & -\dfrac{\sqrt{3}\,a}{2} & 0 \\[2mm] 0 & 0 & \dfrac{\sqrt{3}\,a}{8} \end{bmatrix}$$

$$\|A_1\|_1 = \frac{(\sqrt{3}+1)a}{2} \qquad \|A_1\|_2 = a \qquad \|A_1\|_\infty = \frac{(\sqrt{3}+1)a}{2}$$

$$\|B_1\|_1 = \frac{(2\sqrt{3}+3)a}{8} \qquad \|B_1\|_2 = 0.7106a \qquad \|B_1\|_\infty = \frac{(2\sqrt{3}+3)a}{8} \qquad (5.21)$$

$$\|C_1\|_1 = \frac{\sqrt{3}a}{2} \qquad \|C_1\|_2 = \frac{\sqrt{3}a}{2} \qquad \|C_1\|_\infty = \frac{\sqrt{3}a}{2}$$

It is not difficult to find that the norm of rotation, symmetric deformation, and stretching matrices are all different, except for $\|A\|_2$, under different coordinate frames. A question arises that under which coordinates does the Cauchy−Stokes tensor decomposition give the right stretching (compression) and deformation? The answer is the velocity tensor decomposition must be conducted under the "Principal Coordinate," which has been defined in Chapter 2. Under the principal coordinate, coordinate $\vec{Z}$ is the real eigenvector $\vec{r}$ and $\frac{\partial U_\vartheta}{\partial X} = \frac{\partial V_\vartheta}{\partial Y}$. Performing the velocity gradient decomposition under this coordinate system can easily, clearly, and correctly get the rotation, stretching, and shear deformation.

5.4 Principal coordinates and principal decompositions

Traditionally, the velocity gradient tensor $\nabla \vec{v}$ is decomposed to a symmetric part A representing deformation and an antisymmetric part B representing rotation (the Cauchy−Stokes decomposition). A is called strain rate tensor and B is called vorticity tensor. And the local velocity increment decomposition between two neighboring points can be written as $d\vec{v} = A \cdot d\vec{r} + \frac{1}{2}\left(\nabla \times \vec{v}\right) \times d\vec{r}$. Although vorticity is well defined and considered as a fluid rotation measurement for a long time, it has very weak correlation with fluid rotation.

The problems of vorticity for the identification and visualization of vortical structures prompt Liu et al. to propose an idea that vorticity should be decomposed to a rotational part (a part of vorticity associated with rigid body rotation) and a nonrotational part (a part of vorticity due to shear) in 2016 and 2018, which is called Liutex.

$$\nabla \vec{v} \cdot \vec{r} = \lambda_r \vec{r} \qquad (5.22)$$

where λ_r is the real eigenvalue of the velocity gradient tensor. A coordinate rotation (Q rotation) is used to rotate the original z-axis to the direction of the local rotational axis $\vec{r}$ and the velocity gradient tensor ∇V in the resulting XYZ frame will be given by

$$\nabla \vec{V} = Q\nabla \vec{v} Q^{\mathrm{T}} = \begin{bmatrix} \dfrac{\partial U}{\partial X} & \dfrac{\partial U}{\partial Y} & 0 \\[2ex] \dfrac{\partial V}{\partial X} & \dfrac{\partial V}{\partial Y} & 0 \\[2ex] \dfrac{\partial W}{\partial X} & \dfrac{\partial W}{\partial Y} & \dfrac{\partial W}{\partial Z} \end{bmatrix} \qquad (5.23)$$

where $\boldsymbol{Q}$ is a (proper) rotation matrix and U, V, W represent the velocity components in the XYZ frame. And then, a second rotation (P rotation) is applied to rotate the reference frame around the Z-axis and the corresponding velocity gradient tensor $\nabla\vec{V}_\theta$ can be written as

$$\nabla\vec{V}_\theta = \boldsymbol{P}\nabla\vec{V}\boldsymbol{P}^{-1} \tag{5.24}$$

where $\boldsymbol{P}$ is a (proper) rotation matrix around the Z-axis and can be written as

$$\boldsymbol{P} = \begin{bmatrix} \cos\theta & \sin\theta & 0 \\ -\sin\theta & \cos\theta & 0 \\ 0 & 0 & 1 \end{bmatrix} \tag{5.25}$$

Therefore, the components of 2×2 upper left submatrix of $\nabla\vec{V}_\theta$ are

$$\left.\frac{\partial U}{\partial Y}\right|_\theta = \alpha\sin(2\theta + \varphi) - \beta \tag{5.26a}$$

$$\left.\frac{\partial V}{\partial X}\right|_\theta = \alpha\sin(2\theta + \varphi) + \beta \tag{5.26b}$$

$$\left.\frac{\partial U}{\partial X}\right|_\theta = -\alpha\cos(2\theta + \varphi) + \frac{1}{2}\left(\frac{\partial U}{\partial X} + \frac{\partial V}{\partial Y}\right) \tag{5.26c}$$

$$\left.\frac{\partial V}{\partial Y}\right|_\theta = \alpha\cos(2\theta + \varphi) + \frac{1}{2}\left(\frac{\partial U}{\partial X} + \frac{\partial V}{\partial Y}\right) \tag{5.26d}$$

where

$$\alpha = \frac{1}{2}\sqrt{\left(\frac{\partial V}{\partial Y} - \frac{\partial U}{\partial X}\right)^2 + \left(\frac{\partial V}{\partial X} + \frac{\partial U}{\partial Y}\right)^2} \tag{5.27}$$

$$\beta = \frac{1}{2}\left(\frac{\partial V}{\partial X} - \frac{\partial U}{\partial Y}\right) \tag{5.28}$$

$$\varphi = \begin{cases} a\cos\left(\dfrac{\frac{1}{2}\left(\frac{\partial V}{\partial Y} - \frac{\partial U}{\partial X}\right)}{\alpha}\right), & \frac{\partial V}{\partial X} + \frac{\partial U}{\partial Y} \geq 0 \\[4ex] a\sin\left(\dfrac{\frac{1}{2}\left(\frac{\partial V}{\partial X} + \frac{\partial U}{\partial Y}\right)}{\alpha}\right), & \frac{\partial V}{\partial X} + \frac{\partial U}{\partial Y} < 0, \ \frac{\partial V}{\partial Y} - \frac{\partial U}{\partial X} \geq 0 \\[4ex] a\sin\left(\dfrac{-\frac{1}{2}\left(\frac{\partial V}{\partial X} + \frac{\partial U}{\partial Y}\right)}{\alpha}\right) + \pi, & \frac{\partial V}{\partial X} + \frac{\partial U}{\partial Y} < 0, \ \frac{\partial V}{\partial Y} - \frac{\partial U}{\partial X} < 0 \end{cases} \tag{5.29}$$

(Note: If $\alpha = 0$, we have $\frac{\partial V}{\partial Y} - \frac{\partial U}{\partial X} = 0, \frac{\partial V}{\partial X} + \frac{\partial U}{\partial Y} = 0, \frac{\partial V}{\partial X} = \beta, \frac{\partial U}{\partial Y} = -\beta$ for any θ. In this case, φ is not needed.)

The Liutex magnitude is defined as twice the minimal absolute value of the off-diagonal component of the 2×2 upper left submatrix and is given by

$$R = \begin{cases} 2(\beta - \alpha), & \beta^2 > \alpha^2 \\ 0, & \alpha^2 \geq \beta^2 \end{cases} \tag{5.30}$$

And the vector form of Liutex is obtained by

$$\vec{R} = R\vec{r}$$

$$R = \left(\vec{\omega} \cdot \vec{r}\right) - \sqrt{\left(\vec{\omega} \cdot \vec{r}\right)^2 - 4\lambda_{ci}^2} \text{ and } \vec{\omega} \cdot \vec{r} > 0 \tag{5.31}$$

According to Gao et al. (2018), when $2\theta + \varphi = \pi/2$, the velocity gradient tensor given by Eq. (5.23) becomes

$$\nabla \vec{V}_{\theta \text{ min}} = \begin{bmatrix} \frac{1}{2}\left(\frac{\partial U}{\partial X} + \frac{\partial V}{\partial Y}\right) & -(\beta - \alpha) & 0 \\ \beta + \alpha & \frac{1}{2}\left(\frac{\partial U}{\partial X} + \frac{\partial V}{\partial Y}\right) & 0 \\ \frac{\partial W}{\partial X}\Big|_{\theta \text{ min}} & \frac{\partial W}{\partial Y}\Big|_{\theta \text{ min}} & \frac{\partial W}{\partial Z}\Big|_{\theta \text{ min}} \end{bmatrix} \tag{5.32}$$

In order to simplify the expressions, in the following let us omit the subscript θ min and still use X, Y, Z to represent the resulting reference frame under the second rotation. It should be noted that the following Liutex-based decompositions are performed in the XYZ frame, which is obtained through successive Q rotation and P rotation when the velocity gradient tensor has two complex conjugates and one real eigenvalue. If the decomposition in the original xyz frame is required, the corresponding coordinate transformations will be needed. Let us define that

$$\lambda_r = \frac{\partial W}{\partial Z} \tag{5.33}$$

$$\lambda_{cr} = \frac{1}{2}\left(\frac{\partial U}{\partial X} + \frac{\partial V}{\partial Y}\right) \tag{5.34}$$

$$\beta - \alpha = R/2 \text{ if } \beta > 0 \tag{5.35}$$

$$s = 2\alpha \tag{5.36}$$

$$\xi = \frac{\partial W}{\partial X} \tag{5.37}$$

$$\eta = \frac{\partial W}{\partial Y} \tag{5.38}$$

It can be found that λ_{cr} is the real part of the complex eigenvalues and λ_r is the real eigenvalue.

The first version of the velocity gradient tensor decomposition can be written as

$$\nabla \vec{V} = \begin{bmatrix} \lambda_{cr} & -R/2 & 0 \\ R/2 + \varepsilon & \lambda_{cr} & 0 \\ \xi & \eta & \lambda_r \end{bmatrix} = R + NR \tag{5.39}$$

$$R = \begin{bmatrix} 0 & -R/2 & 0 \\ R/2 & 0 & 0 \\ 0 & 0 & 0 \end{bmatrix} \tag{5.40}$$

$$NR = \begin{bmatrix} \lambda_{cr} & 0 & 0 \\ s & \lambda_{cr} & 0 \\ \xi & \eta & \lambda_r \end{bmatrix} \tag{5.41}$$

where R stands for the rotational part of the local fluid motion, which is called the tensor version of Liutex, and NR the nonrotational part. It is clear that NR has three real eigenvalues, so NR itself implies no local rotation. The R-NR decomposition of velocity gradient tensor is extremely important in fluid mechanics, which is unique, Galilean invariant, well representing physics and should replace the Cauchy–Stokes (Helmholtz) decomposition, which is not unique, Galilean variant and in addition to the fact that vorticity or antisymmetric tensor cannot represent fluid rotation. As UTA is the birthplace of Liutex, the R-NR decomposition of the velocity gradient tensor is then called UTA R-NR decomposition.

The second version of the velocity gradient tensor decomposition can be obtained by

$$\nabla \vec{V} = \begin{bmatrix} \lambda_{cr} & -R/2 & 0 \\ R/2 + \varepsilon & \lambda_{cr} & 0 \\ \xi & \eta & \lambda_r \end{bmatrix} = R + AS + CS + SD \tag{5.42}$$

$$R = \begin{bmatrix} 0 & -R/2 & 0 \\ R/2 & 0 & 0 \\ 0 & 0 & 0 \end{bmatrix} \tag{5.43}$$

$$AS = \begin{bmatrix} 0 & -\dfrac{\varepsilon}{2} & -\dfrac{\xi}{2} \\ \dfrac{\varepsilon}{2} & 0 & -\dfrac{\eta}{2} \\ \dfrac{\xi}{2} & \dfrac{\eta}{2} & 0 \end{bmatrix} \tag{5.44}$$

$$CS = \begin{bmatrix} \lambda_{cr} & 0 & 0 \\ 0 & \lambda_{cr} & 0 \\ 0 & 0 & \lambda_r \end{bmatrix} \tag{5.45}$$

$$SD = \begin{bmatrix} 0 & \dfrac{\varepsilon}{2} & \dfrac{\xi}{2} \\ \dfrac{\varepsilon}{2} & 0 & \dfrac{\eta}{2} \\ \dfrac{\xi}{2} & \dfrac{\eta}{2} & 0 \end{bmatrix} \tag{5.46}$$

where **AS** can be called the antisymmetric shearing part, **CS** the compression-stretching part, and **SD** the symmetric deformation part. According to Eq. (5.42), the simple shear part **SS** can be obtained from the velocity gradient tensor, which is given by

$$SS = \begin{bmatrix} 0 & 0 & 0 \\ \varepsilon & 0 & 0 \\ \xi & \eta & 0 \end{bmatrix} = SD + AS = \begin{bmatrix} 0 & \dfrac{\varepsilon}{2} & \dfrac{\xi}{2} \\ \dfrac{\varepsilon}{2} & 0 & \dfrac{\eta}{2} \\ \dfrac{\xi}{2} & \dfrac{\eta}{2} & 0 \end{bmatrix} + \begin{bmatrix} 0 & -\dfrac{\varepsilon}{2} & -\dfrac{\xi}{2} \\ \dfrac{\varepsilon}{2} & 0 & -\dfrac{\eta}{2} \\ \dfrac{\xi}{2} & \dfrac{\eta}{2} & 0 \end{bmatrix} = \begin{bmatrix} 0 & 0 & 0 \\ \varepsilon & 0 & 0 \\ \xi & \eta & 0 \end{bmatrix} \tag{5.47}$$

Since this decomposition is performed in a principal coordinate, the decomposition shown in Eq. (5.42) is called "Principal Decomposition" of the velocity gradient tensor. If the traditional Cauchy–Stokes decomposition is applied, one can have

$$\nabla \vec{V} = \begin{bmatrix} \lambda_{cr} & -R/2 & 0 \\ R/2 + \varepsilon & \lambda_{cr} & 0 \\ \xi & \eta & \lambda_r \end{bmatrix} = A + B = \begin{bmatrix} \lambda_{cr} & \dfrac{\varepsilon}{2} & \dfrac{\xi}{2} \\ \dfrac{\varepsilon}{2} & \lambda_{cr} & \dfrac{\eta}{2} \\ \dfrac{\xi}{2} & \dfrac{\eta}{2} & \lambda_r \end{bmatrix}$$

$$+ \begin{bmatrix} 0 & -R/2 - \dfrac{\varepsilon}{2} & -\dfrac{\xi}{2} \\ R/2 + \dfrac{\varepsilon}{2} & 0 & -\dfrac{\eta}{2} \\ \dfrac{\xi}{2} & \dfrac{\eta}{2} & 0 \end{bmatrix} \tag{5.48}$$

$$A = \begin{bmatrix} \lambda_{cr} & \dfrac{\varepsilon}{2} & \dfrac{\xi}{2} \\ \dfrac{\varepsilon}{2} & \lambda_{cr} & \dfrac{\eta}{2} \\ \dfrac{\xi}{2} & \dfrac{\eta}{2} & \lambda_r \end{bmatrix} = CS + SD = \begin{bmatrix} \lambda_{cr} & 0 & 0 \\ 0 & \lambda_{cr} & 0 \\ 0 & 0 & \lambda_r \end{bmatrix} + \begin{bmatrix} 0 & \dfrac{\varepsilon}{2} & \dfrac{\xi}{2} \\ \dfrac{\varepsilon}{2} & 0 & \dfrac{\eta}{2} \\ \dfrac{\xi}{2} & \dfrac{\eta}{2} & 0 \end{bmatrix} \quad (5.49)$$

$$B = \begin{bmatrix} 0 & -R/2 - \dfrac{\varepsilon}{2} & -\dfrac{\xi}{2} \\ R/2 + \dfrac{\varepsilon}{2} & 0 & -\dfrac{\eta}{2} \\ \dfrac{\xi}{2} & \dfrac{\eta}{2} & 0 \end{bmatrix} = R + S$$

$$= \begin{bmatrix} 0 & -R/2 & 0 \\ R/2 & 0 & 0 \\ 0 & 0 & 0 \end{bmatrix} + \begin{bmatrix} 0 & -\dfrac{\varepsilon}{2} & -\dfrac{\xi}{2} \\ \dfrac{\varepsilon}{2} & 0 & -\dfrac{\eta}{2} \\ \dfrac{\xi}{2} & \dfrac{\eta}{2} & 0 \end{bmatrix} \quad (5.50)$$

Eq. (5.48) indicates that the antisymmetric part (vorticity tensor) can be further decomposed to the rotational part (the tensor version of Liutex) and the antisymmetric shearing part. This decomposition is called the tensor form of "vorticity R-S decomposition". In addition, from Eq. (5.48), vorticity vector decomposition can be obtained by

$$\vec{\omega} = \nabla \times \vec{V} = \vec{R} + \vec{S} = \begin{bmatrix} 0 \\ 0 \\ R \end{bmatrix} + \begin{bmatrix} \eta \\ -\xi \\ s \end{bmatrix} \quad (5.51)$$

where $\vec{S} = \vec{\omega} - \vec{R}$ can be considered as shearing vector since the components of $\vec{S}$ indicate the strengths of the simple shear along different axes. Vorticity is decomposed to Liutex which represents the fluid rigid rotation and S which is a non-rotational shear. This decomposition is called "vorticity R-S decomposition" which is extremely important to vortex science and turbulence research.

Accordingly, the velocity gradient tensor decomposition given by Eq. (5.42) can be considered as a universal decomposition, from which different forms of decomposition can be obtained (when the velocity gradient tensor has two complex conjugates and one real eigenvalue). Meanwhile, it provides an approach to quantify the local rotational, compression stretching, and shearing effects.

Based on the decomposition given by Eq. (5.42), the local velocity increment can be decomposed as

$$d\vec{v} = \nabla\vec{v}\cdot d\vec{r} = (\boldsymbol{R} + \boldsymbol{S} + \boldsymbol{CS} + \boldsymbol{SD})\cdot\begin{bmatrix} dX \\ dY \\ dZ \end{bmatrix} = d\vec{R} + d\vec{S} + d\vec{CS} + d\vec{SD}$$

(5.52)

$$d\vec{R} = (-R/2dY + R/2dX)\vec{K} \tag{5.53}$$

$$d\vec{S} = \left[-\frac{\varepsilon}{2}dY - \frac{\xi}{2}dZ\right]\vec{I} + \left[\frac{\varepsilon}{2}dX - \frac{\eta}{2}dZ\right]\vec{J} + \left[\frac{\xi}{2}dX + \frac{\eta}{2}dY\right]\vec{K} \tag{5.54}$$

$$d\vec{CS} = \lambda_{cr}\vec{I} + \lambda_{cr}\vec{J} + \lambda_{r}\vec{K} \tag{5.55}$$

$$d\vec{SD} = \left[\frac{\varepsilon}{2}dY + \frac{\xi}{2}dZ\right]\vec{I} + \left[\frac{\varepsilon}{2}dX + \frac{\eta}{2}dZ\right]\vec{J} + \left[\frac{\xi}{2}dX + \frac{\eta}{2}dY\right]\vec{K} \tag{5.56}$$

where $\vec{I}$ represents the unit vector along the X-axis, $\vec{J}$ the unit vector along the Y-axis, and $\vec{K}$ the unit vector along the Z-axis.

A vortex structure in boundary layer transition on a flat plate is studied by Liutex-based decomposition. The direct numerical simulation of boundary layer transition is performed with about 60 million grid points at a free stream Mach number of 0.5. The typical quasi-stream vortices and hairpin vortices are shown in Fig. 5.8, using the Ω criterion with $\Omega = 0.52$. Point A located in the region of quasi-stream vortices

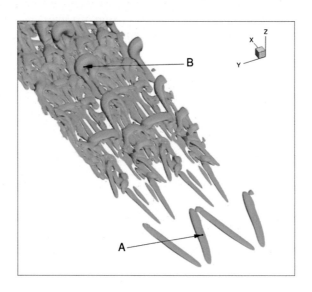

FIGURE 5.8

Quasi-stream vortices and hairpin vortices identified by $\Omega = 0.52$.

Reproduced from Y. Gao and C. Liu, Rortex based velocity gradient tensor decomposition, Phys. Fluids 31 (011704), 2019, with the permission of AIP Publishing.

Table 5.1 The local rotational, compression stretching, and shearing effects of Point A and Point (B).

	A	B
$R/2$	0.0271	0.0207
λ_{cr}	−0.0038	−0.0190
λ_r	0.0073	0.0396
s	0.2013	0.1245
ξ	−0.4764	0.0897
η	0.0035	−0.0608

Reproduced from Y. Gao and C. Liu, Rortex based velocity gradient tensor decomposition, Phys. Fluids 31 (011704), 2019, with the permission of AIP Publishing.

and Point B located in the region of hairpin vortices, as shown in Fig. 5.8, are examined. The corresponding local rotational, compression, stretching, and shearing effects are listed in Table 5.1. According to Table 5.1, it can be found that the local shearing effect ε, ξ are predominant for both Point A and Point B (while the strength of the shearing effect η is relative weak). The local rotational part $R/2$ is much smaller than the strength of the shearing effect ε, which indicates the local rotation is weak, compared to shearing. Accordingly, many eigenvalue-based vortex identification methods are prone to be severely contaminated by shearing. Fig. 5.9 demonstrates the iso-surfaces of vorticity magnitude $\left|\vec{\omega}\right| = 0.541$, shearing magnitude $\left|\vec{S}\right| = 0.517$, and Liutex magnitude $\left|\vec{R}\right| = 0.0543$ (the thresholds of the iso-surfaces are determined by the decomposition of Point A). From Fig. 5.9, it is obvious that the structures identified by vorticity magnitude and shearing

(A)   **(B)**   **(C)**

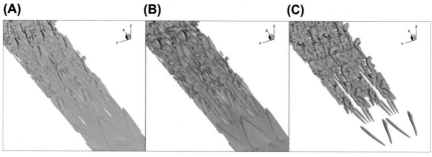

FIGURE 5.9

Iso-surfaces of vorticity magnitude $\left|\vec{\omega}\right| = 0.541$ (A), shearing vector magnitude $\left|\vec{S}\right| = 0.517$ (B), and Liutex magnitude $\left|\vec{R}\right| = 0.0543$ (C).

Reproduced from Y. Gao and C. Liu, Rortex based velocity gradient tensor decomposition, Phys. Fluids 31 (011704), 2019, with the permission of AIP Publishing.

Table 5.2 Vorticity vector decomposition of Point A $\vec{\omega}_A = \vec{R}_A + \vec{S}_A$.

	Vorticity $\vec{\omega}_A$	Liutex $\vec{R}_A$	Shearing vector $\vec{S}_A$
x component	−0.0718	−0.0508	−0.0210
y component	0.5255	0.0188	0.5067
z component	−0.1047	−0.0034	−0.1013
Magnitude	0.5406	0.0543	0.5172

Reproduced from Y. Gao and C. Liu, Rortex based velocity gradient tensor decomposition, Phys. Fluids 31 (011704), 2019, with the permission of AIP Publishing.

Table 5.3 Vorticity vector decomposition of Point B $\vec{\omega}_B = \vec{R}_B + \vec{S}_B$.

	Vorticity $\vec{\omega}_B$	Liutex $\vec{R}_B$	Shearing vector $\vec{S}_B$
x component	−0.0273	0.0143	−0.0416
y component	0.0668	0.0238	0.0430
z component	−0.1846	−0.0307	−0.1538
Magnitude	0.1982	0.0414	0.1650

Reproduced from Y. Gao and C. Liu, Rortex based velocity gradient tensor decomposition, Phys. Fluids 31 (011704), 2019, with the permission of AIP Publishing.

magnitude are very similar, which means that vorticity is shearing-dominant, and the rotational part of vorticity is rather weak, consistent with the analysis of the tensor decomposition. Tables 5.2 and 5.3 provide detailed vorticity vector decompositions of Point A and Point B given by Eq. (5.51), respectively. The corresponding illustrations are shown in Figs. 5.10 and 5.11, respectively. It can be found that the rotation part of vorticity or Liutex only occupies a small part of vorticity and the majority of vorticity represents shearing. In other words, vorticity cannot properly represent

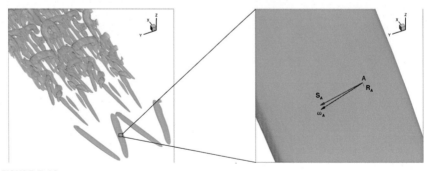

FIGURE 5.10

Illustration of vorticity vector decomposition of Point A.

Reproduced from Y. Gao and C. Liu, Rortex based velocity gradient tensor decomposition, Phys. Fluids 31 (011704), 2019, with the permission of AIP Publishing.

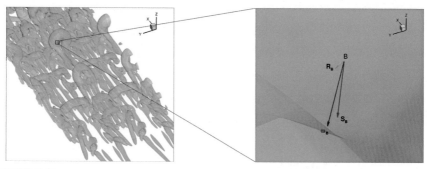

FIGURE 5.11

Illustration of vorticity vector decomposition of Point (B).

Reproduced from Y. Gao and C. Liu, Rortex based velocity gradient tensor decomposition, Phys. Fluids 31
(011704), 2019, with the permission of AIP Publishing.

rotation and vorticity magnitude is not an appropriate measure of the rotation strength. Fig. 5.12 shows the vector lines of vorticity, shearing vector, and Liutex through the vorticity vector decomposition. As can be seen, vorticity lines and shearing vector lines are very similar, but can only represent the ring part of the hairpin vortex. Accordingly, vorticity lines mainly indicate the directions of the local shearing effects, rather than the local rotational axis. In contrast, Liutex lines can provide skeletons of both quasi-streamwise vortex and hairpin vortex.

$$\nabla \vec{V}_A = \begin{bmatrix} -0.0038 & -0.0271 & 0 \\ 0.2285 & -0.00380 & 0 \\ -0.4764 & 0.0035 & 0.0073 \end{bmatrix} = R_A + AS_A + CS_A + SD_A$$

(A)   **(B)**   **(C)**

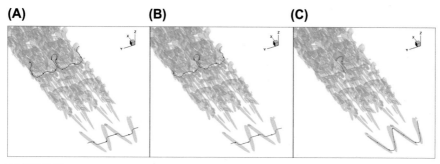

FIGURE 5.12

Vector lines of vorticity (A), shearing vector (B), and Liutex (C) in quasi-stream vortices and hairpin vortices.

Reproduced from Y. Gao and C. Liu, Rortex based velocity gradient tensor decomposition, Phys. Fluids 31
(011704), 2019, with the permission of AIP Publishing.

$$
\boldsymbol{R}_A = \begin{bmatrix} 0 & -0.0271 & 0 \\ 0.0271 & 0 & 0 \\ 0 & 0 & 0 \end{bmatrix}
$$

$$
\boldsymbol{AS}_A = \begin{bmatrix} 0 & -0.1007 & 0.2382 \\ 0.1007 & 0 & -0.0018 \\ -0.2382 & 0.00140 & 0 \end{bmatrix}
$$

$$
\boldsymbol{CS}_A = \begin{bmatrix} -0.0038 & 0 & 0 \\ 0 & -0.0038 & 0 \\ 0 & 0 & 0.0073 \end{bmatrix}
$$

$$
\boldsymbol{SD}_A = \begin{bmatrix} 0 & 0.1007 & -0.2382 \\ 0.1007 & 0 & 0.0018 \\ -0.2382 & 0.0018 & 0 \end{bmatrix}
$$

$$
\nabla \vec{V}_B = \begin{bmatrix} -0.0190 & -0.0207 & 0 \\ 0.1452 & -0.00190 & 0 \\ 0.0897 & -0.0608 & 0.0396 \end{bmatrix} = \boldsymbol{R}_B + \boldsymbol{AS}_B + \boldsymbol{CS}_B + \boldsymbol{SD}_B
$$

$$
\boldsymbol{R}_B = \begin{bmatrix} 0 & -0.0207 & 0 \\ 0.0207 & 0 & 0 \\ 0 & 0 & 0 \end{bmatrix}
$$

$$
\boldsymbol{AS}_B = \begin{bmatrix} 0 & -0.0623 & -0.0448 \\ 0.0623 & 0 & 0.0304 \\ 0.0448 & -0.0304 & 0 \end{bmatrix}
$$

$$
\boldsymbol{CS}_B = \begin{bmatrix} -0.0190 & 0 & 0 \\ 0 & -0.0190 & 0 \\ 0 & 0 & 0.0396 \end{bmatrix}
$$

$$
\boldsymbol{SD}_B = \begin{bmatrix} 0 & 0.0623 & 0.0448 \\ 0.0623 & 0 & -0.0304 \\ 0.0448 & -0.0304 & 0 \end{bmatrix}
$$

5.5 2D velocity gradient tensor decomposition in the original coordinate

5.5.1 Velocity gradient tensor decomposition in the principal coordinate

A general 2D velocity gradient tensor can be decomposed to be a rigid rotation part **R** and a nonrotation part **NR** in the principal coordinate as

$$
\nabla \vec{V} = \begin{bmatrix} \lambda_{cr} & \dfrac{\partial U}{\partial Y} \\[2mm] \dfrac{\partial V}{\partial X} & \lambda_{cr} \end{bmatrix} = \begin{bmatrix} \lambda_{cr} & -R/2 \\ R/2 + \varepsilon & \lambda_{cr} \end{bmatrix} = \begin{bmatrix} 0 & -R/2 \\ R/2 & 0 \end{bmatrix}
$$

$$
+ \begin{bmatrix} \lambda_{cr} & 0 \\ \varepsilon & \lambda_{cr} \end{bmatrix} = R + NR \tag{5.57}
$$

where $R/2 = -\frac{\partial U}{\partial Y} > 0$, and $\frac{\partial V}{\partial X} = R/2 + \varepsilon$ (assume $\frac{\partial V}{\partial X} > 0$ and $\left|\frac{\partial V}{\partial X}\right| > \left|\frac{\partial U}{\partial Y}\right|$).

Eq. (5.57) is called 2D **UTA R-NR** decomposition. The first part on the right side represents the rigid rotation, while the second part has two real eigenvalues and therefore has no fluid rotation itself.

The nonrotation part **NR** in Eq. (5.57) can be further decomposed to a diagonal matrix **CS**, which represents the compression or stretching and a lower triangular matrix **SS** represents the shearing as

$$
NR = \begin{bmatrix} \lambda_{cr} & 0 \\ \varepsilon & \lambda_{cr} \end{bmatrix} = \begin{bmatrix} \lambda_{cr} & 0 \\ 0 & \lambda_{cr} \end{bmatrix} + \begin{bmatrix} 0 & 0 \\ \varepsilon & 0 \end{bmatrix} = CS + SS \tag{5.58}
$$

5.5.2 Velocity gradient tensor decomposition in the original coordinate

The decompositions given by Eqs. (5.57) and (5.58) are performed in the principal coordinate. In practice, sometimes it is better to perform the tensor decomposition in the original coordinate without any explicit coordinate rotation. The explicit expression of Liutex in two-dimensional case can be written as

$$
R = \omega - \sqrt{\omega^2 - 4\lambda_{ci}^2} \tag{5.59}
$$

where ω represents vorticity, which is a scalar in two-dimensional case. So the tensor decomposition given by Eq. (5.57) can be expressed in the original coordinate system as

$$\nabla \vec{v} = \begin{bmatrix} \dfrac{\partial u}{\partial x} & \dfrac{\partial u}{\partial y} \\[2mm] \dfrac{\partial v}{\partial x} & \dfrac{\partial v}{\partial y} \end{bmatrix} = \begin{bmatrix} 0 & -R/2 \\[2mm] R/2 & 0 \end{bmatrix} + \begin{bmatrix} \dfrac{\partial u}{\partial x} & \dfrac{\partial u}{\partial y} + R/2 \\[2mm] \dfrac{\partial v}{\partial x} - R/2 & \dfrac{\partial v}{\partial y} \end{bmatrix} = R + NR$$

(5.60)

Furthermore, the tensor decomposition given by Eq. (5.58) can be expressed in the original coordinate system as

$$NR = \begin{bmatrix} \lambda_{cr} & 0 \\[2mm] 0 & \lambda_{cr} \end{bmatrix} + \begin{bmatrix} \dfrac{\partial u}{\partial x} - \lambda_{cr} & \dfrac{\partial u}{\partial y} + R/2 \\[2mm] \dfrac{\partial v}{\partial x} - R/2 & \dfrac{\partial v}{\partial y} - \lambda_{cr} \end{bmatrix}$$

(5.61)

5.6 3D velocity gradient tensor decomposition in the original coordinate

5.6.1 Velocity gradient tensor decomposition in the principal coordinate

In the 3D principal coordinate, the velocity gradient tensor $\nabla \vec{V}$ can be decomposed as

$$\nabla \vec{V} = \begin{bmatrix} \lambda_{cr} & -R/2 & 0 \\ R/2 + \varepsilon & \lambda_{cr} & 0 \\ \xi & \eta & \lambda_r \end{bmatrix} = \begin{bmatrix} 0 & -R/2 & 0 \\ R/2 & 0 & 0 \\ 0 & 0 & 0 \end{bmatrix}$$

$$+ \begin{bmatrix} \lambda_{cr} & 0 & 0 \\ \varepsilon & \lambda_{cr} & 0 \\ \xi & \eta & \lambda_r \end{bmatrix} = R + NR$$

(5.62)

where R represents the rigid rotation part and NR is the nonrotational part. Since NR has three real eigenvalues as the diagonal elements, there exists no rotation in NR, consistent with the definition of fluid rotation. Eq. (5.62) is the 3D **UTA R-NR** decomposition in the principal coordinate.

Similar to the 2D decomposition, the nonrotational part NR can be further decomposed as

$$NR = \begin{bmatrix} \lambda_{cr} & 0 & 0 \\ \varepsilon & \lambda_{cr} & 0 \\ \xi & \eta & \lambda_r \end{bmatrix} = \begin{bmatrix} \lambda_{cr} & 0 & 0 \\ 0 & \lambda_{cr} & 0 \\ 0 & 0 & \lambda_r \end{bmatrix} + \begin{bmatrix} 0 & 0 & 0 \\ \varepsilon & 0 & 0 \\ \xi & \eta & 0 \end{bmatrix} = CS + SS \quad (5.63)$$

where CS represents stretching or compression along the axes of the principal coordinate, SS the shearing effects along different directions. This decomposition is made only in the principal coordinates and is unique no matter what the original co-ordinates are. On the other hand, the classical Cauchy–Stokes decomposition is closely dependent to the original coordinates, or, in other words, different coordinates will give different Cauchy–Stokes decompositions.

5.6.2 Velocity gradient tensor decomposition in the original coordinate

The rotation part of the velocity gradient tensor in the principal coordinate is:

$$\boldsymbol{R} = \begin{bmatrix} 0 & -R/2 & 0 \\ R/2 & 0 & 0 \\ 0 & 0 & 0 \end{bmatrix} \tag{5.64}$$

Since it is an antisymmetric matrix and the explicit expression of the Liutex vector is available by Eq. (5.31), the matrix representation of the rotation part in the original coordinate can be immediately obtained as

$$\boldsymbol{R} = \frac{1}{2} \begin{bmatrix} 0 & -R_z & R_y \\ R_z & 0 & -R_x \\ -R_y & R_x & 0 \end{bmatrix}, \tag{5.65}$$

where

$$\begin{bmatrix} R_x \\ R_y \\ R_z \end{bmatrix} = R\vec{r} = \left\{ \vec{\omega} \cdot \vec{r} - \sqrt{(\vec{\omega} \cdot \vec{r})^2 - 4\lambda_{ci}^2} \right\} \vec{r}.$$

Then, the 3D **UTA R-NR** decomposition can be written in the original coordinate as

$$\nabla \vec{v} = \begin{bmatrix} \dfrac{\partial u}{\partial x} & \dfrac{\partial u}{\partial y} & \dfrac{\partial u}{\partial z} \\[2mm] \dfrac{\partial v}{\partial x} & \dfrac{\partial v}{\partial y} & \dfrac{\partial v}{\partial z} \\[2mm] \dfrac{\partial w}{\partial x} & \dfrac{\partial w}{\partial y} & \dfrac{\partial w}{\partial z} \end{bmatrix} = \frac{1}{2} \begin{bmatrix} 0 & -R_z & R_y \\ R_z & 0 & -R_x \\ -R_y & R_x & 0 \end{bmatrix}$$

$$+ \begin{bmatrix} \dfrac{\partial u}{\partial x} & \dfrac{\partial u}{\partial y} + \dfrac{1}{2}R_z & \dfrac{\partial u}{\partial z} - \dfrac{1}{2}R_y \\[2mm] \dfrac{\partial v}{\partial x} - \dfrac{1}{2}R_z & \dfrac{\partial v}{\partial y} & \dfrac{\partial v}{\partial z} + \dfrac{1}{2}R_x \\[2mm] \dfrac{\partial w}{\partial x} + \dfrac{1}{2}R_y & \dfrac{\partial w}{\partial y} - \dfrac{1}{2}R_x & \dfrac{\partial w}{\partial z} \end{bmatrix} = R + NR \tag{5.66}$$

5.6.3 Simple way for Liutex magnitude

In general, a velocity gradient tensor can be decomposed in the principal coordinate:

$$
\nabla \vec{V} = \begin{bmatrix} \lambda_{cr} & -\dfrac{1}{2}R & 0 \\ \dfrac{1}{2}R + \varepsilon & \lambda_{cr} & 0 \\ \xi & \eta & \lambda_r \end{bmatrix}
\tag{5.67}
$$

The characteristic equation of $\nabla \vec{V}$ is:

$$
(\lambda - \lambda_r)\left[(\lambda - \lambda_{cr})^2 + \frac{R}{2}\left(\frac{R}{2} + \varepsilon \right) \right] = 0
\tag{5.68}
$$

Thus, the eigenvalues are:

$$
\begin{aligned}
\lambda_1 &= \lambda_r \\
\lambda_2 &= \lambda_{cr} + i\sqrt{R/2(R/2 + \varepsilon)} = \lambda_{cr} + i\lambda_{ci} \\
\lambda_3 &= \lambda_{cr} - i\sqrt{R/2(R/2 + \varepsilon)} = \lambda_{cr} - i\lambda_{ci}
\end{aligned}
\tag{5.69}
$$

Therefore,

$$
\frac{R}{2}\left(\frac{R}{2} + \varepsilon \right) = \lambda_{ci}^2
\tag{5.70}
$$

Note that the orthogonal transformation does not change the eigenvalues. On the other hand,

$$
(R/2 + \varepsilon) - \left(-\frac{R}{2} \right) = \frac{\partial V}{\partial X} - \frac{\partial U}{\partial Y} = \omega_z = \vec{\omega} \cdot \vec{r},
\tag{5.71}
$$

where $\vec{\omega}$ is vorticity and ω_z is the third component of vorticity. There are two equations for two unknowns.

Solve R and ε from Eqs. (5.70) and (5.71), one can get

$$
R = \left(\vec{\omega} \cdot \vec{r} \right) - \sqrt{\left(\vec{\omega} \cdot \vec{r} \right)^2 - 4\lambda_{ci}^2}
\tag{5.72}
$$

$$
\varepsilon = \frac{\lambda_{ci}^2}{\dfrac{R}{2}} - \frac{R}{2}
\tag{5.73}
$$

This provides a simple way to get the explicit formula of Liutex magnitude. There is no need to do any coordinate rotation to find these elements except for ξ and η under the principal coordinate indeed.

5.7 Vorticity tensor and vector decomposition
5.7.1 2D vorticity tensor decomposition

As previously discussed, the traditional Cauchy–Stokes decomposition cannot provide a real rotation part of the velocity gradient tensor. Therefore, vorticity should be further decomposed. In the principal coordinate, the Cauchy–Stokes decomposition can be written as

$$
\nabla \vec{V} = \begin{bmatrix} \dfrac{\partial U}{\partial X} & \dfrac{\partial U}{\partial Y} \\[2mm] \dfrac{\partial V}{\partial X} & \dfrac{\partial V}{\partial Y} \end{bmatrix} = \begin{bmatrix} \lambda_{cr} & -\dfrac{1}{2}R \\[2mm] \dfrac{1}{2}R + \varepsilon & \lambda_{cr} \end{bmatrix} = \begin{bmatrix} \lambda_{cr} & \varepsilon \\[2mm] \varepsilon & \lambda_{cr} \end{bmatrix}
$$

$$
+ \begin{bmatrix} 0 & -\dfrac{1}{2}(R+\varepsilon) \\[2mm] \dfrac{1}{2}(R+\varepsilon) & 0 \end{bmatrix} = A + B \tag{5.74}
$$

where A is a symmetric tensor and B is an antisymmetric tensor, which is usually considered as a vorticity tensor. To obtain the real rotation part from the vorticity tensor, the 2D vorticity tensor can be further decomposed as

$$
B = \begin{bmatrix} 0 & -\dfrac{1}{2}(R+\varepsilon) \\[2mm] \dfrac{1}{2}(R+\varepsilon) & 0 \end{bmatrix} = \begin{bmatrix} 0 & -\dfrac{1}{2}R \\[2mm] \dfrac{1}{2}R & 0 \end{bmatrix} + \begin{bmatrix} 0 & -\dfrac{1}{2}\varepsilon \\[2mm] \dfrac{1}{2}\varepsilon & 0 \end{bmatrix} = R + S
$$

$$\tag{5.75}$$

where R is the real rigid rotation part and S is the antisymmetric shear. Eq. (5.75) is called 2D R-S decomposition. It should be noted that although both R and S are antisymmetric matrices, their physical meanings are totally different: R is the real rotation part of the vorticity tensor while S represents antisymmetric shearing, which will be balanced by the symmetric part A.

The 2D R-S decomposition can be expressed in the original coordinate system as

$$
B = \frac{1}{2}\begin{bmatrix} 0 & -\omega \\ \omega & 0 \end{bmatrix} = \frac{1}{2}\begin{bmatrix} 0 & -R \\ R & 0 \end{bmatrix} + \frac{1}{2}\begin{bmatrix} 0 & -\omega+R \\ \omega-R & 0 \end{bmatrix} = R + S \tag{5.76}
$$

where R is obtained through Eq. (5.59).

5.7.2 3D vorticity tensor and vector decomposition

The Cauchy–Stokes decomposition in 3D principal coordinate can be written as

$$
\nabla \vec{V} =
\begin{bmatrix}
\dfrac{\partial U}{\partial X} & \dfrac{\partial U}{\partial Y} & 0 \\[2ex]
\dfrac{\partial V}{\partial X} & \dfrac{\partial V}{\partial Y} & 0 \\[2ex]
\dfrac{\partial W}{\partial X} & \dfrac{\partial W}{\partial Y} & \dfrac{\partial W}{\partial Z}
\end{bmatrix}
=
\begin{bmatrix}
\dfrac{\partial U}{\partial X} & \dfrac{1}{2}\left(\dfrac{\partial V}{\partial X}+\dfrac{\partial U}{\partial Y}\right) & \dfrac{1}{2}\dfrac{\partial W}{\partial X} \\[2ex]
\dfrac{1}{2}\left(\dfrac{\partial V}{\partial X}+\dfrac{\partial U}{\partial Y}\right) & \dfrac{\partial V}{\partial Y} & \dfrac{1}{2}\dfrac{\partial W}{\partial Y} \\[2ex]
\dfrac{1}{2}\dfrac{\partial W}{\partial X} & \dfrac{1}{2}\dfrac{\partial W}{\partial Y} & \dfrac{\partial W}{\partial Z}
\end{bmatrix}
$$

$$
+
\begin{bmatrix}
0 & -\dfrac{1}{2}\left(\dfrac{\partial V}{\partial X}-\dfrac{\partial U}{\partial Y}\right) & -\dfrac{1}{2}\dfrac{\partial W}{\partial X} \\[2ex]
\dfrac{1}{2}\left(\dfrac{\partial V}{\partial X}-\dfrac{\partial U}{\partial Y}\right) & 0 & -\dfrac{1}{2}\dfrac{\partial W}{\partial Y} \\[2ex]
\dfrac{1}{2}\dfrac{\partial W}{\partial X} & \dfrac{1}{2}\dfrac{\partial W}{\partial Y} & 0
\end{bmatrix}
= A + B \qquad (5.77)
$$

Similar to the 2D case, the vorticity tensor B can be decomposed as

$$
B =
\begin{bmatrix}
0 & -\dfrac{1}{2}\left(\dfrac{\partial V}{\partial X}-\dfrac{\partial U}{\partial Y}\right) & -\dfrac{1}{2}\dfrac{\partial W}{\partial X} \\[2ex]
\dfrac{1}{2}\left(\dfrac{\partial V}{\partial X}-\dfrac{\partial U}{\partial Y}\right) & 0 & -\dfrac{1}{2}\dfrac{\partial W}{\partial Y} \\[2ex]
\dfrac{1}{2}\dfrac{\partial W}{\partial X} & \dfrac{1}{2}\dfrac{\partial W}{\partial Y} & 0
\end{bmatrix}
$$

$$
=
\begin{bmatrix}
0 & -\dfrac{R}{2} & 0 \\[2ex]
\dfrac{R}{2} & 0 & 0 \\[2ex]
0 & 0 & 0
\end{bmatrix}
+
\begin{bmatrix}
0 & -\dfrac{1}{2}\left(\dfrac{\partial V}{\partial X}-\dfrac{\partial U}{\partial Y}\right)+\dfrac{R}{2} & -\dfrac{1}{2}\dfrac{\partial W}{\partial X} \\[2ex]
\dfrac{1}{2}\left(\dfrac{\partial V}{\partial X}-\dfrac{\partial U}{\partial Y}\right)-\dfrac{R}{2} & 0 & -\dfrac{1}{2}\dfrac{\partial W}{\partial Y} \\[2ex]
\dfrac{1}{2}\dfrac{\partial W}{\partial X} & \dfrac{1}{2}\dfrac{\partial W}{\partial Y} & 0
\end{bmatrix}
$$

$$
= R + S
$$

$$(5.78)$$

where R is the real rigid rotation part and S is the antisymmetric shear. Eq. (5.78) is 3D R-S decomposition. Also, both R and S are antisymmetric matrices, but only R represents the real rotation.

In the original coordinate system, **R-S** decomposition is expressed as

$$
B = \frac{1}{2}
\begin{bmatrix}
0 & -\omega_z & \omega_y \\
\omega_z & 0 & -\omega_x \\
-\omega_y & \omega_x & 0
\end{bmatrix}
$$

$$
= \frac{1}{2}
\begin{bmatrix}
0 & -R_z & R_y \\
R_z & 0 & -R_x \\
-R_y & R_x & 0
\end{bmatrix}
+ \frac{1}{2}
\begin{bmatrix}
0 & -\omega_z + R_z & \omega_y - R_y \\
\omega_z - R_z & 0 & -\omega_x + R_x \\
-\omega_y + R_y & \omega_x - R_x & 0
\end{bmatrix}
= R + S
$$

$$(5.79)$$

where $\omega_x = \frac{\partial w}{\partial y} - \frac{\partial v}{\partial z}$, $\omega_y = \frac{\partial u}{\partial z} - \frac{\partial w}{\partial x}$, $\omega_z = \frac{\partial v}{\partial x} - \frac{\partial u}{\partial y}$.

According to Eq. (5.79), the velocity gradient tensor in the original coordinate system can be decomposed to three parts

$$
\nabla \vec{v} =
\begin{bmatrix}
\frac{\partial u}{\partial x} & \frac{\partial u}{\partial y} & \frac{\partial u}{\partial z} \\
\frac{\partial v}{\partial x} & \frac{\partial v}{\partial y} & \frac{\partial v}{\partial z} \\
\frac{\partial w}{\partial x} & \frac{\partial w}{\partial y} & \frac{\partial w}{\partial z}
\end{bmatrix}
=
\begin{bmatrix}
\frac{\partial u}{\partial x} & \frac{1}{2}\left(\frac{\partial u}{\partial y} + \frac{\partial v}{\partial x}\right) & \frac{1}{2}\left(\frac{\partial u}{\partial z} + \frac{\partial w}{\partial x}\right) \\
\frac{1}{2}\left(\frac{\partial v}{\partial x} + \frac{\partial u}{\partial y}\right) & \frac{\partial v}{\partial y} & \frac{1}{2}\left(\frac{\partial v}{\partial z} + \frac{\partial w}{\partial y}\right) \\
\frac{1}{2}\left(\frac{\partial w}{\partial x} + \frac{\partial u}{\partial z}\right) & \frac{1}{2}\left(\frac{\partial w}{\partial y} + \frac{\partial v}{\partial z}\right) & \frac{\partial w}{\partial z}
\end{bmatrix}
$$

$$
+ \frac{1}{2}
\begin{bmatrix}
0 & -R_z & R_y \\
R_z & 0 & -R_x \\
-R_y & R_x & 0
\end{bmatrix}
+ \frac{1}{2}
\begin{bmatrix}
0 & -\omega_z + R_z & \omega_y - R_y \\
\omega_z - R_z & 0 & -\omega_x + R_x \\
-\omega_y + R_y & \omega_x - R_x & 0
\end{bmatrix}
$$

$$
= A + R + S \qquad\qquad (5.80)
$$

The component of **S** in the original coordinate system only represents one part of the shearing effect, because part of the shearing effect will appear in the diagonal terms of the velocity gradient tensor after the coordinate transformation (rotation). If the exact part of the shearing effect in the original coordinate system is required, the explicit coordinate transformation (rotation) should be applied.

The **R-S** decomposition of the antisymmetric tensor or vorticity tensor will lead to an extremely important vector decomposition, which is the vorticity vector decomposition.

Theorem 5.1. Vorticity vector can be decomposed to a Liutex vector and a nonrotational shear vector, i.e.,

$$
\nabla \times \vec{V} = \vec{R} + \vec{S}. \qquad\qquad (5.81)
$$

Proof. Assume $\vec{dl}$ is an arbitrarily selected real vector,

$$2\boldsymbol{B}\cdot d\vec{l} = -d\vec{l} \times \left(\nabla \times \vec{V} \right) = \boldsymbol{R}\cdot d\vec{l} + \boldsymbol{S}\cdot d\vec{l} = -d\vec{l} \times \vec{R} - d\vec{l} \times \vec{S}\, d\vec{l}$$

$$\times \left(\nabla \times \vec{V} \right) = d\vec{l} \times \vec{R} + d\vec{l} \times \vec{S} = d\vec{l} \times \left(\vec{R} + \vec{S} \right) \qquad (5.82)$$

Since $d\vec{l}$ is arbitrarily selected,

$$\nabla \times \vec{V} = \vec{R} + \vec{S} \qquad (5.83)$$

Theorem 5.1 is called R-S decomposition for vorticity vector, which is illustrated in Fig. 5.13.

5.8 Summary

In this chapter, several decompositions are introduced, including **UTA R-NR** decomposition for velocity gradient tensor and **R-S** decomposition for vorticity tensor and vector, based on the core idea that only Liutex can represent the real rotation part of fluid motion rather than vorticity. Through **UTA R-NR** decomposition, the velocity gradient tensor will be decomposed to a rotational part, i.e., the tensor of Liutex, and a nonrotational part including stretching, compression, and shearing effects. The **UTA R-NR** tensor decomposition can be performed in the principal coordinate or the original coordinate. But for 3-D case, if the detailed shearing part is required, an explicit coordinate rotation is necessary to do the decomposition in the principal coordinate. **R-S** decomposition for vorticity tensor indicates the vorticity tensor consists of two parts: one is the rotational part, i.e., Liutex and the other is the antisymmetric shearing part. It leads to the corresponding decomposition of the vorticity vector $\nabla \times \vec{V} = \vec{R} + \vec{S}$. Only these decompositions can correctly interpret the local fluid motion. Several conclusions can be made:

(1) Vorticity is not vortex. From the principal decomposition of vorticity, it is observed that vorticity is actually a rigid rotation plus shear. For rigid body,

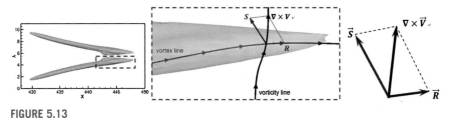

FIGURE 5.13

The decomposition of vorticity vector to a Liutex vector and a nonrotational vector.

Reproduced from Y. Gao and C. Liu, Rortex based velocity gradient tensor decomposition, Phys. Fluids 31 (011704), 2019, with the permission of AIP Publishing.

shear deformation is assumed to be zero. Only in such a special case, vorticity can be considered as a measurement of rotation. Vorticity should be decomposed to Liutex and shear or, in other words, $\nabla \times \vec{V} = \vec{R} + \vec{S}$, which is given by Liu et al. (2016, 2018a).

(2) Liutex is a physical quantity to accurately represent the fluid rotation or vortex.

(3) The Cauchy–Stokes decomposition itself is not Galilean invariant, so it is uncertain under which coordinates one can get the right physics such as rotation, shear, deformation, and stretching (compression). In addition, the antisymmetric part of the velocity gradient tensor only represents vorticity, but vorticity cannot represent vortex or fluid rotation.

(4) A new and unique coordinate called "Principal Coordinate" is defined based on the velocity gradient tensor. Under this coordinate, the fluid motion can be decomposed uniquely to the rotation, shearing, and stretching parts corresponding to the real physics, which is called principal decomposition of vorticity and velocity gradient tensor.

(5) Principal decomposition not only has a clear and correct physical meaning but also overcomes the Galilean variant drawbacks of the Cauchy–Stokes decomposition, which also mistreats the antisymmetric tensor as fluid rotation.

(6) The traditional Helmholtz decomposition and its corresponding Cauchy–Stokes tensor decomposition in general are not a right way for decomposition of fluid motion and they should be replaced by Liutex-based **UTA R-NR** velocity gradient tensor decomposition and Liutex-based **RS** decomposition of vorticity, which could become a foundation of fluid kinematics for further analysis in fluid dynamics.

Liutex and third generation of vortex identification methods

Since Liutex is defined as a vector with the local rotation axis as its direction and twice the local angular speed as its strength, which will have several ways to describe the vortex structure. These methods are introduced in this chapter.

6.1 Liutex iso-surface

Since Liutex magnitude is also a scalar, its iso-surface could be applied to show the vortex structure, which is similar to the iso-surface of Q, for example, but for rigid rotation only (pure rotation) without stretching, compression and shearing contamination.

In Chapter 4, the definition of Liutex and relation with Q and λ_{ci} have been described, which is briefly discussed here.

Liutex is defined as the rigid rotation part of fluid motion

$$\vec{R} = R\,\vec{r} \text{ and } \vec{\omega} \cdot \vec{r} > 0$$

$$R = \left(\vec{\omega} \cdot \vec{r}\right) - \sqrt{\left(\vec{\omega} \cdot \vec{r}\right)^2 - 4\lambda_{ci}^2} \tag{6.1}$$

Here, $\vec{r}$ is the real eigenvector of $\nabla \vec{v}$ and the restriction of $\nabla \times \vec{v} \cdot \vec{r} > 0$ is to keep Liutex definition unique and consistent when the fluid motion is pure rotation. In Eq. (6.1) $\vec{\omega} = \nabla \times \vec{v}$ is vorticity and λ_{ci} is the imaginary part of conjugate complex eigenvalues of $\nabla \vec{v}$.

Q is the second invariant of the characteristic equation of $\nabla \vec{v}$, i.e., $\lambda^3 + P\lambda^2 + Q\lambda + \tilde{R} = 0$

$$Q = \lambda_1\lambda_2 + \lambda_2\lambda_3 + \lambda_3\lambda_1 = (\lambda_{cr} + i\lambda_{ci})(\lambda_{cr} - i\lambda_{ci}) + (\lambda_{cr} - i\lambda_{ci})\lambda_r + \lambda_r(\lambda_{cr} + i\lambda_{ci})$$

$$= \lambda_{cr}^2 + \lambda_{ci}^2 + 2\lambda_{cr}\lambda_r = \lambda_{cr}^2 + \frac{R}{2}\left(\frac{R}{2} + \varepsilon\right) + 2\lambda_{cr}\lambda_r \tag{6.2}$$

$$4\lambda_{ci}^2 = R(R + \varepsilon) \quad \text{or} \quad \lambda_{ci} = \sqrt{\frac{R}{2}\left(\frac{R}{2} + \varepsilon\right)}, \tag{6.3}$$

Liutex and Its Applications in Turbulence Research. https://doi.org/10.1016/B978-0-12-819023-4.00007-0
Copyright © 2021 Elsevier Inc. All rights reserved.

where R is twice the rigid rotation angular speed, λ_r is the real eigenvalue, λ_{cr} is the real part of the conjugate complex eigenvalues of the velocity gradient tensor $\nabla \vec{v}$, and ε is the nonrotational shear part of $\nabla \vec{v}$.

From the aforementioned equations, it is clearly shown that Liutex or $\vec{R}$ represents the rigid rotation part of the fluid motion. However, λ_{ci} is seriously contaminated by shear (treats shear as part of fluid rotation) and Q is more seriously contaminated by both shear and stretching as λ_r and λ_{cr} really represent the stretching (or compression). In the principal coordinate $(X_\vartheta, Y_\vartheta, Z)$, $\frac{\partial U_\vartheta}{\partial X_\vartheta} = \lambda_{cr}$, $\frac{\partial V_\vartheta}{\partial Y_\vartheta} = \lambda_{cr}$, $\frac{\partial W_\vartheta}{\partial Z} = \lambda_r$, which represent stretching (or compression) in X_ϑ, Y_ϑ, Z directions, respectively. Although all second generation of vortex identification methods can claim capable to show vortex structures, they really cannot represent the vortex rotation strength as they are contaminated by shear and stretching (or compression), as shown in Eqs. (6.2) and (6.3). In addition, Q cannot be applied for high speed as the Q-criterion assumes the first invariant $P = 0$.

Like the Q-criteria, a threshold must be selected and all points on the iso-surface should have same angular speed as $R = R_{threshold}$ (see Fig. 6.1). Of course, since the scalar of Liutex is used, vortex structure is still strongly dependent on the threshold. Different threshold will give different vortex structure for the same data set. Note that the iso-surface of Liutex and λ_{ci} (or Q) are not identical as Liutex magnitude represents local angular speed, but λ_{ci} and Q contain shear (See Fig. 6.2).

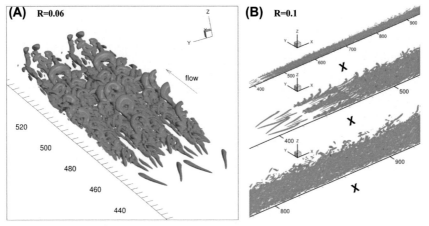

FIGURE 6.1

Liutex iso-surfaces with R $=$ 0.06 and R $=$ 0.1 for natural flow transition.

Reproduced from Y. Gao and C. Liu, Rortex and comparison with eigenvalue-based vortex identification criteria, Phys. Fluids 30 (085107), 2018, with the permission of AIP Publishing.

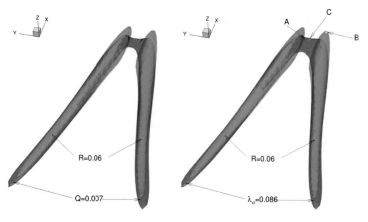

FIGURE 6.2

Iso-surfaces of R = 0.06 (green), Q = 0.007 (blue) and λ_{ci} = 0.086 (blue).

Reproduced from Y. Gao and C. Liu, Rortex and comparison with eigenvalue-based vortex identification criteria, Phys. Fluids 30 (085107), 2018, with the permission of AIP Publishing.

6.2 Liutex vectors, lines, and tubes

In Chapter 1, it has been addressed that vorticity cannot be applied to represent vortex and they are not correlated. For example, the vorticity magnitude is very large near the wall of a laminar boundary layer, but there is no fluid rotation or vortex found near the wall. According to Helmholtz first vortex theory, the strength Γ of the vortex is constant all along its length, which indicates vortex tube can never end inside the flow field except for the wall or infinitely far field boundaries. However, almost all observed vortices are ended inside the flow field, which is just the opposite of the Helmholtz theories.

Therefore, one cannot use vorticity tubes to represent vortices. However, Liutex is a mathematical vector definition, which can exactly represent fluid rotation. Therefore, the Liutex vector field, Liutex line, which has its tangent aligned with Liutex vector everywhere, and the Liutex tube can be applied to visualize vortices and vortical structures. Figs. 6.3–6.9 show the vortex structures for early flow transition in a boundary layer visualized by Liutex vectors, Liutex lines, Liutex tubes, and Liutex iso-surface of $|R| = 0.1$.

6.3 Liutex-Ω vortex identification method

It is a tradition of the second-generation vortex identification methods to use iso-surfaces of different magnitudes to represent vortices. This is also true for Liutex magnitude, as shown in Figs. 6.9 and 6.10, in which the typical vortical structures of boundary layer transition, Λ and hairpin vortices, are well captured. A major advantage of the Liutex magnitude iso-surface over the second-generation methods is that no contamination by shearing and/or stretching is contained in vortex

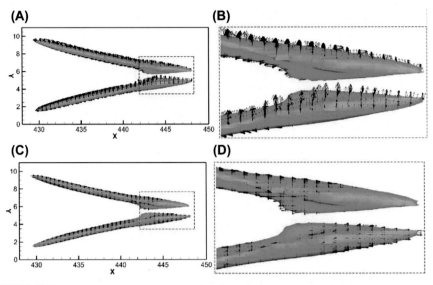

FIGURE 6.3

Comparison of vorticity vector field and Liutex vector field for the vortex legs: (A) vorticity vector field and the Liutex iso-surface, (B) the local enlargement of vorticity vector field near the head part, (C) Liutex vector field and the Liutex iso-surface, and (D) the local enlargement of Liutex vector field near the head part.

Reproduced from Y. Gao and C. Liu, Rortex and comparison with eigenvalue-based vortex identification criteria, Phys. Fluids 30 (085107), 2018, with the permission of AIP Publishing.

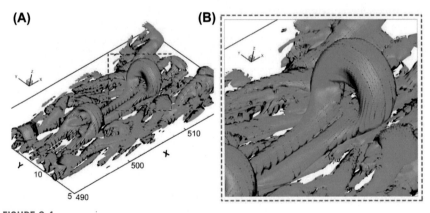

FIGURE 6.4

Liutex vector field and the Liutex iso-surface for the first vortex ring: (A) global view and (B) local enlargement of the marked area.

Reproduced from Y. Gao and C. Liu, Rortex and comparison with eigenvalue-based vortex identification criteria, Phys. Fluids 30 (085107), 2018, with the permission of AIP Publishing.

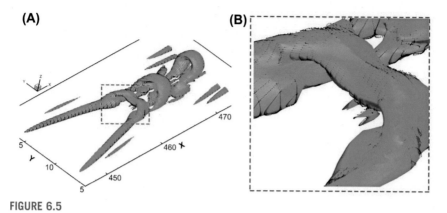

FIGURE 6.5

Liutex vector field and the Liutex iso-surface for the later vortex rings: (A) global view and (B) local enlargement of the marked area.

Reproduced from Y. Gao and C. Liu, Rortex and comparison with eigenvalue-based vortex identification criteria, Phys. Fluids 30 (085107), 2018, with the permission of AIP Publishing.

structures, while those previous methods are prone to be contaminated by shearing to different extent as detailed in Gao and Liu (2018). As a popular method, Q is even worse by treating the stretching (or compression) as part of the vortex strength, which will lead to more discrepancy for compressible flows.

It should be acknowledged that, however, the threshold issue of the second-generation methods persists as the nature of all iso-surface-based methods. With a smaller threshold of Liutex magnitude $R = 0.005$, as shown in Fig. 6.10A, the T-S vortices are captured while the Λ and hairpin vortices are however smeared and blurred in the early transition region. With $R = 0.02$, the Λ vortex around $x = 385$ is clearly visualized but the T-S waves are lost in Fig. 6.10B. An even larger threshold of $R = 0.1$ as shown in Fig. 6.10C leads to a clearer representation of hairpin vortices. However, both T-S vortices and Λ vortices are lost. In addition, the legs of hairpin chain around $x = 340$ are disconnected with hairpin rings. This inconsistency of threshold problem roots from the fact that vortices have different rotational strengths which leads to that almost every iso-surface-based method has the threshold problem. It is thus concluded that iso-surfaces are not adequate to describe the vortical structure in a flow field.

6.3.1 Ω method

The Ω *method* has been introduced in Chapter 1. The problem that vorticity is not vortex and the threshold issue of the second generation of vortex identification methods prompts Liu et al. (2016) to introduce the Ω method. Rather than capturing the swirling strength, the Ω parameter is a measure of local vorticity density or fluid rigidity, which can be viewed as relative vortex strength,

$$\Omega = \frac{\|\boldsymbol{B}\|_F}{\|\boldsymbol{A}\|_F + \|\boldsymbol{B}\|_F + \varepsilon} \tag{6.4}$$

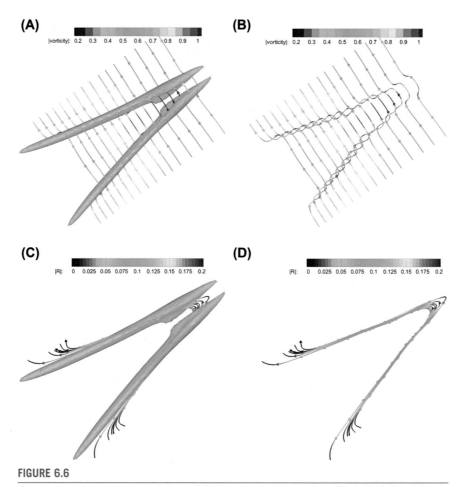

FIGURE 6.6

Comparison of vorticity lines and Liutex lines for the lambda vortex: (A) vorticity lines with the Liutex iso-surface visible, (B) vorticity lines with the Liutex iso-surface hidden, (C) Liutex lines with the Liutex iso-surface visible, and (D) Liutex lines with the Liutex iso-surface hidden.

Reproduced from Y. Gao and C. Liu, Rortex and comparison with eigenvalue-based vortex identification criteria, Phys. Fluids 30 (085107), 2018, with the permission of AIP Publishing.

where A and B are the symmetric and antisymmetric part from the Cauchy−Stokes decomposition, ε is a small positive number introduced to avoid division by zero. Dong et al. (2018c) suggests that ε could be determined at each time step by

$$\varepsilon = 0.001\left(||B||_F - ||A||_F\right)_{max} \tag{6.5}$$

The maximum in Eq. (6.5) is achieved over the whole flow domain. According to the definition, the parameter Ω ranges from 0 to 1 and thus has been normalized. It

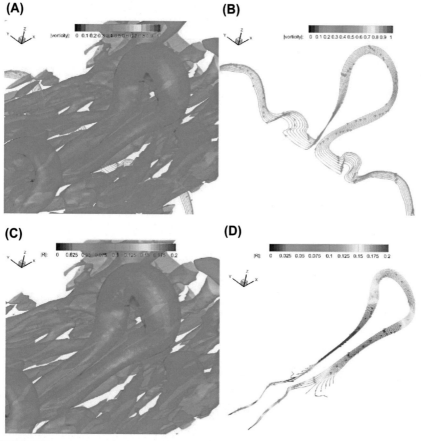

FIGURE 6.7

Comparison of vorticity lines and Liutex lines for the ring-like vortex: (A) vorticity lines with the Liutex iso-surface visible, (B) vorticity lines with the Liutex iso-surface hidden, (C) Liutex lines with the Liutex iso-surface visible, and (D) Liutex lines with the Liutex iso-surface hidden.

Reproduced from Y. Gao and C. Liu, Rortex and comparison with eigenvalue-based vortex identification criteria, Phys. Fluids 30 (085107), 2018, with the permission of AIP Publishing.

has been proved by many users (Zhang et al., 2019; Wang L., et al., 2019; Wang Y., et al., 2019) that the Ω method with a fixed threshold of 0.52 could well capture both strong and weak vortices in transient flows without adjustment of the threshold and thus could be viewed as a reliable, robust, threshold-insensitive, and easy-to-use vortex identification method in practice or the best method among the second generation of vortex identification methods. This means the Omega method is better than any of the second generation vortex identification methods in many aspects. Note that Eq. (6.5) is empirical and may need to adjust, case by case. The main reason to introduce ε is to avoid a division by an extremely small number.

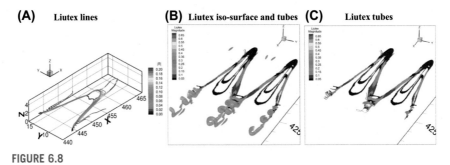

FIGURE 6.8

Liutex lines and Liutex tubes for vortex structure in early transition.

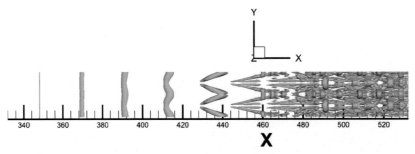

FIGURE 6.9

Liutex iso-surface for the vortex structure in early transition.

6.3.2 Liutex omega vortex identification method

In order to introduce the idea of the Liutex-Ω method, the Liutex method should be briefly described. In the Liutex method, the velocity gradient tensor $\nabla \vec{v}$ has complex conjugate eigenvalues $\lambda_{cr} \pm \lambda_{ci} i$ in the region with rotation, and the real unit eigenvector $\vec{r}$ is the direction of the local rotational axis of vortex. Hence, we have $\nabla \vec{v} \cdot \vec{r} = \lambda_r \vec{r}$ where λ_r denotes the real eigenvalue and indicates the rate of change on the axis of rotation. In addition, the definition of local rotation axis means there is no cross-velocity gradient on the local rotation axis. In order to obtain the rotation strength, the first coordinate rotation Q_r is used so that the rotation axis is the $Z-$ axis. Then

$$\nabla \vec{V} = Q_r \nabla \vec{v} Q_r^T = \begin{bmatrix} \dfrac{\partial U}{\partial X} & \dfrac{\partial U}{\partial Y} & 0 \\[2mm] \dfrac{\partial V}{\partial X} & \dfrac{\partial V}{\partial Y} & 0 \\[2mm] \dfrac{\partial W}{\partial X} & \dfrac{\partial W}{\partial Y} & \dfrac{\partial W}{\partial Z} \end{bmatrix} \tag{6.6}$$

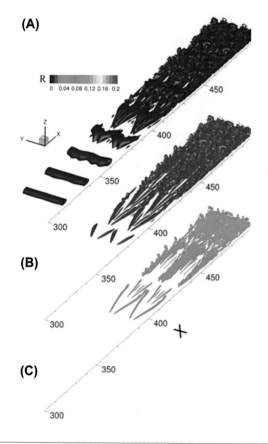

FIGURE 6.10

Iso-surface of (A) $R = 0.005$, (B) $R = 0.02$, (C) $R = 0.1$ in a boundary layer transition on a flat plate.

Hence, in the new XYZ frame, the coordinate $Z-$ axis is the local rotation axis. From the definition of Liutex vector, the rotation strength is obtained by a second coordinate rotation $\boldsymbol{P}_r$ in the $XY-$ plane to make $\left|\frac{\partial U}{\partial Y}\right|$ or $\left|\frac{\partial V}{\partial X}\right|$ take the minimum value. After turning θ angle, the velocity gradient tensor will become

$$\nabla \vec{V}_\theta = \boldsymbol{P}_r \nabla \vec{V} \boldsymbol{P}_r^T \tag{6.7}$$

Therefore, the terms in the 2×2 upper left submatrix of $\nabla \vec{V}_\theta$ are the following

$$\left.\frac{\partial U}{\partial Y}\right|_\theta = \alpha \sin(2\theta + \phi) - \beta,$$

$$\left.\frac{\partial V}{\partial X}\right|_\theta = \alpha \sin(2\theta + \phi) + \beta,$$

$$\left.\frac{\partial U}{\partial X}\right|_\theta = -\alpha \cos(2\theta + \phi) + \frac{1}{2}\left(\frac{\partial U}{\partial X} + \frac{\partial V}{\partial Y}\right),$$

$$\left.\frac{\partial V}{\partial Y}\right|_\theta = \alpha \cos(2\theta + \phi) + \frac{1}{2}\left(\frac{\partial U}{\partial X} + \frac{\partial V}{\partial Y}\right),$$

(6.8)

where

$$\alpha = \frac{1}{2}\sqrt{\left(\frac{\partial V}{\partial Y} - \frac{\partial U}{\partial X}\right)^2 + \left(\frac{\partial V}{\partial X} + \frac{\partial U}{\partial Y}\right)^2},$$

(6.9)

$$\beta = \frac{1}{2}\left(\frac{\partial V}{\partial X} - \frac{\partial U}{\partial Y}\right),$$

(6.10)

The detailed production and the expression of φ can be referred to Liu et al. (2019). In fact, β in (6.10) is the vorticity in the XY frame and independent of the P_r rotation. The Liutex vector is defined by

$$\vec{R} = R\vec{r}$$

(6.11)

And the magnitude of Liutex is defined as twice the minimal absolute value of the off-diagonal component of the 2×2 upper left submatrix and can be given by

$$R = \begin{cases} 2(\beta - \alpha), & \alpha^2 - \beta^2 < 0 \\ 0, & \alpha^2 - \beta^2 \geq 0 \end{cases}$$

(6.12)

In the Liutex- Omega vortex identification method, Ω_R is defined as the ratio of β squared over the sum of β squared and α squared and can be written as

$$\Omega_R = \frac{\beta^2}{\alpha^2 + \beta^2 + \varepsilon}$$

(6.13)

where ε is a small parameter introduced in the denominator of Ω_R to remove the noises caused by computer rounding errors, which could be case-dependent. It is defined as a function of the maximum of the term of $\beta^2 - \alpha^2$. In this book, ε is proposed as follows:

$$\varepsilon = b * \left(\beta^2 - \alpha^2\right)_{max}$$

(6.14)

where b is a small positive number around 0.001—0.002. ε is case-related since noises in different cases have different dimensions. However, if the case is set up, b is a constant (0.001) and the term $\left(\beta^2 - \alpha^2\right)_{max}$ is a fixed number for each time step, which can be easily obtained, the manual adjustment of ε for each case can therefore be avoided. Thus, in such a situation, usually the iso-surface of $\Omega_R = 0.52$ can be chosen to visualize the vortex structures and to indicate the region where the vorticity overtakes the principal strain rate on the plane perpendicular to the local rotation axis.

In the following, two examples are provided to examine the Liutex-Omega method. As the first example, the data from direct numerical simulation of boundary layer transition on a flat plate (Liu et al., 2014) are used. Fig. 6.11 shows vortex structures at four time steps of the early stage of the boundary layer transition obtained by the iso-surfaces of $\Omega_R = 0.52$. Based on Eq. (6.14), the values of ε in Fig. 6.11A—D are 0.00064, 0.0006, 0.00048, and 0.00044 respectively, but b in Eq. (6.14) is fixed as 0.001 and no adjustment is needed. As can be found, the vortex structures are well captured in different time steps without noises appearing in the upper side of the computation domain. Furthermore, one doesn't need to adjust the threshold of Ω_R for different time steps. The iso-surface of $\Omega_R = 0.52$ performs well for accurately capturing the vortex structures in each time step. The vortex structures at the late stage of the boundary layer transition are also shown by the iso-surfaces of $\Omega_R = 0.52$ in Fig. 6.12. The values of ε are 0.0006 and 0.0003, respectively. It can be seen that the iso-surface of $\Omega_R = 0.52$ can also visualize the vortex structures at the late stage of the flow transition very clearly.

The SWBLI (shock wave and boundary layer interaction) in a supersonic ramp flow with micro-vortex generator (MVG) control at Mach 2.5 is selected as another test case (Li et al., 2010). Fig. 6.13 shows the iso-surfaces of $\Omega_R = 0.52$ at (A) $t = 1284T^*$ (B) $t = 1984T^*$ (T^* is the characteristic time in MVG case). According to Eq. (6.14), ε is equal to 1.54 and 1.35, respectively. It is clearly shown that, at different time steps, both strong vortex structures such as vortex rings and weak structures such as turbulent boundary layers are well captured without any noise in the upper side of the computation domain.

In the aforementioned two different cases, although the values of ε are quite different, the iso-surface of $\Omega_R = 0.52$ can always capture both strong and weak vortex structures well. In addition, the current method is valid for both compressible and incompressible flows.

6.3.3 Summery on Ω_R

The new Ω_R method has the following advantages:

(1) Ω_R is a normalized function from 0 to 1, which can be further used in statistics and correlation analysis as a physical quantity just like pressure, density, and so on.
(2) Compared with many vortex identification methods, which require case-dependent thresholds to capture the vortex structures, it is quite robust and can be always set as 0.52 to capture both weak and strong vortex structures simultaneously.
(3) Ω_R is applicable to both compressible and incompressible flows.

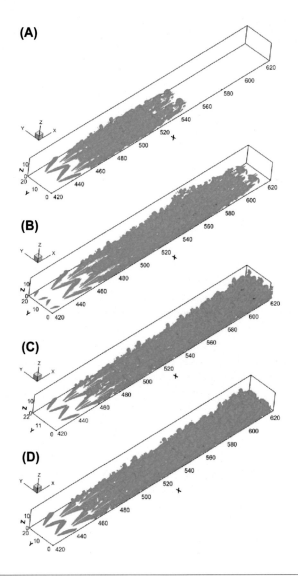

FIGURE 6.11

ISO-surface of $\Omega_R = 0.52$ at different time steps of the early stage of boundary layer transition: (A) $t = 7.1T$ (B) $t = 8.2T$ (C) $t = 9.5T$ (D) $t = 11.0T$ (T is the period of T-S wave).

6.4 Modified Liutex-Omega method

In the last section, the original normalized Liutex/vortex identification method Ω_R is defined by Dong et al. (2019b)

$$\Omega_R = \frac{\beta^2}{\alpha^2 + \beta^2 + \varepsilon} \tag{6.15}$$

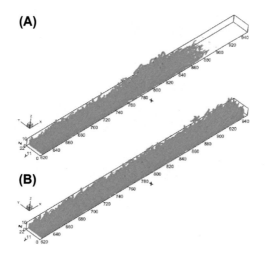

FIGURE 6.12

Iso-surfaces of $\Omega_R = 0.52$ at different time steps of the late stage of boundary layer transition: (A) $t = 15.2T$ (B) $t = 19.5T$ (T is the period of T-S wave).

where a small parameter ε is used to prohibit the computational noise caused by an extremely small divisor, which is proposed as $\varepsilon = b_0\left(\beta^2 - \alpha^2\right)_{\max}$ where b_0 is a small positive number around $0.001 \sim 0.002$. In subsection 6.3.3, although there are many advantages in the vortical identification by the original method (6.15), the iso-surfaces formed by it are slightly not smooth and many bulges appeared on the iso-surfaces. From the definition of the original Ω method (Liu et al., 2016), given the denotations of the symmetric tensor $A = \frac{1}{2}\left(\nabla \vec{v} + \nabla \vec{v}^T\right)$ and the antisymmetric spin tensor $B = \frac{1}{2}\left(\nabla \vec{v} - \nabla \vec{v}^T\right)$, Ω method can be formulated as

$$\Omega = \frac{b}{a + b + \varepsilon} \tag{6.16}$$

where $a = \|A\|_F^2$, $b = \|B\|_F^2$, and $\|\cdot\|_F$ represents the Frobenius norm. For two-dimensional flow, the Liutex decomposition and Galilean invariance of Ω method (Liu et al., 2019) revealed that

$$\nabla \vec{V}_{\theta\min} = \begin{bmatrix} \lambda_{cr} & -(\beta - \alpha) \\ \beta + \alpha & \lambda_{cr} \end{bmatrix} = \begin{bmatrix} \lambda_{cr} & \alpha \\ \alpha & \lambda_{cr} \end{bmatrix} + \begin{bmatrix} 0 & -\beta \\ \beta & 0 \end{bmatrix} = A + B \tag{6.17}$$

and

$$\Omega_{2D} = \frac{b}{a + b + \varepsilon} = \frac{\beta^2}{\beta^2 + \alpha^2 + \lambda_{cr}^2 + \varepsilon} \tag{6.18}$$

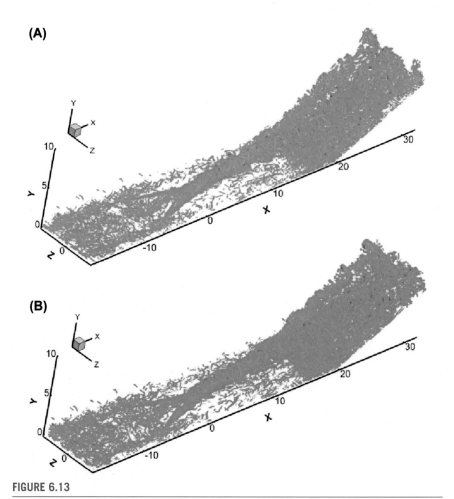

FIGURE 6.13

Iso-surfaces of $\Omega_R = 0.52$ at different time steps of shock and boundary layer interaction: (A) $t = 1284T^*$ (B) $t = 1984T^*$ (T* is the characteristic time).

Similarly, for three-dimensional flow,

$$\nabla \vec{V}_{\theta\min} = \begin{bmatrix} \lambda_{cr} & -(\beta - \alpha) & 0 \\ \beta + \alpha & \lambda_{cr} & 0 \\ \xi & \eta & \lambda_r \end{bmatrix} = \begin{bmatrix} \lambda_{cr} & \alpha & \dfrac{1}{2}\xi \\ \alpha & \lambda_{cr} & \dfrac{1}{2}\eta \\ \dfrac{1}{2}\xi & \dfrac{1}{2}\eta & \lambda_r \end{bmatrix} + \begin{bmatrix} 0 & -\beta & -\dfrac{1}{2}\xi \\ \beta & 0 & -\dfrac{1}{2}\eta \\ \dfrac{1}{2}\xi & \dfrac{1}{2}\eta & 0 \end{bmatrix}$$

$$= A + B$$

$$(6.19)$$

and

$$\Omega_{3D} = \frac{\beta^2 + \frac{1}{4}\left(\xi^2 + \eta^2\right)}{\beta^2 + \frac{1}{2}\left(\xi^2 + \eta^2\right) + \alpha^2 + \lambda_{cr}^2 + \frac{1}{2}\lambda_r^2 + \varepsilon} \tag{6.20}$$

Hence, the Ω method given by Eqs. (6.18) and (6.20) is different from the normalized Liutex/vortex identification method Ω_R (6.15). Considering the idea of the original Ω method, the following new modified Liutex-Omega identification method was recommended

$$\widetilde{\Omega}_R = \frac{\beta^2}{\beta^2 + \alpha^2 + \lambda_{cr}^2 + \frac{1}{2}\lambda_r^2 + \varepsilon} \tag{6.21}$$

And $\widetilde{\Omega}_R$ requires a parameter greater than 0.5, which is same as Ω or Ω_R methods. In practice, $\widetilde{\Omega}_R = 0.52$ is feasible. Here, for the convenience of explanation, α and β are used to represent the modified $\widetilde{\Omega}_R$ method. Then an explicit formula is given that does not require the previous Q_r and P_r rotations to get α and β. It will greatly simplify the solution process.

At present, the Q criterion is a very popular vortex identification method used in engineering. The mathematical relationship between the new modified $\widetilde{\Omega}_R$ method and the popular Q criterion can be derived. In fact, from Eq. (6.21), neglect the small ε, the vortex domain should be governed by

$$\widetilde{\Omega}_R = \frac{\beta^2}{\beta^2 + \alpha^2 + \lambda_{cr}^2 + \frac{1}{2}\lambda_r^2} > 05 \tag{6.22}$$

i.e.,

$$\beta^2 - \alpha^2 > \lambda_{cr}^2 + \frac{1}{2}\lambda_r^2 \tag{6.23}$$

Assume the flow is incompressible, then $\lambda_r = -2\lambda_{cr}$. Hence, Eq. (6.23) becomes

$$\beta^2 - \alpha^2 > 3\lambda_{cr}^2 \tag{6.24}$$

From Eq. (6.19), the relation of $\beta^2 - \alpha^2 = \lambda_{ci}^2$ can be obtained by a simple algebra calculation. Therefore, it is satisfied by $\lambda_{ci}^2 > 3\lambda_{cr}^2$. That is

$$\left(\frac{\lambda_{cr}}{\lambda_{ci}}\right)^2 < \frac{1}{3} \tag{6.25}$$

In addition, the second invariant Q of the velocity gradient tensor for incompressible flow can be explicitly written as

$$Q = \lambda_{ci}^2\left(1 - 3\left(\frac{\lambda_{cr}}{\lambda_{ci}}\right)^2\right) \tag{6.26}$$

$Q > 0$ criterion requires $\left(\frac{\lambda_{cr}}{\lambda_{ci}}\right)^2 < \frac{1}{3}$. Therefore, the mathematical vortex boundaries for $\widetilde{\Omega}_R > 0.5$ and $Q > 0$ are same completely, and all regions described by them are a subset of the region defined by $\lambda_{ci} > 0$. But the new modified quantity $\widetilde{\Omega}_R$ is dimensionless from 0 to 1. Furthermore, from Eq. (6.15), $\Omega_R = \frac{\beta^2}{\alpha^2+\beta^2} > 0.5$ and $\beta^2 - \alpha^2 = \lambda_{ci}^2$, the mathematical vortex boundaries defined by $\Omega_R > 0.5$ and $\lambda_{ci} > 0$ are same. Although the region defined by $\widetilde{\Omega}_R > 0.5$ is a subset defined by $\Omega_R > 0.5$, Eq. (6.22) will avoid regions of strong outward spiraling given by $\left(\frac{\lambda_{cr}}{\lambda_{ci}}\right)^2 > \frac{1}{3}$. Furthermore, the new modified $\widetilde{\Omega}_R$ method can keep smooth and get rid of the bulges on the iso-surfaces.

Although Eq. (6.21) is very simple, the calculations of α and β are not trivial. Recently, Wang et al. (2019a) gave an explicit formula of Liutex vector to simplify the calculations of β and α. Then the explicit expressions of β and α are

$$\beta = \frac{1}{2}\vec{\omega} \cdot \vec{r} \tag{6.27}$$

$$\alpha = \frac{1}{2}\sqrt{\left(\vec{\omega} \cdot \vec{r}\right)^2 - 4\lambda_{ci}^2} \tag{6.28}$$

Hence, the new modified Liutex-Omega identification method (6.21) can be reformulated as

$$\widetilde{\Omega}_R = \frac{\left(\vec{\omega} \cdot \vec{r}\right)^2}{2\left[\left(\vec{\omega} \cdot \vec{r}\right)^2 - 2\lambda_{ci}^2 + 2\lambda_{cr}^2 + \lambda_r^2\right] + \varepsilon} \tag{6.29}$$

The new expression (6.29) does not require the coordinate rotations Q_r and P_r, and the calculation is greatly simplified.

In order to show the modified Liutex-Omega identification method (6.21) or (6.29) can get rid of bulging phenomenon, which can be caused by original Ω_R method, the flow data calculated by high-order direct numerical simulation (DNS) of boundary layer transition on a flat plate at Mach number 0.5 and Reynolds number 1000 are chosen to compare. The iso-surfaces shown in Fig. 6.14 reflect the different results by the two normalized Liutex-Omega vortex identification methods. As shown in Fig. 6.14A, by the original Ω_R method (6.15), the iso-surfaces lose smoothness and the emergence of the bulging phenomenon is obvious. But, by the new modified $\widetilde{\Omega}_R$ method, as is shown in Fig. 6.14B, the effectiveness of the new method is demonstrated. Furthermore, $\widetilde{\Omega}_R$ is a dimensionless relative quantity from 0 to 1. The iso-surface formed by $\widetilde{\Omega}_R$ method is not sensitive to the threshold. Generally, for different examples, the value of $\widetilde{\Omega}_R$ can always be 0.52 and it has the shape-preserving feature as the threshold increases. However, for the popular Q criterion, the determination of the threshold is unclear in advance and needs to be determined according to the problem, case by case, sometimes for some special cases as large as

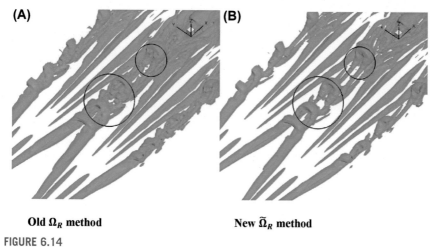

(A)　(B)

Old Ω_R method　**New $\widetilde{\Omega}_R$ method**

FIGURE 6.14

Iso-surfaces of hairpin vortex structures.

*Reproduced from J. Liu and C. Liu, Modified normalized Rortex/vortex identified method, Phys. Fluids 31
(061704), 2019, with the permission of AIP Publishing.*

10^8, but the current example can only take a small value. As a demonstration, the results of the iso-surfaces obtained by Q criterion are also provided. The results obtained by Q criterion at different value of $Q = 0.005, 0.02, 0.04, 0.05$ are shown in Figs. 6.15A, C, E, and G. The vortex structures obtained by $\widetilde{Q}_R$ method at $\widetilde{Q}_R = 0.52, 0.6, 0.7, 0.8$ are shown in Figs. 6.15B, D, F, and H. The graph of the DNS data shows that as the threshold of the Q criterion increases, many vortex structures disappear in the direction of flow, especially the ring structure. However, the $\widetilde{Q}_R$ method still maintains the vortex structures as the threshold increases. Furthermore, the vortex intensity expressed by the Q criterion threshold has been greatly reduced in the flow direction, and the relative vortex strength is still relatively large by the $\widetilde{Q}_R$ or Liutex-Omega method.

A modified normalized Liutex/vortex (Liutex-Omega) identification method named $\widetilde{Q}_R$ is a modification to the original Ω_R method, which can alleviate the bulging phenomenon on the iso-surfaces. The new modified $\widetilde{Q}_R$ method retains the advantages of original Ω_R method. $\widetilde{Q}_R$ is a dimensionless and relative quantity from 0 to 1, which can be used to do statistics and correlation analysis directly. $\widetilde{Q}_R$ can distinguish vortex from high vorticity concentration due to high shear in the boundary layers. In general, $\widetilde{Q}_R$ is robust and can always be set as 0.52 to visualize the vortex structures. As a relative quantity, $\widetilde{Q}_R$ has the capability of capturing both strong and weak vortices simultaneously. With respect to the results by Q criterion, the $\widetilde{Q}_R$ method can retain the ring structures when the threshold increases and in many places, the relative strength of vortex structures is high. Therefore, for the iso-surface-based vortex identification methods, the modified Liutex-Omega method or $\widetilde{Q}_R$ is the best iso-surface-based vortex identification method so far according to the practical experience by many users.

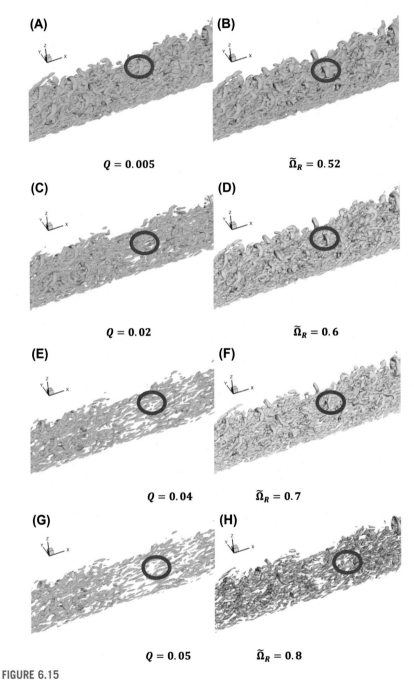

(A)

$$Q = 0.005$$

(B)

$$\widetilde{\Omega}_R = 0.52$$

(C)

$$Q = 0.02$$

(D)

$$\widetilde{\Omega}_R = 0.6$$

(E)

$$Q = 0.04$$

(F)

$$\widetilde{\Omega}_R = 0.7$$

(G)

$$Q = 0.05$$

(H)

$$\widetilde{\Omega}_R = 0.8$$

FIGURE 6.15

Iso-surfaces of vortex structures obtained by Q criterion and $\widetilde{\Omega}_R$ method.

Reproduced from Wang, Y., et al., Liutex theoretical system and six core elements of vortex identification, Journal of Hydrodynamics, 32(2), 2020, with the permission of Journal of Hydrodynamics.

6.5 Objective Liutex method

6.5.1 Short review

Previously, researchers used streamlines to represent vortices (Lugt, 1979). The biggest shortcoming of using streamlines to define vortices is that the streamline is neither Galilean nor rotation-invariant. Streamlines obtained from different observers will show different vortex patterns, and hence, are unreliable for right vortex identification. For a long time, the first generation of vorticity-based methods, such as vorticity tubes, is also applied to represent vortices and the vortex is regarded as the region with high concentration of vorticity. But as an immediate counter-example, the laminar boundary layer has large vorticity magnitude without any rotational motion. Hence, the vorticity is misused and is unable to distinguish the vortex region from the shear layer (Epps, 2017). Furthermore, the direction of the local vorticity and the vortex core direction of the vortex structure cannot align, especially in the wall-bounded turbulent flow (Zhou et al., 1999; Pirozzoli et al., 2008; Gao et al., 2011).

The vortex is intuitively recognized as the rotational motion of fluids around an axis. Accordingly, it should be associated with a rotating axis and a rotational angular velocity. Recently, Liutex is proposed by Liu et al. (2018a, 2019), Gao et al. (2018) to define the local rigid rotational axis and angular speed. This method decomposes the vorticity vector into a pure rigid body rotational vector (the Liutex vector) and a nonrotational vector called shear vector. The local rotational axis defined by Liutex is determined by the real eigenvector of the velocity gradient tensor, and the rotational strength is given by the minimal rotational angular speed calculated in the plane perpendicular to the local rotational axis. The Liutex lines, tubes, and magnitude can all be used to visualize turbulent flow and coherent vortex structures. Furthermore, the Liutex lines are aligned with the vortex core orientation, while most of the vorticity lines are somewhat misaligned with the vortex core. Although the proposed Liutex shows potentially important advantages in the research of turbulent flow and vortex identification, the definition of Liutex is not objective (Liu et al., 2019b).

The method keeping objectivity requires that the rotational direction and magnitude of Liutex vector should be independent of the observer (Epps, 2017; Haller et al., 2016; Liu et al., 2019b; Martins et al., 2016). Furthermore, the vortex structures visualized by the identification method should also be consistent regardless of arbitrary time variable translation or rotation of the observer. Nonobjective method will yield different vortex structures and contaminate the visualized flow field due to the rotation of the observer. Hence, it cannot be used in a time variable rotating reference frame directly. However, most of popular vortex identification methods ignored the objectivity. In order to make people pay attention to objectivity of vortex identification, Haller (2016) once again started the discussion of objectivity and provided an objective version of Lagrangian vortex identification method. Objectivity is mainly influenced by the time variable rotating reference system, which changes the velocity gradient tensor. In order to keep objectivity, the impact of the rotational motion of reference frames must be offset. At the present time, there are mainly three kinds of balance methods in literature (Epps, 2017). For example, Drouot (1976) used rotation rate of the eigenvector of the strain-rate tensor to define a strain rotation tensor to balance the impact of the Eulerian rotational reference frame change. A rigid vorticity tensor is defined by Wedgewood (1999) to offset the impact of the motion reference frame, but the process is too complicated to easily be used. Recently, with the help of spatially averaged vorticity, Haller (2016) defined objective coherent vortices by the Lagrangian-averaged vorticity deviation. Martins et al. (2016) used the relative rate-of-rotation tensor to define the objective Q criterion, the objective Δ criterion, the objective λ_2 criterion, and the objective λ_{ci} criterion. In this section, with the help of the spatially averaged vorticity to form a spatially averaged spin tensor, a net/relative velocity gradient tensor is given to define an objective Liutex vector. The defined objective Liutex vector will not change its direction and magnitude under Eulerian time variable rotational frame (Liu et al., 2019b). The detailed mathematical derivation is described and the numerical examples are presented to demonstrate the developed method in this section.

6.5.1.1 Objective extension of the Liutex vortex vector

The universal vortex vector definition should be objective. Otherwise, the vortex structures obtained by the moving observer at different states are different. Therefore, it is impossible to specifically reflect the essential characteristics of the vortex. In other words, the obtained vortex structures are not objective.

Objectivity requires that the definition of vortex is independent of the observer. The structure of the vortices should maintain its shape regardless of arbitrary time variable translation or rotation of the observer. Based on the research work by Haller (2016), vorticity is not objective. Hence, any method directly defined by vorticity to express vortex is not objective. In fact, the vorticity is not enough to express the flow rotation and define the vortex. Objectivity requires that the vortex vector should retain unchanged under the moving reference frame. The moving frame is generally defined by following Euclidean transformation

$$
\begin{bmatrix} x \\ y \\ z \end{bmatrix} = \boldsymbol{Q}_{or}(t) \begin{bmatrix} x^* \\ y^* \\ z^* \end{bmatrix} + \overrightarrow{b}(t), \tag{6.30}
$$

where x, y, z and x^*, y^*, z^* represent the different coordinates in the original and moving reference frame respectively, and $\boldsymbol{Q}_{or}(t)$ is a time-dependent orthogonal rotation matrix with $\boldsymbol{Q}_{or}^T \boldsymbol{Q}_{or} = \boldsymbol{Q}_{or} \boldsymbol{Q}_{or}^T = \boldsymbol{I}$, and $\overrightarrow{b}(t)$ is a time-dependent translation. The vector and tensor are called objective vector and objective tensor, if they remain unchanged under the Euclidean transformation (6.30). From the concept of objectivity, at first, the definitions of the object vector and tensor are given.

Definition 6.5.1. A vector $\vec{l}$ is called objective vector under the Euclidean transformation (6.30), if the new vector $\vec{l}^{\,*}$ satisfies the transformation

$$\vec{l}^{\,*} = Q_{or}^T(t)\,\vec{l}. \tag{6.31}$$

Definition 6.5.2. A tensor P is called objective tensor under the Euclidean transformation (6.30), if the new tensor P^* satisfies the transformation

$$P^* = Q_{or}^T(t)PQ_{or}(t). \tag{6.32}$$

In Euclidean transformation (6.30), both $Q_{or}(t)$ and $\vec{b}(t)$ are time-dependent. If $Q_{or}(t)$ is independent of time and $\vec{b}(t) = \vec{c}_1 t$, Eq. (6.30) become Galilean transformation. Wang et al. (2018) have proved that the Liutex vector is invariant under the Galilean transformation. From Eq. (6.30), the velocity vector in the new reference frame can be written as

$$\vec{v}^{\,*} = \begin{bmatrix} u^* \\ v^* \\ w^* \end{bmatrix} = Q_{or}^T(t) \begin{bmatrix} u \\ v \\ w \end{bmatrix} + \dot{Q}_{or}^T(t) \begin{bmatrix} x \\ y \\ z \end{bmatrix} - \frac{d}{dt}\left(Q_{or}^T(t)\,\vec{b}(t)\right) \tag{6.33}$$

where u, v, w denote the velocity components in the original reference frame, u^*, v^*, w^* denote the velocity components in the moving reference frame, and $\dot{Q}_{or}(t) = \frac{d}{dt}(Q_{or}(t))$. And the velocity gradient tensor in the moving reference frame can be expressed by

$$\nabla \vec{v}^{\,*} = Q_{or}^T(t)\nabla \vec{v} \, Q_{or}(t) + \dot{Q}_{or}^T(t)Q_{or}(t). \tag{6.34}$$

From Definition 6.5.2, $\nabla \vec{v}^{\,*}$ is not objective. Hence, if the velocity gradient tensor $\nabla \vec{v}^{\,*}$ is directly used to define the Liutex vector, the final Liutex vector is not objective. This conclusion will be verified in the section of numerical examples.

In order to describe the objective Liutex conveniently, the following lemmas and definitions are given.

Lemma 6.5.1. If $Q_{or}(t) \in \mathbb{R}^{3\times 3}$ is an orthogonal rotation matrix, then

$$\dot{Q}_{or}^T(t)Q_{or}(t) + Q_{or}^T(t)\dot{Q}_{or}(t) = 0 \tag{6.35}$$

Proof. As an orthogonal rotation matrix, from the fact $Q_{or}^T Q_{or} = Q_{or}Q_{or}^T = I$, it can be obtained that

$$\frac{d}{dt}\left(Q_{or}^T(t)Q_{or}(t)\right) = \dot{Q}_{or}^T(t)Q_{or}(t) + Q_{or}^T(t)\dot{Q}_{or}(t) = 0.$$

this finishes the proof of Lemma 6.5.1.

Lemma 6.5.2. The antisymmetric spin tensor in the new reference frame x^*, y^*, z^* can be formulated as

$$B^* = Q_{or}^T(t)BQ_{or}(t) - Q_{or}^T(t)\dot{Q}_{or}(t), \tag{6.36}$$

here B is the antisymmetric spin tensor in the original frame.

Proof. In the new reference frame, the antisymmetric spin tensor can be uniquely calculated by

$$B^* = \frac{1}{2}\left[\nabla\vec{v}^* - (\nabla\vec{v}^*)^T\right]$$

According to Eq. (6.34), it can be obtained by

$$B^* = \frac{1}{2}\left[Q_{or}^T(t)\nabla\vec{v}\,Q_{or}(t) + \dot{Q}_{or}^T(t)Q_{or}(t) - Q_{or}^T(t)\left(\nabla\vec{v}\right)^T Q_{or}(t) - Q_{or}^T(t)\dot{Q}_{or}(t)\right]$$

$$= \frac{1}{2}Q_{or}^T(t)\left[\nabla\vec{v} - \left(\nabla\vec{v}\right)^T\right]Q_{or}(t) + \frac{1}{2}\left[\dot{Q}_{or}^T(t)Q_{or}(t) - Q_{or}^T(t)\dot{Q}_{or}(t)\right]$$

$$= Q_{or}^T(t)\,BQ_{or}(t) + \frac{1}{2}\left[\dot{Q}_{or}^T(t)Q_{or}(t) - Q_{or}^T(t)\dot{Q}_{or}(t)\right]$$

By Lemma 6.5.1, the conclusions can be made immediately.

Lemma 6.5.3. Given the original symmetric rate-of-strain tensor $A = \frac{1}{2}\left(\nabla\vec{v} + \left(\nabla\vec{v}\right)^T\right)$, from Eqs. (6.34)–(6.36), the symmetric rate-of-strain tensor in the new reference frame can be obtained by

$$A^* = \nabla\vec{v}^* - B^* = Q_{or}^T(t)\,AQ_{or}(t). \tag{6.37}$$

For Lemma 6.5.3, the conclusion is straightforward, and the proof is omitted here. From the definition of the objectivity of a tensor, the strain-rate tensor is objective. In order to define an objective Liutex vector, an instantaneous spatially averaged vorticity is defined.

Definition 6.5.3. An instantaneous spatially averaged vorticity $\overline{\vec{\omega}}$ over $Vol(t)$ (volume of the fluid domain) is defined as

$$\overline{\vec{\omega}}(t) = \frac{1}{Vol(t)}\int_{Vol(t)}\vec{\omega}\left(\vec{x},t\right)dV, \tag{6.38}$$

where $\vec{\omega} = \nabla\times\vec{v}$.

From Eq. (6.36), an antisymmetric tensor $\overline{B}$ can easily be obtained by

$$\overline{B}\vec{e} = \frac{1}{2}\overline{\vec{\omega}}\times\vec{e}, \ \forall\ \vec{e}\in\mathbb{R}^3. \tag{6.39}$$

Due to the action of the moving reference frame, the vorticity will also change, and the new vorticity $\vec{\omega}^*$ will become

$$\vec{\omega}^* = Q_{or}^T(t)\vec{\omega} + \dot{\vec{q}}(t), \tag{6.40}$$

where the new defined vector $\dot{\vec{q}}(t)$ satisfied $Q_{or}^T(t)\dot{Q}_{or}(t)\vec{e} = -\frac{1}{2}\left(\dot{\vec{q}}\times\vec{e}\right)$ for all $\vec{e}\in\mathbb{R}^3$. Taken spatial average over Eq. (6.40), the spatially averaged vorticity in the new reference frame can be formulated by

$$\overline{\vec{\omega}}^* = Q_{or}^T(t)\overline{\vec{\omega}} + \dot{\vec{q}}(t). \tag{6.41}$$

From the aforementioned equation, a unique antisymmetric tensor $\overline{B}^*$ can be obtained, which satisfied

$$\overline{B}^* \overrightarrow{e} = \frac{1}{2}\overrightarrow{\overline{\omega}}^*$$

$$\times \overrightarrow{e} = \frac{1}{2}\left(Q_{or}^T(t)\overrightarrow{\overline{\omega}} + \overrightarrow{\dot{q}}(t)\right) \times \overrightarrow{e} = \frac{1}{2}Q_{or}^T(t)\overrightarrow{\overline{\omega}} \times \overrightarrow{e} - Q_{or}^T(t)\dot{Q}_{or}(t)\overrightarrow{e}. \quad (6.42)$$

Thus,

$$\overline{B}^* = Q_{or}^T(t)\overline{B}Q_{or}(t) - Q_{or}^T(t)\dot{Q}_{or}(t). \quad (6.43)$$

In fact, Eq. (6.43) can also be derived directly by taking spatial average to Eq. (6.36).

In order to obtain an objective Liutex vector and offset the impact of moving observers, a net velocity gradient tensor is defined.

Definition 6.5.4.   A net velocity gradient tensor is defined by

$$\widehat{\nabla \overrightarrow{v}} = \nabla \overrightarrow{v} - \overline{B}. \quad (6.44)$$

Then in the moving reference frame, the net velocity gradient tensor is

$$\widehat{\nabla \overrightarrow{v}}^* = \nabla \overrightarrow{v}^* - \overline{B}^*. \quad (6.45)$$

From Eqs. (6.34), (6.35), (6.37), and (6.43), in fact, the following equation can be achieved

$$\widehat{\nabla \overrightarrow{v}}^* = \nabla \overrightarrow{v}^* - \overline{B}^* = Q_{or}^T(t)\widehat{\nabla \overrightarrow{v}}Q_{or}(t). \quad (6.46)$$

In order to obtain the objective rotation axis and rotation strength, all of the following discussion is based on the net velocity gradient tensor. From the definition of Liutex vector, in the first step, it is needed to determine the eigenvalues and the corresponding eigenvectors of the net velocity gradient tensor $\widehat{\nabla \overrightarrow{v}}$. If the net velocity gradient tensor $\widehat{\nabla \overrightarrow{v}}$ has only one real eigenvalue λ_r, and two complex eigenvalues, it can be confirmed that it has local rotation and defines the corresponding real eigenvector $\widehat{r}$ as the rotational axis. Otherwise, there is no local rotation at this point. For the objective Liutex vector, the following theorem can be given (Liu et al., 2019b).

Theorem 6.5.1.   Based on the definition of the net velocity gradient tensor (6.44), the Liutex vector is objective.

Proof. From Eq. (6.46), the rotational axis, in the new time-dependent moving frame, can be calculated by

$$\widehat{\overrightarrow{r}}^* = Q_{or}^T(t)\widehat{\overrightarrow{r}} \quad (6.47)$$

Here, $\hat{\vec{r}}$ is the real eigenvector of $\widehat{\nabla v}$. Recalling Definition 6.5.1, the rotational axis is invariant in the time-dependent moving frame.

For net velocity gradient tensor $\widehat{\nabla v}$., the calculation of the rotation strength needs two coordinate rotations: $\boldsymbol{Q_r}, \boldsymbol{P_r}$ (see previous sections). After the $\boldsymbol{Q_r}$ rotation, the rotation axis $\hat{\vec{r}}$ will become $\vec{Z} = [0, 0, 1]^T$ as

$$\boldsymbol{Q_r}\hat{\vec{r}} = [0, 0, 1]^T \tag{6.48}$$

and the net velocity gradient tensor becomes

$$\widehat{\nabla V} = \boldsymbol{Q_r}\widehat{\nabla \hat{v}}\boldsymbol{Q_r^T} = \begin{bmatrix} \dfrac{\partial \widehat{U}}{\partial X} & \dfrac{\partial \widehat{U}}{\partial Y} & 0 \\[2ex] \dfrac{\partial \widehat{V}}{\partial X} & \dfrac{\partial \widehat{V}}{\partial Y} & 0 \\[2ex] \dfrac{\partial \widehat{W}}{\partial X} & \dfrac{\partial \widehat{W}}{\partial Y} & \dfrac{\partial \widehat{W}}{\partial Z} \end{bmatrix} \tag{6.49}$$

In the time-dependent moving frame, in order to rotate $\hat{\vec{r}}^*$ to $\vec{Z}^* = [0, 0, 1]^T$, by considering Eqs. (6.47) and (6.48), it can be obtained that

$$\vec{Z}^* = [0, 0, 1]^T = \boldsymbol{Q_r Q_{or}}(t)\hat{\vec{r}}^* \tag{6.50}$$

In addition, with the $\boldsymbol{P_r}$ rotation at a rotation angle θ_0 to take the minimal absolute value of the off-diagonal components of the 2×2 upper left submatrix of $\widehat{\nabla V}_\theta$, from Eqs. (6.46) and (6.49), one can get

$$\widehat{\nabla V}_{\theta_0} = \boldsymbol{P_r}\widehat{\nabla V}\boldsymbol{P_r^T} = \boldsymbol{P_r Q_r}\widehat{\nabla \hat{v}}\boldsymbol{Q_r^T P_r^T} = \boldsymbol{P_r Q_r Q_{or}}(t)\widehat{\nabla v}^*\boldsymbol{Q_{or}^T}(t)\boldsymbol{Q_r^T P_r^T}$$

$$= (\boldsymbol{P_r Q_r Q_{or}}(t))\widehat{\nabla v}^*(\boldsymbol{P_r Q_r Q_{or}}(t))^T \tag{6.51}$$

Based on the uniqueness of the minimal absolute value, it is concluded that

$$\widehat{\nabla V}_{\theta_0}^* = \widehat{\nabla V}_{\theta_0} = (\boldsymbol{P_r Q_r Q_{or}}(t))\widehat{\nabla v}^*(\boldsymbol{P_r Q_r Q_{or}}(t))^T \tag{6.52}$$

Hence, the rotation strength $\widehat{R}$ of Liutex is invariable. Thus, the proof of Theorem 6.5.1 is finished.

In order to identify objective vortex structures, the iso-surfaces of the magnitude of objective Liutex vector can be directly used to visualize vortex structures. And the Liutex lines can be used to represent the direction of the vortex.

6.5.2 Numerical examples

In this section, three examples will be presented to verify and compare the original Liutex vortex vector with the objective version. Objectivity requires that the moving observer can get same vortex structures. The moving reference frame can be described by a time-dependent rotation matrix $Q_{or}(t)$ combined by three basic rotations around three coordinate axes as

$$Q_{or}(t) = Q_{cx}(\gamma_1 t)Q_{cy}(\gamma_2 t)Q_{cz}(\gamma_3 t)$$

$$= \begin{bmatrix} 1 & 0 & 0 \\ 0 & \cos \gamma_1 t & \sin \gamma_1 t \\ 0 & -\sin \gamma_1 t & \cos \gamma_1 t \end{bmatrix} \begin{bmatrix} \cos \gamma_2 t & 0 & -\sin \gamma_2 t \\ 0 & 1 & 0 \\ \sin \gamma_2 t & 0 & \cos \gamma_{2t} \end{bmatrix} \begin{bmatrix} \cos \gamma_3 t & \sin \gamma_3 t & 0 \\ -\sin \gamma_3 t & \cos \gamma_3 t & 0 \\ 0 & 0 & 1 \end{bmatrix},$$

$$(6.53)$$

where γ_1, γ_2, and γ_3 are the corresponding rotational angular frequencies around the coordinate axes.

(1) Sullivan vortex

As a prototype example, the Sullivan vortex (Sullivan, 1959; Wang Y., et al., 2018; Liu et al., 2019b), an analytical two-cell vortex solution to the Navier–Stokes equations, is first examined. This special analytical flow is often used to depict a strong tornado with a central downdraft flow. In general, the formula of the Sullivan vortex is formulated in cylinder coordinate system by

$$u_r = -ar + \frac{6\nu}{r}\left[1 - \exp\left(-\frac{ar^2}{2\nu}\right)\right],$$

$$u_\theta = \frac{\Gamma}{2\pi r} \frac{H\left(\frac{ar^2}{2\nu}\right)}{H(\infty)}, \qquad (6.54)$$

$$u_z = 2az\left[1 - 3\exp\left(-\frac{ar^2}{2\nu}\right)\right],$$

where $H(\eta) = \int_0^\eta \exp\left(-s + 3\int_0^s \frac{1-e^{-\tau}}{\tau} d\tau\right) ds$ and thus $H(\infty) = 37.905$. In this special example, a Sullivan vortex with $a = 1$, $\nu = 0.02$, and $\Gamma = 5$ is considered. Wang Y., et al. (2018) have shown that the streamlines are not Galilean invariant while Liutex is Galilean invariant. In order to verify the present method is objective, as a sample, angular frequencies are randomly selected, $\gamma_1 = 0.35$, $\gamma_2 = 2.25$, and $\gamma_3 = 1.50$ and the translation of $\overrightarrow{b}(t) = \left[1, t^2, 0.5\right]^T$ in Eqs. (6.30) and (6.53), which would not lose generality. After time $t = 0.5$, the motion frame will reach a new position and has a speed of movement. The contours of the nonobjective Liutex magnitudes in the original reference frame and in the new reference frame are shown in Fig. 6.16. The figure shows that the original Liutex has changed greatly in size and orientation under the old and new reference frames. Hence, the original Liutex definition is not objective. Using the objective Liutex vortex vector defined by the net velocity gradient tensor (6.44), the size and direction of the Liutex vector are given in Fig. 6.17. The visualization results show that the new objective Liutex vector can keep its size and direction unchanged. Therefore, the use of nonobjective Liutex method in the moving frame does not result in a consistent vortex structure.

6.5.3 Unsteady ABC-type flow

ABC-type flow is used by many people to study the objectivity of vortex identification methods in literatures by Haller (2005), Haller et al. (2016), and Martins et al. (2016). ABC-type flow is an unstable analytical solution to the inviscid Euler system with high-frequency instabilities. The unsteady velocity components in general are

$$u(\boldsymbol{x}, t) = A(t)\sin z + C \cos y,$$
$$v(\boldsymbol{x}, t) = B \sin x + A(t)\cos z, \qquad (6.55)$$
$$w(\boldsymbol{x}, t) = C \sin y + B \cos x,$$

where $A(t) = A_0 + (1 - e^{-qt})\sin \omega t$ represents the effect of a growing and saturating unstable mode (Haller, 2005). As a sample example, it is set that

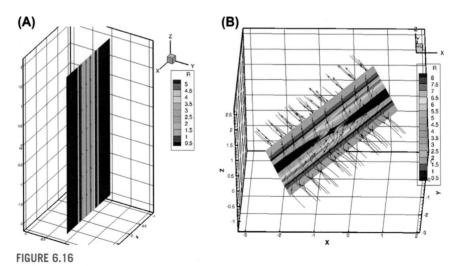

(A)   **(B)**

FIGURE 6.16

Slices of the nonobjective Liutex in the original reference frame (A) and in the new reference frame (B).

Reproduced from J. Liu, Y. Gao and C. Liu, An objective version of the Rortex vector for vortex identification, Phys. Fluids 31 (065112), 2019, with the permission of AIP Publishing.

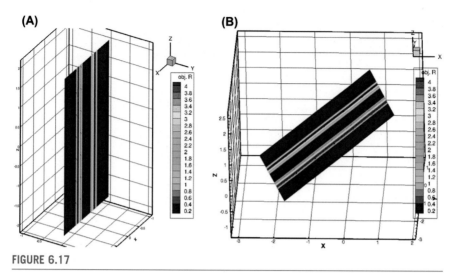

FIGURE 6.17

Slices of the objective Liutex in the original reference frame (A) and in the new reference frame (B).

Reproduced from J. Liu, Y. Gao and C. Liu, An objective version of the Rortex vector for vortex identification, Phys. Fluids 31 (065112), 2019, with the permission of AIP Publishing.

$A_0 = \sqrt{3}$, $q = 0.1$, $\omega = 2\pi$, $B = \sqrt{2}$ and $C = 1$. The time-dependent rotation matrix $\boldsymbol{Q}_{or}(t)$ and the translation $\overrightarrow{b}(t)$ are completely same as the ones in the Sullivan vortex example. The contours of the nonobjective Liutex strength in different reference frames are reported in Fig. 6.18. Due to the influence of the moving observer's reference frame, the vortex strength by the nonobjective Liutex at the new position is very different from the original strength. In Fig. 6.19, the contours of the vortex strength by the new developed objective Liutex vector are shown in different frames, where no visible differences can be found by the naked eyes. To further investigate the value of the vortex vector, the vortex strengths at the original point $P(1.202641, 2.429826, 0.957204)$ and the new point $P^*(-0.797187, 2.082218, 0.174931)$ at the new reference frame are chosen. For the Liutex strength at the point P in the original reference frame, the nonobjective Liutex strength is $R = 1.489584$ and the objective Liutex strength is $\widehat{R} = 1.479505$. In the new reference frame, the point P becomes P^*, and the nonobjective Liutex strength is $R = 5.873281$ and the objective Liutex strength is $\widehat{R} = 1.479505$. Nonobjective Liutex strengths are quite different under different moving reference frames. More seriously, the nonobjective Liutex method would mistakenly detect the region with no rotation as a rotational region in some place. The present objective Liutex still retains its real strength. This also shows that the use of objective Liutex vector is important in the moving reference frames.

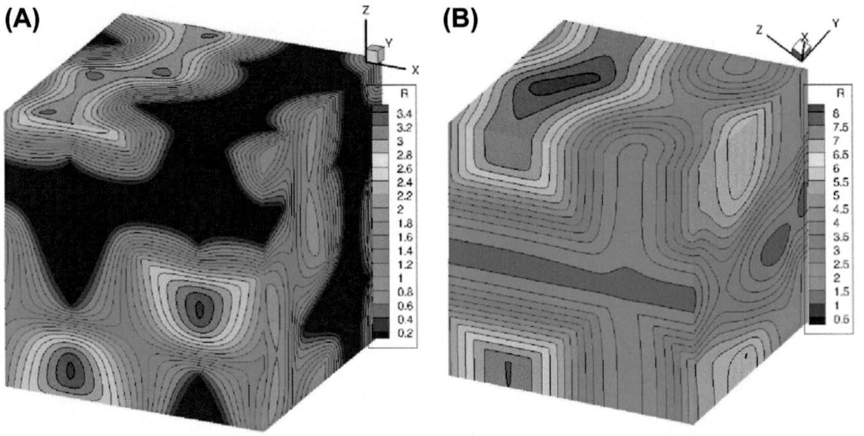

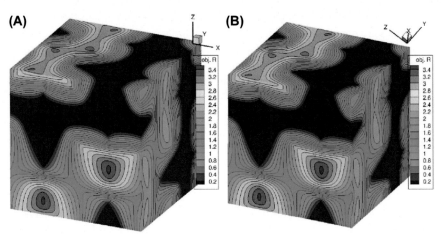

FIGURE 6.18

Contours of the nonobjective Liutex strength in the original reference frame (A) and in the new reference frame (B).

Reproduced from J. Liu, Y. Gao and C. Liu, An objective version of the Rortex vector for vortex identification, Phys. Fluids 31 (065112), 2019, with the permission of AIP Publishing.

FIGURE 6.19

Contours of the objective Liutex strength in the original reference frame (A) and in the new reference frame (B).

Reproduced from J. Liu, Y. Gao and C. Liu, An objective version of the Rortex vector for vortex identification, Phys. Fluids 31 (065112), 2019, with the permission of AIP Publishing.

6.5.4 Objective vortex structures behind micro-vortex generator (MVG)

In order to show the importance of the objectivity in practical flows, the data obtained by implicit large-eddy simulation (ILES) method for a supersonic flow (Mach 2.5) with the flow control by a micro-vortex generator (MVG) are also studied by the present objective Liutex method. The simulation is carried on by a high-order WENO finite difference method on a body-fitted structural mesh with about 40 million grids, and the detailed settings for the computation can refer to Liu et al. (2018b).

The specified time-dependent reference frame in this case is same as the one used in the example of Sullivan vortex. For this special case, the Liutex vector can be used to visualize the complex vortex structures generated by the MVG. The iso-surfaces of the nonobjective Liutex strength with $R = 2$ in the original reference frame are shown in Fig. 6.20. In the new reference frame, due to the motion of the reference frame, the overall polluted iso-surfaces by the nonobjective Liutex method are shown in Fig. 6.21. Hence, the nonobjective method is not suitable for the moving reference frame. In Figs. 6.22 and 6.23, the iso-surfaces of the present objective Liutex strength with $\widehat{R} = 2$ in the original and new reference frame are demonstrated, and the vortex structures are retained by the objective Liutex method. This example again demonstrates that the method is capable of maintaining vortex structures between moving reference frames.

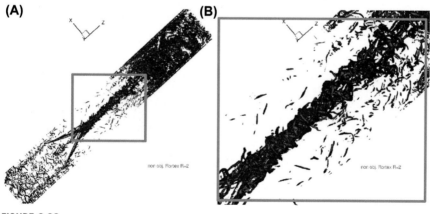

(A)   **(B)**

non obj. Rortex R=2

non obj. Rortex R=2

FIGURE 6.20

Iso-surfaces of vortex vector strength with $R = 2$ by nonobjective Liutex method in the original reference frame. (A) Full view of iso-surfaces. (B) Local enlargement.

Reproduced from J. Liu, Y. Gao and C. Liu, An objective version of the Rortex vector for vortex identification, Phys. Fluids 31 (065112), 2019, with the permission of AIP Publishing.

(A) **(B)**

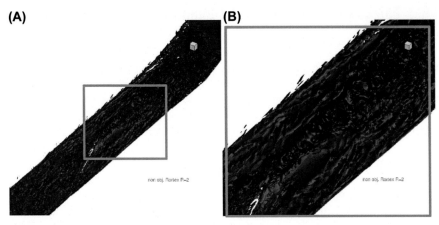

FIGURE 6.21

Iso-surfaces of vortex vector strength with $R = 2$ by nonobjective Liutex method in the new reference frame. (A) Full view of iso-surfaces. (B) Local enlargement.

Reproduced from J. Liu, Y. Gao and C. Liu, An objective version of the Rortex vector for vortex identification, Phys. Fluids 31 (065112), 2019, with the permission of AIP Publishing.

6.5.5 Summary

In this section, the objective Liutex vector is defined by a net velocity gradient tensor. The newly defined objective vector can keep the size and direction of the vortex vector unchanged under the moving reference frame, so it can be used to

(A) **(B)**

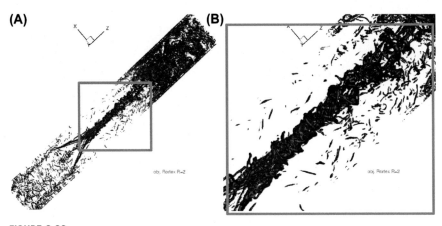

FIGURE 6.22

Iso-surfaces of vortex vector strength with $\widehat{R} = 2$ by objective Liutex method in the original reference frame. (A) Full view of iso-surfaces. (B) Local enlargement.

Reproduced from J. Liu, Y. Gao and C. Liu, An objective version of the Rortex vector for vortex identification, Phys. Fluids 31 (065112), 2019, with the permission of AIP Publishing.

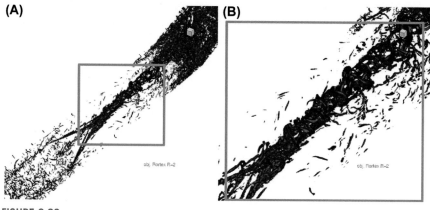

FIGURE 6.23

Iso-surfaces of vortex vector strength with $\widehat{R} = 2$ by objective Liutex method in the new reference frame. (A) Full view of iso-surfaces. (B) Local enlargement.

Reproduced from J. Liu, Y. Gao and C. Liu, An objective version of the Rortex vector for vortex identification, Phys. Fluids 31 (065112), 2019, with the permission of AIP Publishing.

visualize the vortex structure in the time-dependent reference frame. The objective method is carefully verified by several examples including a practical turbulent flow simulated by ILES method. From the results, the objectivity is very important for the moving reference frame. Otherwise, the nonobjective method will pollute the vortex structures. From the results, it is concluded that the objective Liutex method should be used in the moving reference frames. Of course, there are many other ways to make Liutex to be objective and here is just one of them.

6.6 Summary

In this chapter, several Liutex-based vortex identification methods, which are called the third generation of vortex identification methods, are introduced including Liutex vector, Liutex line, Liutex tube, Liutex isosurface, Liutex-omega, modified Liutex-Omega, objective Liutex methods. As a vector accurately representing fluid rotation strength and direction, Liutex is overwhelmingly superior over all first and second generations of vortex identification methods and should replace all the second generation methods, which are scalar and contaminated by shears, to identify the vortices and visualize the vortex structures. These methods should be widely applied in vortex science and turbulence research in many disciplines.

LXC-core line method and R-S decomposition for vortex dynamics

7.1 Comments on vortex identification methods

According to Liu et al. (2019), there exist, so far, three generations of vortex identification (VI) methods in the long history of the vortex-related studies, which well reflected and represented the efforts made by the research community of fluid mechanics in a historically time-sequential clue and context.

The first-generation (1st-G) VI theory and method are represented by the well-known classical vorticity vector, defined as the curl of velocity originally given by Helmholtz (1858) dated back to 1858. Within the framework of 1st-G VI theory, the vorticity was believed in as a physical entity appropriate for representing vortex, which includes (1) the vorticity scalar, i.e., the magnitude of velocity curl; (2) the vorticity vector, i.e., the vector of velocity curl; and (3) the vorticity tensor, i.e., the antisymmetric portion of the Cauchy—Stokes decomposition of velocity gradient tensor.

As well known, the vorticity has been popularly accepted as a synonym to vortex for a long period of time by fluid mechanics community and has been well recorded by many classical fluid mechanics teaching books (Truesdell, 1954; Warsi, 2005). In the meantime, numerous vortex-related researches have been conducted under the theoretical framework of the vorticity set by Helmholtz (Saffman, 1992; Fuenteso, 2007; Wu et al., 2005). Within the context, one of the famous foundations was the vorticity transport equation, which was historically well known as the Helmholtz equation (Helmholtz, 1858). Numerous so-called vortex dynamics were studied based on the equation with vorticity as the dependent variable and therefore, from the nowadays third-generation (3rd-G) VI theory, should be more strictly and critically called vorticity dynamics.

Moreover, the vorticity was later found incapable of properly representing the vortex behaviors (Liu et al., 2019) with the major drawbacks in three aspects:

(1) the scalar defined by vorticity magnitude is actually not a quantity truly representing the fluid particle rotation and therefore is not suitably applied for measuring vortex rotation strength (Liu et al., 2019);

Liutex and Its Applications in Turbulence Research. https://doi.org/10.1016/B978-0-12-819023-4.00010-0
Copyright © 2021 Elsevier Inc. All rights reserved.

(2) the vector defined by velocity curl is not a directional quantity for vortex rotational direction, which, from the perspective of 3rd-G VI, is a fatal drawback;

(3) the antisymmetrical tensor resulting from the Cauchy–Stokes decomposition of velocity gradient is not a suitable quantity for the analysis of fluid rotational motion.

Therefore, searching for more appropriate quantities and ways to identify vortex was launched as the campaign of the second-generation (2nd-G) VI, which are nowadays well-known by the so-called vortex criteria including the Q criterion, the Δ criterion, the λ_2 criterion and the λ_{ci} criterion, and lately the Ω criterion, and so on (Hunt et al., 1988; Chong et al., 1990; Jeong et al., 1995; Zhou et al., 1999; Liu et al., 2016). Looking back retrospectively on these 2nd-G VI approaches, it is not difficult to find that most of them are the Eulerian local region-type VI based on the characteristic quantities of local velocity gradient tensor. More specifically, these criteria are exclusively determined by the eigenvalues or invariants of the velocity gradient tensor and thereby are suitably classified as the eigenvalue-based criteria (Gao et al., 2018) with the VI quantity being a scalar in nature. Obviously, the variety of these criteria clearly indicates that these 2nd-G VI methods are essentially the on-trial based experiments in anticipation to obtaining a proper characteristic quantity in velocity gradient tensor for the VI purpose. Because of their scalar property, these VI approaches are visualized, without any exception, by the scalar iso-surfaces to represent vortex structures with a fairly arbitrary threshold. Consequently, the entire 2nd-G VI campaign can be summarized as the trial-and-error approaches with a variety of conjectures and threshold arbitrarinesses, which does not stand up to even the most superficial scrutiny due to the lack of sufficient and solid grounds in terms of both mathematics and physics. However, the most valuable lesson that can be learned from the researches of 2nd-G VI is that the target quantity in VI research became clearer and more focused, which enlightens the community eventually locking on to the local velocity gradient tensor as the most promising and potential candidate target from which a quantity is expected to be extracted out that can truly represent vortex motions in the next-generation VI.

Within the context, the 3rd-G VI method was pioneered by Liu's group at UTA around 2015 and was publicly recognized in 2018 (Liu et al., 2018a). The research was started by rethinking the most distinctive and universal characteristics in vortex motions and their mathematical descriptions and definitions. Their major arguments include: (1) intuitively the fundamental element in vortex motions is rotation and logically the quantity of fluid rotation needs to be mathematically defined; (2) to rigorously define the rotation, the rotational axis has to be first determined, which must be a vector quantity in three-dimensional space. With the logical thinking, through a careful study of velocity gradient tensor, the rigid rotation axis was, for the first time, successfully detected and discovered as the real eigenvector of the gradient tensor. Thereby, the Liutex vector was born enlighteningly with a physically meaningful and mathematically rigorous definition. Subsequently, the magnitude of Liutex vector was found being a quantity representing twice the local angular speed of fluid-particle rigid rotation, which was taken out from the gradient tensor as the most promising quantity pursued by the VI research community over many decades for the VI purpose. Therefore, the 3rd-G VI theoretical framework is established (Liu et al., 2018a, 2019; Dong et al., 2019b; Gao et al., 2018, 2019a; Wang Y. et al., 2018, 2019a) as limpidly and definitively expressed by the following series of definitions, theorems, and proofs.

7.1.1 Liutex magnitude and vector definition

Definition 7.1. The direction of Liutex vector is defined as the local rotation axis and the magnitude of Liutex vector is defined as the twice local angular speed of fluid motion, i.e.,

$$\vec{R} = R\vec{r} \tag{7.1}$$

According to Gao and Liu (2018), $\vec{r}$ is the unit eigenvector of the velocity gradient tensor ∇V, namely

$$\nabla V \cdot \vec{r} = \lambda_r \vec{r} \tag{7.2}$$

Wang Y. et al. (2019a) further gave an explicit formula to calculate R, i.e.,

$$R = \vec{\omega} \cdot \vec{r} - \sqrt{\left(\vec{\omega} \cdot \vec{r}\right)^2 - 4\lambda_{ci}^2} \tag{7.3}$$

and Liu et al. (2018a) provided the decomposition of $\vec{\omega}$ by

$$\vec{\omega} = \vec{R} + \vec{S} \tag{7.4}$$

where $\vec{\omega}$ is the local vorticity vector, λ_{ci} is the imaginary part of the complex eigenvalue of ∇V, and $\vec{S}$ is the local shearing vector. The definition is provided in Chapter 3 and is here reiterated for the following analyses.

7.1.2 Liutex (vortex) rotation axis line definition

Definition 7.2. A Liutex (vortex) rotation axis line is defined as the local maxima of Liutex, which is a line instead of iso-surface.

Theorem 7.1. Any small element of a line on the Liutex iso-surface must be orthogonal to the gradient of Liutex magnitude scalar if $\nabla R \neq 0$.

Proof. On the Liutex iso-surface, $dR = \nabla R \cdot d\vec{l} = 0$ if $\nabla R \neq 0$, where $d\vec{l}$ is an infinitesimal line element on the Liutex iso-surface. Therefore, ∇R, $d\vec{l}$ are orthogonal.

Theorem 7.2. If $\nabla R \times d\vec{l} = 0$ at a point, the point must be located on the Liutex rotation axis.

Proof. Any point that is not located in the Liutex rotation core axis has to be in some Liutex iso-surface. If $d\vec{l}$ is on a Liutex iso-surface, $dR = \nabla R \cdot d\vec{l} = 0$ must hold and then $\nabla R \times d\vec{l} \neq 0$. If $\nabla R \times d\vec{l} = 0$, $d\vec{l}$ must not be on any Liutex iso-surface, which is the Liutex rotation axis because according to Definition 7.2, the Liutex rotation axis is a local maxima and has no Liutex iso-surface.

Definition 7.3. The vortex core line is defined as a special Liutex line, which passes the points satisfying the condition of

$$\nabla R \times \vec{r} = 0, \ \vec{r} \neq 0 \tag{7.5}$$

where $\vec{r}$ represents the unit direction of the Liutex vector. Definition 7.3 is used to find the Liutex (vortex) rotation core lines in flow field, which is uniquely defined without any threshold requirement. Therefore, the Liutex core rotation axis lines with the Liutex strength are derived and are believed the only entity that is capable of cleanly and unambiguously representing the vortex structures.

Due to the Liutex completeness for vortex expression, the 3rd-G VI methods can be broadly classified into two types of vortex representations with the first type being the scalar-based, such as the scalars of Liutex magnitude or Liutex-Ω given in (Liu et al., 2018a, 2019; Dong et al., 2019b; Gao et al., 2018, 2019a; Wang Y. et al., 2019a) and the second type being the Liutex-core-based using the LXC-core vector line (Gao et al., 2019b; Xu et al., 2019; Wang et al., 2020a; Wang, 2020b), which are well demonstrated by Fig. 7.1 for a zero-pressure-gradient flat plate turbulent boundary layer (ZPGFP TBL). Among these 3rd VI visualizations, the new physical quantity of Liutex is the key element at their foundation, which fundamentally addresses the rotation issue in vortex and thereby successfully depicts its rotational characteristics. Liutex is a systematic definition of the local fluid rigid rotation, including the scalar, vector, and tensor interpretations. The scalar version or the magnitude of Liutex represents the local rotational strength, i.e., twice the local angular rotation speed, which is suitable to be visualized by scalar-based method using iso-surfaces. The direction of Liutex vector, determined by the real eigenvector of the velocity gradient tensor, represents the local rotation axis, from which the Liutex-core line was, for the first time, defined and extracted by Gao et al. (2019b). The tensor form of Liutex, rather than the vorticity tensor, stands for the rigid rotational part of the velocity gradient tensor, which can be used for the decomposition of the velocity gradient tensor (Gao et al., 2019a).

Given the fact that the more detailed Liutex theory and its mathematical framework along with the development are provided in the previous chapters of the book, the purpose of current chapter is presenting and demonstrating the new findings and key achievements, so far, in applying these two types of innovative 3rd-G VIs, specifically the Liutex-scalar-based and LXC-core-line-based approaches, to explore the vortex-related fluid-mechanics challenges and mechanisms. These include:

(1) Section-2 provides the visualization techniques in the Liutex scalar-based and core-line-based 3rd-G VI;

(2) Section-3 introduces the automatic LXC-core-line identification and massive visualization techniques using computer, which opens the door for 3rd-G VI being applied in a variety of practical vortex studies;

(A)

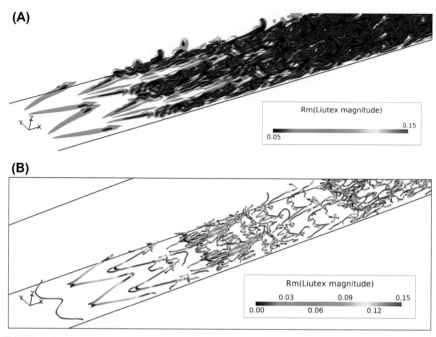

(B)

FIGURE 7.1

Vortex structure effected by the 3rd-G VI of a ZPGFP TBL transition using the Liutex-scalar-based method and the LXC-core-line-based approach. (A) Vortex structure using the Liutex-scalar-based method; (B) vortex structure using the LXC-core-line-based approach.

Reproduced from Wang, D., et al., Law-of-the-wall analytical formulations for Type-A turbulent boundary layers, Journal of Hydrodynamics 32 (2), 2020, with the permission of Journal of Hydrodynamics

(3) Section-4 summarizes the important findings and key knowledge points in the 3rd-G VI.

(4) Section-5 demonstrates the importance of R-S decomposition applied in the vortex dynamics study for the transition of dual cylinders in tandem.

7.2 Visualizations based on Liutex iso-surfaces and core lines

Over the long history of VI research, the essential goal and the major effort have been concentrated on searching for an approach that was imagined to be able to universally represent or ubiquitously describe the vortex phenomena. Specifically, the studies are fundamentally seeking the answers for two questions, firstly what is the physical quantity that is capable of truly representing vortex and secondly how to visualize the quantity to vividly see and mimic the vortex phenomena so that the "seeing is believing" can come true.

To answer the first question, the VI research has gone through the three generations of endeavors with a variety of vortex-related quantities being attempted and tested, such as the vorticity based on the Cauchy–Stokes decomposition in 1st-G VI, the variety of velocity-gradient eigenvalue-based criteria in 2nd-G VI and up to more recently the Liutex-based theory and approach in 3rd-G VI. For the second question, the graphical representation of vortex structures has long been dominated by using the iso-surfaces for three-dimensional vortices or the contour lines in two-dimension cases, which is essentially an approach to graphically visualize the so-called vortex-related scalar with certain "ad hoc" selected thresholds.

7.2.1 Visualization using Liutex iso-surface

As aforementioned, the vorticity theory initiated by Helmholtz prevailed in VI research for a long period of time. When looking back nowadays from the standpoint of 3rd-G VI, it is interesting to note that the vorticity theory possesses every elementary form that is essentially needed to depict a vortex motion, namely the scalar, the vector, and the tensor forms. Although, for the 1st-G VI, the scalar (the magnitude of vorticity), the vector (curl of velocity), and the tensor (antisymmetric tensor in the Cauchy–Stokes decomposition) are found unable to truly represent vortex motion as mentioned before, it does suggest that a complete depiction for vortex behaviors has to contain these three elementary forms of vortex-related scalar, vector, and tensor quantities, and moreover, these forms of quantities have to possess the Galilean-invariant property.

Within the context, as a milestone in vortex identification (VI) research, the introduction of Liutex opens the door to rigorously, quantitatively, and systematically study complex vortex phenomena, since the Liutex theory provides the clear definitions of complete VI quantities in the forms of scalar, vector, and tensor, which can satisfactorily reflect and address the core issues in the vortex description in terms of the vortex strength, direction, core, and boundaries, and so on.

R-S decomposition in Eq. (7.2) accurately and precisely suggests that Liutex, i.e., the rigid rotational speed of fluid particles, eliminates the contamination of shearing motion implicitly contained in the traditional vorticity and thus be capable of quantifying the local strength of pure rigid rotation, which were clearly demonstrated for the incompressible flow by (Liu et al., 2019a,b). Although more accurate in terms of getting rid of the shearing contaminations comparing to the 1st and 2nd-G VIs, a user-specified threshold of Liutex scalar is still required and needs to be selected for identifying and visualizing the vortical structures in the iso-surfaces-based approach. Therefore, the idea of combining the Ω method, a relative strength introduced in Liu et al. (2016), with the Liutex was proposed as the Liutex-Ω in (Dong et al., 2019b; Liu et al., 2019), which was considered as a potential solution to address the issue of threshold determination. The method was found capable of more cleanly representing the true vortical structures in Fig. 6.11 and more effectively easing the arbitrariness of threshold issue to some extent by narrowing the selections of threshold values to around $\Omega = 0.52$. However, in principle, the awkwardness of threshold selection and the related multivortical structures caused by the threshold arbitrariness was still lingering on in the identification until the discovery of the LXC-core lines in (Gao et al., 2019b; Xu et al., 2019) that totally eliminates the need to specify the iso-surface threshold and therefore satisfactorily resolve the issue.

Within this regard, the Liutex vector field and lines, on top of the Liutex scalar-based VI approach, can be utilized and applied to study the vortical structures. Aside from the Liutex scalar-related magnitude standing for the absolute or relative rotational strength, the 3rd-G VI possesses the Liutex vector representing the local rotational axis direction, which is the major advantage over the existing 2nd-G VI using the eigenvalue-based criteria. Definitions 7.1, 7.2, and 7.3 along with Theorems 7.1 and 7.2 provide the theoretical foundations for the exploration of LXC-core lines and their identification criteria, based on which a manual identification of Liutex core line was first introduced by Gao et al. (2019b), and then the automatic generation of massive LXC-core lines was further developed by Xu et al. (2019).

7.2.2 Manual identification and generation of Liutex core lines

The theoretical foundation laid out by the Liutex-related definitions, theorems, and proofs, as aforementioned, enlightens and proves the existence of Liutex core line, which gives rise to a possibility to express the vortex structures by Liutex core line instead of iso-surface. The Liutex vector property permits the total elimination of threshold issue and therefore, nowadays, is considered as the ultimate method in VI to uniquely and perfectly depict and exhibit vortex structures. As a matter of fact, the aforementioned Liutex-related definitions, theorems, and proofs suggest that a special group of Liutex core lines be extracted out from the Liutex vector field, which possess the threshold-free property and the structural uniqueness to represent vortex core structures or vortex skeletons.

Specifically, for the rotational axes of vortices, their rotational strengths, i.e., the Liutex magnitudes, reach a local maximum in the plane normal to the direction of Liutex vector. The requirement is well reflected by Eq. (7.3) of $\nabla R \times \vec{r} = 0$ and $R > 0$, which can be interpreted as that the gradient of the Liutex magnitude ∇R is in the same direction as the local Liutex vector direction of $\vec{r}$ and the second condition of $R > 0$ ensures the point with the local maximum Liutex magnitude locates inside the vortex. The logics of idea is that on the iso-surfaces of R, ∇R is perpendicular to any small line element $d\vec{l}$ that is laid on the iso-surface. As R becomes larger and larger until the iso-surface reaches the vortex core, where ∇R would be in the same direction as $\vec{r}$. The condition of Eq. (7.3) is actually used to detect the seeding points through which the Liutex core lines are identified and are integrated to generate the Liutex core lines representing a vortex structure.

Within the context, the manual generation of vortex-core lines was first done by Gao et al. (2019b), demonstrated by the direct numerical simulation (DNS) data in the early transition of ZPGFP TBL, as seen in Fig. 7.3. The vortex structures in the

(A)   (B)   (C)   (D)

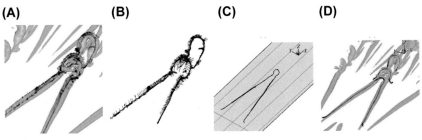

FIGURE 7.2

Procedures of manually generating Liutex core lines in the early transition of ZPGFP TBL.

Reproduced from [H. Xu et al., "Liutex (vortex) core definition and automatic identification for turbulence vortex structures", Journal of Hydrodynamics, 2019, 31(5)], with the permission of "Liutex (vortex) core definition and automatic identification for turbulence vortex structures", Journal of Hydrodynamics, 2019, 31(5)], with the permission of Journal of Hydrodynamics)

stage of early transition are distinctively characterized by the sharp hairpin pattern, which remains at a quite large scale and has yet developed into the turbulent multi-scale stage. As a result, the Liutex core line was identified pretty easily and cleanly with the procedures including: (1) calculating the Liutex vector $\vec{r}$ and Liutex magnitude gradient ∇R with R being the magnitude of Liutex scalar; (2) plotting the iso-surface of R and finding the points on the iso-surface as illustrated by Fig. 7.2A; (3) plotting the integral curves of ∇R based on the points of the iso-surface R and deriving the limiting curve to which all the integral curves are approaching and eventually collapsed as exhibited in Fig. 7.2B; and (4) finding a

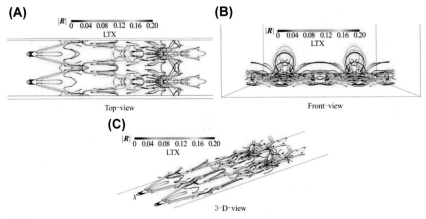

FIGURE 7.3

(Color online) Automatic generation of the LXC-core lines as the representation of vortex structures in the early transition of ZPGFP TBL.

Reproduced from Xu, H., et al., Liutex (vortex) core definition and automatic identification for turbulence vortex structures, Journal of Hydrodynamics 2019, 31 (5), with the permission of Journal of Hydrodynamics.

point on the limiting curve and integrating the Liutex vector $R\vec{r}$ based on the point to obtain a special Liutex core line as seen in Fig. 7.2C. By repeating the same procedures for the other iso-surfaces one by one, the Liutex core lines at other locations in flow field can thereby be generated, visualized, and tracked individually, as seen in Fig. 7.2D. The entire procedures, as demonstrated in Fig. 7.2, first proved the existence of Liutex core line and graphically presented the relations and connection between the core line and Liutex iso-surface.

7.3 Automatic LXC-core line identification using computer

Although quite successful in identifying the Liutex core line in (Gao et al., 2019b), the manual procedures were recognized only valuable for demonstrating the concept of Liutex core lines and were not able to be applied in any meaningful 3rd-G VI practices, such as the VI research in turbulence. The major drawback was caused by the fact that the theoretical foundation of the Liutex core line identification, i.e., the aforementioned definitions and theorems, was not fully and adequately implemented in the manual operation steps. As explained before, the automatic Liutex core line identification was integrally achieved by fully implementing the definitions and theorems into a computer program by Xu et al. (2019) and therefore, satisfactorily overcame the drawback in the manual procedures. These Liutex core lines were named by LXC-core lines (Xu et al., 2019) as a historical memory of joint efforts committed by Profs. Liu, Xu, and Cai for initiating, engaging, and effecting the fruitful research with a Liutex-core-line computing program being successfully developed by Xu.

Based on the theoretical foundation laid out in (Gao et al., 2019b), the automatic generation of Liutex core lines was developed by fully implementing the definitions and theorems into a computer program through the proper algorithms. The chapter makes use of the DNS data for two types of typical TBLs, namely the Type-A (ZPGFP) (Liu et al., 2014) and -B (PGDSAD) (Wang D. et al., 2019) TBLs, to demonstrate the automation capability that the current 3rd-G VI approach can uniquely achieved. To fully implement the advantage of 3rd-G VI offered by the Liutex vector property into a computing code, the special group of Liutex core lines satisfying $\nabla R \times \vec{r} = 0$ and $R > 0$, i.e., the LXC-core lines, need to be first identified and then extracted out from the Liutex vector field. The procedure to identify the LXC-core line consists of first cutting through the vortices with a slice plane, then detecting the seeding points using the condition set by Eq. (7.3), and finally generating the Liutex core lines by plotting the vector line, i.e., the streamline based on Liutex vector, using the identified seeding points as the start locations for the vector-line integration.

As presented by Fig. 7.4, the Liutex core lines are massively generated by the automatic 3G VI method based on the DNS data at the early stage of the natural transition in the ZPGFP TBL. These strings exhibit a clearly symmetrical pattern

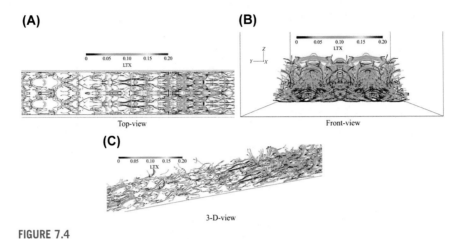

FIGURE 7.4

(Color online) Automatic generation of the LXC-core lines as the representation of vortex structures in the late transition of ZPGFP TBL, namely the Type-A TBL in Cao and Xu (2018).

Reproduced from Xu, H., et al., Liutex (vortex) core definition and automatic identification for turbulence vortex structures, Journal of Hydrodynamics 2019, 31 (5), with the permission of Journal of Hydrodynamics.

representing the formation of horse-shoe and the Λ-shape vortex evolution. Comparing to the traditional iso-surface-based VI methods, these structures are limpidly identified by the strings of LXC core lines instead of the iso-surfaces with an "ad hoc" threshold and therefore lead to the uniquely defined structure depicting the vortical characteristics of fluid motions, i.e., the vortex cores with their tangential direction standing for the local rigid rotating direction. These strings are colored by the Liutex magnitude R, which distinctively represent the vortex local rotational strengths, i.e., the local angular rotation speeds. It can be seen evidently in Fig. 7.3A—C that the vortex rotational strengths tend to be very non-uniformly distributed along these strings, which makes the iso-surface methods with a specific threshold being only capable of approximately, partially, and evasively reflecting and representing the vortex structures. The three typical views from different space view angles, i.e., the top, front, and 3-D views in Fig. 7.4, vividly exhibit the spatial patterns and distributions of LXC-core lines and thereby produce the limpid vortex structures for the first time in the VI history. Only with these LXC-core lines can the vortices along with their swirling strength information be well captured and uniquely presented, which satisfactorily resolve the long-debated and puzzled issues of threshold.

Needless to say, the 3rd-G VI approach based on the LXC-core lines provides the research community with a unique capability of depicting the vortex structure definitively, systematically, and completely. Comparing Fig. 7.3A—C to Fig. 7.4A—C, it can be seen evidently that with the temporal evolution, the vortex structures develop progressively toward the more refined-scale structures from the early to the late transition. Consequently, the automatic identification has to be developed to investigate

(A)

(B)

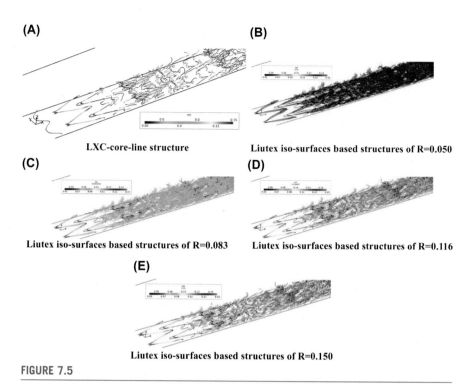

LXC-core-line structure

Liutex iso-surfaces based structures of R=0.050

(C)

Liutex iso-surfaces based structures of R=0.083

(D)

Liutex iso-surfaces based structures of R=0.116

(E)

Liutex iso-surfaces based structures of R=0.150

FIGURE 7.5

Comparative study of vortex structures with (A) LXC-core-line structure and Liutex iso-surfaces-based structures of (B) R = 0.050, (C) R = 0.083, (D) R = 0.116 and (E) R = 0.150. (A) LXC-core-line structure (B) Liutex iso-surfaces-based structures of R = 0.050. (C) Liutex iso-surfaces-based structures of R = 0.083 (D) Liutex iso-surfaces-based structures of R = 0.116. (E) Liutex iso-surfaces-based structures of R = 0.150.

the more complex turbulences with these progressively refined scales. Within the context, the current 3rd-G VI opens the door to make use of the LXC-core lines to more quantitatively and rigorously study the highly twisted and tangled vortex-relevant coherent structures.

More detailed comparative study was carefully conducted to intuitively visualize the difference between the iso-surface-based approaches and the LXC-core-line method, from which the advantage of LXC-core line can be limpidly exhibited and be well understood in terms of the vortex structure presentation. For this purpose, the Liutex iso-surfaces structures with the sequential thresholds at R = 0.050, 0.083, 0.116, and 0.150 are plotted and presented in Fig. 7.5B–E, which are overlapped with the corresponding LXC-core lines. These core-line strings are colored by the Liutex magnitude R, which distinctively represent the local fluid rotational strengths, i.e., the local angular spinning speeds.

Comparing to Fig. 7.5A as a benchmark, the spatial distribution relationships can be clearly seen between the iso-surfaces and their corresponding core lines. The iso-surfaces with the sequential thresholds of R = 0.050, 0.083, 0.116, and 0.150 enwrap their corresponding core lines by a certain portion of space encircling around the core lines. The core-line topology actually forms the skeletons for these iso-surface enclosed spaces. The iso-surface with smaller threshold tends to occupy a larger space around its core line and the one with larger threshold enwraps only a small portion of the core line. Given the fact that the core lines with their tangential direction standing for the fluid local spinning direction and the color representing the rotational strengths, it is not difficult to conclude that the group of iso-surfaces with a specific threshold can only approximately, partially, and evasively represent a portion of the core-lines and, indeed, is incapable of uniformly and comprehensively reflecting the entire topology exhibited by the core-line strings in Fig. 7.5A.

After successfully being applied to the transition of ZPGFP TBL (Type-A TBL in Wang et al. 2020a), the automatic 3rd-G VI approach was further pushed to the Type-B TBL in Wang D. et al. (2019), namely the fully developed pressure-gradient-driven turbulent boundary layer (PGDSAD, TBL). The DNS data were presented in Xu (2009) and later were extended to the thermal turbulence in Wang D. et al. (2019), which received a reliable validation from various experiments and other related DNS data. The 3rd-G VI approach based on the LXC-core lines was conducted, for the first time, in the fully developed TBL. Qualitatively, the visualizations of these core-line strings were very reminiscent for the existing knowledges of the wall-bounded turbulence, such as the near-wall streak structures and the ejection or sweeping characteristic motions of a wall turbulence, but were refreshingly represented and exhibited by the LXC-core lines as seen in Fig. 7.6. For example, the vortex strings near wall, as seen in Fig. 7.7A, tend to be generally lifted-up from the wall at the downstream end and to be suppressed down toward the wall at the upstream end, indicating an ejection and sweeping must exist around the two ends of the strings, respectively. The strings are all elongated in the streamwise direction, giving rise to the nature of the streak or coherent structures mentioned in many existing coherent-structure studies. Moreover, with the automatic approach demonstrated, the LXC core lines uniquely generated and massively visualized in Fig. 7.6 in the fully developed turbulence provide the true skeleton of the vortex structures at a specific time instant.

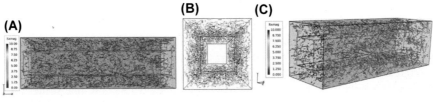

FIGURE 7.6

Automatic generation of the LXC-core lines as the representation of vortex structures in the fully-developed turbulence in PGDSAD TBL, namely the Type-B TBL in Wang, D., et al. (2019).

It can be seen clearly in Figs. 7.5 and 7.6 that the rotational strengths tend to be very nonuniformly distributed along these strings, which makes the iso-surface methods with a specific threshold being only capable of approximately, partially, and evasively reflecting and representing the core-line topology that is considered as the true skeleton of vortex structures. Within the context, it can be concluded that the iso-surface-based approaches generally are not suitable, at least not rigorous, to be applied to represent vortex structures due to their threshold dependency, which evidently causes the nonuniqueness and incompleteness issues of vortex structure depiction. With that being concluded, it is not exaggerated to claim that the LXC-core-line technique is the ultimate way to satisfactorily resolve the issue of rigorously and quantitatively depicting vortex phenomena in terms of the preciseness, uniqueness, and completeness.

In order to detect and study the detailed mechanisms behind the observations presented in Figs. 7.3–7.6, the temporal evolutions of these LXC-core-line strings have to be time-sequentially generated and transiently visualized so that the dynamic processes of these strings can be investigated. In this regard, the definition of Liutex provides a quantity permitting a quantitative interrogation of the vortex dynamics as represented by the evolution of the vortex core lines, and apparently, the automatic 3rd-G VI capability definitely plays a vitally important role in the research community. To achieve the goal, the temporally sequenced images were generated based on the current existing DNS data at the website: http://simplefluids.fudan.edu.cn/publications/book-elsevier/. Despite the successful representation of vortical structures, the automatic approach has a difficulty to distinguish those vortex cores within a physically close distance and therefore the faked ones maybe generated by the numerically related errors. Therefore, an appropriate algorithm to addres the issue still remains for future development. However, it can be definitively recognized that the LXC-core-line method is by far the best method to cleanly extract the vortex structures from a flow field, and the approach uniquely defines the vortex-related skeletons without the traditional threshold issue, which provides an evident advantage over the iso-surface-based methods that are all inevitably threshold-dependent.

7.4 Summary and future of 3rd-G VI techniques

In summary, Sections 7.1, 7.2, and 7.3 elaborate the key points of Liutex theory and the visualization capability of the 3rd-G VI approaches, including both Liutex-scalar and LXC-core-line-based methods, which evidently demonstrated the drawbacks in the 1st-G and 2nd-G VIs and the advantages of the 3rd-G VI techniques.

Regarding the vortex visualization based on a scalar, the vortex structure can only be generated and exhibited by an iso-surface, which is, therefore, inevitably dependent on the associated threshold for almost all the current VI approaches. However, the determination of a threshold is arbitrary, resulting in vortex structures being threshold-dependent and losing their uniqueness. Actually, whether an appropriate threshold exists is unknown before the Liutex core line was detected, generated, and visualized in (Gao et al., 2019b). As analyzed in Section 7.3 through the comparisons of the iso-surfaces with different thresholds and their connections with the LXC-core lines, the role of threshold playing in a VI can be clearly seen and concluded that a larger threshold truly causes the loss of some important structure patterns qualitatively associated to the weak vortices represented by LXC-core lines and a smaller threshold can make the visualized vortex structures vague, blurred, or smeared due to the enlarged space enwrapped by the iso-surfaces. Therefore, the vortex structures exhibited by the LXC-core lines colored by the vortex rotation strength in Figs. 7.3–7.6 are conclusively and more scientifically appropriate in terms of their uniqueness, clarity, and threshold independency.

In this regard, although many efforts were made in (Strawn et al., 1999; Banks and Singer, 1994; Levy et al., 1990; Zhang and Choudhury, 2006; Miura and Kida, 1997; Kida and Miura, 1998; Linnick and Fasel, 2005; Sujudi and Haimes, 1995; Roth, 2000; Zhang et al., 2018; Epps, 2017), only very limited progress and success were achieved until the Liutex core line was successfully derived and introduced in (Gao et al., 2019b; Xu et al., 2019). The Liutex core line in (Gao et al., 2019b) was defined as a special Liutex vector line where $\nabla R \times \vec{r} = 0$ satisfies. However, the Liutex core line generated manually and tracked individually in (Gao et al., 2019b) has very limited application value, since the vortex structures can be very complicated for almost all kinds of flow phenomena including the transition and turbulence as demonstrated by Figs. 7.3–7.6. Consequently, it is not realistic and practical to track each individual Liutex core line manually in any applications.

Within the context, the Liutex core-line identification for 3rd-G VI becomes critically important to be implemented into a computing program to automatically generate the core line. The vortex-related research communities can thereby be significantly benefited by the capability of LXC-core line generation and visualization to explore the true vortex evolution and dynamics, for example the turbulence coherent structures. In this aspect, Xu et al. (2019) introduced an automatic vortex core-line identification method, later named as LXC-core line, and for the first time, applied the method to the representative turbulences resolved by DNS, including the data for the natural transition in a ZPGFP, i.e., the Type-A TBL in (Liu and Liu, 1995; Liu et al., 2014; Liu and Cai, 2017) and the fully-developed turbulence in a square annular duct (SAD) or the pressure-gradient-driven SAD (PGDSAD) TBL, i.e., the Type-B TBL (Yao et al., 2015; Xu, 2009; Cao and Xu, 2018). The vortex structures represented by these LXC-core lines symbolized a milestone in the 3rd-G VI based on Liutex system being truly applied to turbulence research. The VI approach based on the vortex cores is expected to be capable of profoundly uncovering the vortex natures and preserving the vortex-structure uniqueness with a distinctive threshold independency, as demonstrated in the Section 7.5.

The LXC core-line practices applied to the two typical TBLs, as presented in the previous sections, raise a serious question of why many existing research efforts were fruitless in searching for the vortex core lines for a long time. The question can be satisfactorily answered by these practices that a proper vector quantity is missing to represent a vortex and the Liutex vector is the key quantity that people are looking for. The automatic core-line generation actually creates an algorithm to efficiently search for the seeding point on each Liutex core line and then integrate to form the vector line based on the Liutex local magnitude and direction. Moreover, the capability of automatically identifying these LXC-core lines permits to massively visualize large number of these core lines in an evolutionary process, which opens the door to explore the intrinsic dynamic natures of the centennial puzzle of turbulence and other vortex-related phenomena, as demonstrated by the DNS data analyses in Section 7.5.

Given the fact that vortices are essentially the building blocks of turbulence, the existing turbulence theories, indeed, lack a serious research with the regard to a quantitative VI so far. Within the context, the new findings and the potential research directions pointed out and laid out in the current chapter are expected to lead the fluid mechanics into a new era of quantitatively studying turbulence vortical natures using these 3rd-G VI techniques, the most important and key portion of which can be highlighted as following:

(1) Definitions of Liutex scalar and vector given by $R = \vec{\omega} \cdot \vec{r} - \sqrt{\left(\vec{\omega} \cdot \vec{r}\right)^2 - 4\lambda_{ci}^2}$ and $\vec{R} = R\vec{r}$, $\vec{\omega} \cdot \vec{r} > 0$, which provides an effective way to extract the Liutex field out from a flow field;

(2) Relation of Liutex and shear decomposition, i.e., $\vec{\omega} = \vec{R} + \vec{S}$ and the vorticity transport equation, which give rise to a technical roadmap to investigate the dynamic process and evolution of Liutex that truly means the dynamics of vortex, as demonstrated in Section 7.5;

(3) Definition of Liutex core line and the massive generation of LXC-core lines based on $\nabla R \times \vec{r} = 0$, $R > 0$, which permits uniquely and quantitatively representing the essential skeleton of vortex structures and further interrogates the dynamics of these structures.

7.5 Preliminary vortex dynamics enabled by R-S decomposition and LXC-core lines

7.5.1 Research background on transition of cylinder flow

Flow past a cylinder is a classical configuration studied by the fluid-mechanics community for many decades. A variety of research topics relating to a bunch of flow phenomena have been touched, which includes the stagnated vortex, flow separation, von Karman vortex shedding, transitions in the wake of cylinder as reported by Spivack (1946) and Wille (1960) etc. Among them, the secondary instability in wake is a recent topic of interest mainly focused on studying the dynamic evolution of three-dimensional (3-D) structures and their interactions related to the vortices, which represents another type of fundamentals in the wake turbulence illustrated by Leweke and Williamson (1998), instead of the wall-bounded turbulence as discussed in Section 7.1. Regarding the specific configurations for the study, the physical domains typically involve a single cylinder or two identical cylinders in tandem being placed in either a confined or an open space. Although simple in terms of the geometry, the flow phenomena in these configurations are fraught with abundance of dynamics, which contains not only a lot of attracting vortex activities, but also a variety of intriguing wake instabilities and transitions.

The existing researches include a variety of investigations on the transition processes from the two-dimensional (2-D) instability, or the first instability giving rise to von Karman vortex streets, to the 3-D secondary instability leading to a variety of the 3-D vortical structures in an unconfined single-cylinder wake, as given by Roshko (1954), Williamson (1988, 1992, 1996), Braza et al. (2001), and Behara and Mittal (2010). Among these researchers, Roshko (1954) was the pioneer to study the flow past a cylinder, particularly the drag and shedding frequency in the wake transition. Subsequently, Williamson (1988) made the classification for the distinctive instabilities in a cylindrical transition, which gave the three types of "Mode A", "Mode B", and "Vortex dislocations". The "Mode A" instability takes place when the spanwise vortices loses their stability. Within the instability, the streamwise vortices with the relatively large scales come into formation characterized by the wavelengths at roughly three or four times of the cylinder diameters, which gives rise to a sharp transformation in the configuration loads due to the migration of wake patterns.

On the other hand, "Mode B" instability is featured by the relatively small-scale streamwise vortex structures with the wavelength at about one diameter of the cylinder. With the wake being evolved toward the fully-developed state, the vortex dislocations eventually arise in the spanwise along with its distinct frequencies of oscillation, which is claimed due to the low-frequency disturbances reported in Williamson (1992) as another mode in the wake transition.

In the traditional studies of cylinder wake, the instability modes are distinguished by the two distinctive frequencies or the related Strouhal numbers. The first mode relates to the transition from the 2-D spanwise von Karman vortices to the 3-D vortices of Mode A at its characteristic frequencies. With the Reynolds number (Re) being increased, the second mode comes into appearance, which corresponds to Mode (B). Williamson (1996) first took the measurements of velocity and pressure in the wake transitions and thereby studied the characteristics of these measurements. Moreover, based on the wavelengths and wake patterns, the elliptic instability in the near-wake vortex cores and the hyperbolic instability in the shear-layer braid were found responsible for Mode A and Mode B, respectively.

Regarding to the multi-cylinder configurations with more than one cylinder being located close to each other, the effect of interferences gives rise to even more complex vortex shedding and the related instability behaviors. Within the context, it is demanding to interrogate the interacting mechanisms among these cylinders in terms of their instability modes. As one of the representative types of interferences, the wake interactions were widely and intensively investigated by the configuration with double cylinders placed in tandem. For the configuration, both the flow criterion number of Re and the geometry parameter of center-to-center distance (L_x) between the cylinders are the important model characteristics.

The distinctive features of integrated forces for such configuration were reported in some early papers, which included the sharp transformation in the variation of forces changing with L_x and the Re-dependency of flow, and so on. Thomas and Kraus (1964) found that some particular L_x might induce quite special behaviors of the wake vortices under the condition of an identical Re. The performance table of the integrated loads was obtained by Zdravkovich (1972), which was found dependent on the parameter matrix of both Re and L_x. The rear cylinder behaved like a stabilizer of the flow, which imposed a delay to the turbulence transition taking place at a specific Re if the center-to-center distance was kept sufficiently short, i.e., smaller than the drag inversion separation.

Subsequent researches were more concentrated on classifying the flow into different regimes, which were found, for example, in Slaouti and Stansby (1992), Meneghini et al. (2005), Mizushima and Suehiro (2005), Deng et al. (2006), Carmo and Meneghini (2006), and Papaioannou et al. (2006). These investigations were highlighted by qualitatively predicting the regimes and providing more transition details in the process migrating to 3-D flow. The effects of 3-D structures were studied by Carmo and Meneghini (2006), which distinguished the results based on the 2-D structures from the data obtained by the 3-D configuration for a L_x close to the drag inversion separation with Re ranging in $160 < Re < 320$.

The stabilizing or destabilizing effects of the rear cylinder were studied in Papaioannou et al. (2006) by their dependency on the center-to-center distance. Carmo and Meneghini (2006) and Meneghini et al. (2001) conducted the linear and nonlinear stability analyses, which investigated in detail about the early transition of the 3-D wake in different geometrical configurations. The research found four distinctive regimes with each of the regimes being typically selected to represent the scenarios with L_x varying from 1.2 to 10 of the cylinder diameter in an unconfined domain. On the other hand, for the cylinders confined in between parallel walls, both studies from experiment and computation were scarce, in particular for the 3-D vortex structures at a low Re as commented by Kanaris et al. (2011).

However, the configurations in a confined space can be found widely applied in industry, where the enhanced mixing and heat transfer are highly demanded. The variety of applications include the turbulence promoters, the dividers in polymer processing, and the pin-fin cooling structures in aero-engine turbine blade, and so on. The confined single-cylinder wake was found in some existing studies. With the channel height being denoted by G, the blockage ratio can then be defined as $B = D/G$ with D being the cylinder diameter. For the single cylinder case with a blockage ratio of $1/3$ in unconfined space, Rehimi et al. (2008) found the typical secondary instability with both Mode A and B patterns. Kanaris et al. (2011) reported the similar results from the numerical studies with a larger blockage ratio of $1/2$. Moreover, Camarri and Giannetti (2010) conducted the Floquet stability analysis for such a flow geometry.

Based on the aforementioned comments, two topics still remain open for investigation. The first one is on the topic of the 3-D vortex structures and instability analyses of double cylinders in tandem confined by two parallel walls. Secondly, the

Floquet stability analysis was applied in most of the existing studies by successfully finding the critical Re for the 3-D flow evolution with various bifurcations. The major advantage of the approach is the relatively lower amount of computation. However, it is only applicable for the late transition without being able to extend to the entire evolution process and therefore, the structures of streamwise vortices and their formation mechanisms are limited, albeit these streamwise vortices are vital for the industrial application.

To address the aforementioned issues, the DNS was applied to study the secondary instability of the flow around double cylinders in tandem confined by a channel. The computations were conducted by a blockage ratio at $B = 1/4$ in between the ratios used by Kanaris et al. (2011) ($B = 1/5$) and Camarri and Giannetti (2010) ($B = 1/3$). Consequently, the wake vortex shedding in the near field was not severely interfered by the channel wall, which made the results comparable for the unconfined cases to some extent. The center-to-center distance between the cylinders was given at 2.5, which made the flow in the AG shedding regime of Carmo and Meneghini (2010) with the rear cylinder sitting neither entirely immersed in the shear layer from the front cylinder nor fully exposed to the von Karman wake.

Traditionally, the vorticity was commonly applied to represent the vortices with its transport equation, or the Helmholtz equation being popularly used for the vortex-dynamics study. However, the drawbacks of the 1st-G VI based on the vorticity and the shortcomings of the 2nd-G VI represented by a number of so-called criteria of Q, λ_{ci}, or Ω, and so on severely hamper the capability of applying these approaches to obtain the real vortex physics as aforementioned. Within the context, the current section, for the first time, applies the 3rd-G VI based on Liutex, recently introduced by Liu et al. (2018a) and Gao et al. (2018), as the physical quantity to represent the vortex. Moreover, the vortex dynamics are more rigorously and quantitatively analyzed by the Liutex-Shear (R-S) decomposition of vorticity given by Liu et al. (2016, 2018a). In this regard, based on the Liutex dynamics implicitly contained in the vorticity transport equation and the R-S decomposition of vorticity, the formation mechanisms of the streamwise vortices is analyzed in detail based on the available DNS data as reported in Wang (2020b). The Liutex spontaneous behaviors and the adjoint shearing effects are investigated profoundly in both linear and nonlinear evolution phases. A number of intriguing transition and instability phenomena, including the spanwise vortex-bulge structures, the vortex pairs induced by Mode B instability, and the vortex interactions due to the confined walls, are systematically and carefully studied. Moreover, the distinctive hairpin vortices and their formation mechanisms are successfully captured and identified in detail, for the first time, by the LXC-core line in the stage of secondary instability.

7.5.2 Physical model and flow conditions of the DNS data

A number of the existing studies are referenced to set the current physical model, computational domain, and flow conditions. Carmo and Meneghini (2010) investigated the instabilities in the double cylinder at $L_x = 2.3$ in an unconfined domain, in which the Floquet stability analysis suggested that the wake was neutrally stable at $Re = 250 \pm 1$ with the wavenumber of $\beta D \approx 1.366$ ($L_z/D \approx 4.6$) and no obvious variance was found when Re increases to 350. Williamson (1988), Carmo and Meneghini (2010), and Kanaris et al. (2011) found that the range of Mode A unstable wavelength was slightly wider compared to those in the single-cylinder case. In addition, a spanwise length of $L_z/D = 8$ in the confined single-cylinder flow was found by Kanaris et al. (2011) sufficient for developing the unstable modes.

Therefore, to maintain computation at a reasonable cost and to allocate sufficient grids to resolve the streamwise vortices in the near field of wake, a slightly shorter length of $L_z = 2\pi$ is applied in spanwise for the current computation. The streamwise length behind the rear cylinder is set at $7.5D$ for the wake vortices in the near field, which is capable of containing about two complete von Karman vortices as shown in Fig. 7.7A. Comparing to the existing studies, the current computation domain has relatively smaller dimensions so that a sufficient resolution can be obtained in the near field to guarantee the wake vortices being fully resolved and thereby, the vorticity being accurately calculated and the Liutex dynamics equation being reliably analyzed.

7.5.3 Vortex dynamics analyses enabled by R-S decomposition

With the advent of the 3rd-G VI of Liutex introduced by Liu et al. (2018a), Gao et al. (2018), and Wang Y. et al. (2020), as elucidated in the other chapter of the book, vortex can be quantitatively represented by the scalar and vector forms of Liutex, which unambiguously stands for the local rigid-rotation strength and direction and consequently provides a more appropriateness or suitability to quantitatively study the vortex dynamics. More specifically, the introduction of Liutex straightforwardly leads to the $R - S$ decomposition of vorticity given in Liu et al. (2016, 2018a), which provides the feasibility to quantitatively study the vortex dynamics as implicitly contained in the DNS data presented in the current study. Therefore, as an in-depth analysis of the flow instability for the double cylinder in tandem, the vortex formation mechanisms and the associated vortex dynamics are presented in the following discussions, which, for the first time, demonstrate that dynamics of Liutex, indeed, truly reflect the vortex behaviors in the transition in terms of the first and second instability and the associated modes of Mode A and B. Via making use of the visualization of Liutex iso-surfaces and core lines, the instability processes, particularly the secondary instability, are interrogated in both linear and nonlinear phases in the evolution. The dynamic evolutions and the vortex formations are quantitatively studied by the DNS data under the guidance of the Liutex transport equation enabled by the $R - S$ decomposition.

The vorticity transport equation, or Helmholtz equation, for the unsteady incompressible flow is written as

$$\frac{D\vec{\omega}}{Dt} = (\nabla v) \cdot \vec{\omega} + \frac{1}{Re} \Delta \vec{\omega} \tag{7.6}$$

(A)

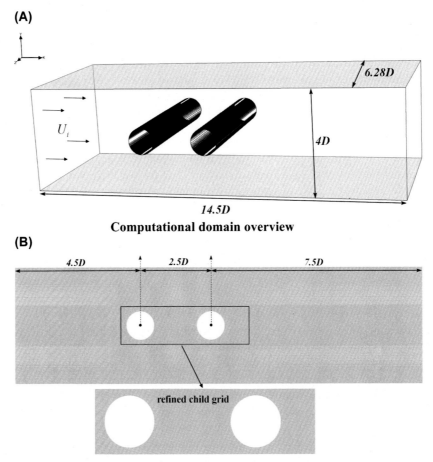

Computational domain overview

(B)

Child grid and mesh distributions near cylinders

FIGURE 7.7

Diagram for the flow configuration with dual cylinders in tandem. (A) Computational domain overview. (B) Child grid and mesh distributions near cylinders.

with the vorticity vector $\vec{\omega}$ being defined as the curl of velocity i.e., $\vec{\omega} = \nabla \times \vec{V}$.

Based on the *R-S* decomposition proposed by Liu et al. (2018a), the vorticity vector $\vec{\omega}$ then satisfies the relation of $\vec{\omega} = \vec{R} + \vec{S}$, if $\vec{R}$ and $\vec{S}$ denote the Liutex vector, i.e., the rigid rotation, and the shear vector, i.e., shearing motion, respectively.

As explained before, the local velocity gradient tensor, i.e., $\nabla \vec{v}$, usually possesses one real eigenvalue λ_r and two conjugated complex eigenvalues $\lambda_{cr} \pm i\lambda_{ci}$ in a vortical region. The eigenvector $\vec{r}$ corresponding to the real eigenvalue λ_r is then defined as the Liutex vector. Consequently, the vector of Shear can be determined by $\vec{S} = \vec{\omega} - \vec{R}$.

Due to the incompressible assumption, the trace of velocity gradient, or equivalently the velocity divergence, has to satisfy $trace\left(\nabla\vec{v}\right) = \lambda_r + 2\lambda_{cr} = 0$. Therefore, if the eigenvectors of $\nabla\vec{v}$ are normalized as l_r and $\bar{l}_c$ for λ_r and $\lambda_{cr} + i\lambda_{ci}$, respectively, and $\bar{l}_c$ stands for the conjugated-counterpart eigenvector of $\lambda_{cr} - i\lambda_{ci}$, $\vec{S}$ can correspondingly be decomposed by $\vec{S} = c_r l_r + c_c l_c + \overline{c_c l_c}$. Eq. (7.6) can be further simplified and rewritten into the following form:

$$
\frac{D\vec{R}}{Dt} = \nabla v \cdot \left(\vec{R} + \vec{S}\right) + \frac{1}{Re}\Delta\left(\vec{R} + \vec{S}\right) - \frac{D\vec{S}}{Dt}
$$

$$
= \lambda_r\vec{R} + \nabla v \cdot \left(c_r l_r + c_c l_c + \overline{c_c l_c}\right) + \frac{1}{Re}\Delta\left(\vec{R} + \vec{S}\right) - \frac{D\vec{S}}{Dt}
$$

$$
= \lambda_r\vec{R} + \lambda_r c_r l_r + (\lambda_{cr} + i\lambda_{ci})c_c l_c + (\lambda_{cr} - i\lambda_{ci})\overline{c_c l_c} + \frac{1}{Re}\Delta\left(\vec{R} + \vec{S}\right) - \frac{D\vec{S}}{Dt}
$$

$$
= \lambda_r\vec{R} + \lambda_r c_r l_r - \lambda_r Real(c_c l_c) - 2\lambda_{ci}Imag(c_c l_c) + \frac{1}{Re}\Delta\left(\vec{R} + \vec{S}\right) - \frac{D\vec{S}}{Dt} \quad (7.7)
$$

with $Real(c_c l_c)$ and $Imag(c_c l_c)$ being the real and image parts of $c_c l_c$, respectively, and Δ being the Laplacian operator.

Considering the flow being inviscid, or assuming the Reynolds number Re being infinitely large, the viscous term $\frac{1}{Re}\Delta\left(\vec{R} + \vec{S}\right)$ is then negligible and Eq. (7.7) can be further simplified and rewritten as

$$
\frac{D\vec{R}}{Dt} = \lambda_r\vec{R} + \underbrace{\lambda_r c_r l_r - \lambda_r Real(c_c l_c) - 2\lambda_{ci}Imag(c_c l_c) - \frac{D\vec{S}}{Dt}}_{\text{Shearing term}} \quad (7.8)
$$

Although the general analytical solution of Eq. (7.8) is difficult to be obtained, the first-order differential equation can provide the governing terms for analyzing the generation and dynamics of Liutex, if considering the right-hand-side term $g = \lambda_r c_r l_r - \lambda_r Real(c_c l_c) - 2\lambda_{ci}Imag(c_c l_c) - \frac{D\vec{S}}{Dt}$ as the source term under the Eulerian framework. Moreover, if $\tilde{f}, f = \vec{R}, \lambda_r, \lambda_c, g$ are written under the Lagrange framework, i.e., $\tilde{f}(X_0, t) = f(x(X_0, t), y(X_0, t), z(X_0, t), t)$ with the vector X_0 representing the location at an initial time t_0, Eq. (7.9) can be used to provide the governing equation for the dynamics analysis of $\tilde{R}$

$$
\frac{\partial\tilde{R}}{\partial t} = \tilde{\lambda}_r\tilde{R} + \underbrace{\tilde{\lambda}_r\tilde{c}_r\tilde{l}_r - \tilde{\lambda}_r Real\left(\tilde{c}_c\tilde{l}_c\right) - 2\tilde{\lambda}_{ci}Imag\left(\tilde{c}_c\tilde{l}_c\right) - \frac{D\tilde{S}}{Dt}}_{\text{Shearing term}} \quad (7.9)
$$

and the solution of Eq. (7.9) can generally be integrated as

$$\widetilde{R}(X_0, t) = \left(\widetilde{R}(X_0, t_0) + \int_{t_0}^{t} \widetilde{g}(X_0, \tau) e^{-\int_{t_0}^{\tau} \widetilde{\lambda}_r(X_0, \xi) d\xi} d\tau \right) e^{\int_{t_0}^{t} \widetilde{\lambda}_r(X_0, \tau) d\tau}$$

$$= \underbrace{\widetilde{R}(X_0, t_0) e^{\int_{t_0}^{t} \widetilde{\lambda}_r(X_0, \tau) d\tau}}_{\text{spontaneous term}} + \underbrace{\int_{t_0}^{t} \widetilde{g}(X_0, \tau) e^{\int_{\tau}^{t} \widetilde{\lambda}_r(X_0, \xi) d\xi} d\tau}_{\text{shearing term}} \qquad (7.10)$$

The solution of Eq. (7.9) suggests that the formation or generation of Liutex can be attributed to two parts of contributions, with the first part being $\widetilde{R}(X_0, t_0) e^{\int_{t_0}^{t} \widetilde{\lambda}_r(X_0, \tau) d\tau}$ and the second part being $\int_{t_0}^{t} \widetilde{g}(X_0, \tau) e^{\int_{\tau}^{t} \widetilde{\lambda}_r(X_0, \xi) d\xi} d\tau$ interpreted as the effect of shearing S. The physical meanings of λ_r and λ_{ci} have to be profoundly discussed and understood so that the Liutex generation terms, i.e., $\widetilde{R}(X_0, t_0) e^{\int_{t_0}^{t} \widetilde{\lambda}_r(X_0, \tau) d\tau}$ and $\int_{t_0}^{t} \widetilde{g}(X_0, \tau) e^{\int_{\tau}^{t} \widetilde{\lambda}_r(X_0, \xi) d\xi} d\tau$, being applied to the current transition and vortex dynamics analyses.

Considering an arbitrary location X in space with its neighborhood point in the real eigenvector direction being defined and expressed as $X + \Delta x l_r$, the spatial difference of the two points is then in line with the eigenvector direction of l_r representing the rotating axis. The corresponding velocity difference is written as $\nabla \vec{v}(X) \cdot \Delta x l_r = \lambda_r \Delta x l_r$. In a time increment of Δt, the neighborhood point $X + \Delta x l_r$ moves along l_r with a distance of $\lambda_r \Delta x \Delta t$, and still stays in the rotation direction. Thereby the λ_r actually stands for, or is interpreted as, the relative speed at the point in the rotating direction measuring the stretching or compressing strength of the local vortex. As explained by Wang Y. et al. (2019a), the physical interpretation for λ_{ci} is the pseudo time-averaged angular rotation speed, which reflects the mean local spinning effect.

Eq. (7.10) suggests that $\vec{R}$ be generated by the spontaneous and shearing terms, respectively. To analyze the spontaneous mechanism of $\vec{R}$, g can first be considered negligible in Eq. (7.10), which corresponds to the late developing stage of Mode B vortex pairs when the vortices are stretched and satisfy $\lambda_r > 0$. The spontaneous terms, although not being generated autonomously, can lead to a vortex deformation when λ_r is not distributed uniformly in space. If the two points of X_1 and X_2 are on the same iso-surface of Liutex scalar at the initial time instant of t_0 i.e., $\left\| \widetilde{R}(X_1, t_0) \right\|_2 = \left\| \widetilde{R}(X_2, t_0) \right\|_2$ with $\|\cdot\|_2$ denoting the L_2 norm and the shearing effects being ignored, $\left\| \widetilde{R}(X_1, t_0 + \Delta t) \right\|_2 = \left\| \widetilde{R}(X_1, t_0) \right\|_2 e^{\int_{t_0}^{t_0 + \Delta t} \widetilde{\lambda}_r(X_1, \tau) d\tau}$ and $\left\| \widetilde{R}(X_2, t_0 + \Delta t) \right\|_2 = \left\| \widetilde{R}(X_2, t_0) \right\|_2 e^{\int_{t_0}^{t_0 + \Delta t} \widetilde{\lambda}_r(X_2, \tau) d\tau}$ can be calculated at the time of $t + \Delta t$. Thereby, the nonuniformity of λ_r, i.e., the nonuniform vortex stretching,

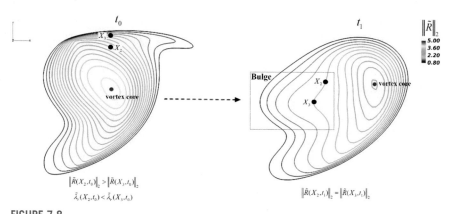

FIGURE 7.8

Diagram of vortex-bulge generation on $\left\|\widetilde{R}\right\|_2$ iso-surface.

can give rise to $\left\|\widetilde{R}(X_1, t_0 + \Delta t)\right\|_2 \neq \left\|\widetilde{R}(X_2, t_0 + \Delta t)\right\|_2$, resulting in the deformation or reshaping of the vortex vertical to its rotating direction. The phenomenon takes place where the local maximum of $\widetilde{\lambda}_r$ exists around the vortex. As shown by the diagram in Fig. 7.8 where the vortex is 2-D with the rotating direction in Z, $\left\|\widetilde{R}\right\|_2$ tends to be larger near the vortex core and $\left\|\widetilde{R}(X_2, t_0)\right\|_2 > \left\|\widetilde{R}(X_3, t_0)\right\|_2$ at the time instant of t_0.

The shearing term given by $\int_{t_0}^{t} \widetilde{g}(X_0, \tau) e^{\int_{\tau}^{t} \widetilde{\lambda}_r(X_0, \xi) d\xi} d\tau$ is even more complicated, which essentially involves three mechanisms as suggested in Eq. (7.9). The first mechanism relates to the existence of vortex stretching, which is expressed by the term of $\widetilde{\lambda}_r \widetilde{c}_r \widetilde{l}_r - \widetilde{\lambda}_r Real\left(\widetilde{c}_c \widetilde{l}_c\right)$ in Eq. (7.9) with $\widetilde{\lambda}_r$ being the multiplication factor. The term can be neglected if the stretching strength is trivial and $\widetilde{\lambda}_r \approx 0$. The second mechanism is represented by the term of $-2\lambda_{ci} Imag(c_c l_c)$ and obviously $\widetilde{\lambda}_{ci}$ is a multiplication factor, which generally satisfies $\widetilde{\lambda}_{ci} \neq 0$ in vortical region. The term can be induced by the directional separation of $\overrightarrow{R}$ and $\overrightarrow{S}$. The third mechanism is described by the shearing allocation effect of $-\frac{D\overrightarrow{S}}{Dt}$. These terms in Eq. (7.9) and their related streamwise-vortex generations can be directly investigated if a DNS data is available. Moreover, the term of $-\frac{1}{Re}\Delta\left(\overrightarrow{R} + \overrightarrow{S}\right)$ in Eq. (7.7) cannot be neglected for a viscous flow, which, in general, causes the diffusion of Liutex. Within the context, the relationships between the viscous term and the generation of streamwise vortices can be interrogated in detail based on the DNS data discussed in the following sections.

7.5.4 Vortex dynamics and secondary instability in transition

7.5.4.1 Comments on vortex dynamics in transition of dual cylinders in tandem

The center-to-center gap distance is set at $2.5D$ between the two cylinders in tandem, similar to the $2.3D$ used in Carmo and Meneghini (2010). The 3-D transition occurs first in the gap between the cylinders. The front cylinder acts like a destabilizer, which makes the transition time for a secondary instability remarkably shortened comparing to the single-cylinder wake. The instability developed in the gap originates from a pair of counter-rotating vortices with an elliptic nature in a cooperative manner. The comprehensive descriptions of 3-D elliptic stability were reported in Billant et al. (1999) based on the studies of a counter-rotating vortex pair. The essential characteristics of elliptic-instability with the cooperative manner in the linear phase is presented by the LXC-core lines given by Gao et al. (2019b) and Xu et al. (2019). A sinusoidal oscillation in the core lines was originated from the instability inside the vortex cores, as seen in Fig. 7.19A, which are exhibited by the tangled core lines, meanderingly concave or convex relative to the center plane of $y = 0$.

The instability wavelengths tend to be longer in the gap regime than those in the wake. The vortex pair is constrained by the two cylinders and the shear layers in between the gap, which plays an important role in the gap instability. The similarity and difference were explained in Carmo and Meneghini (2010) for the elliptic instabilities in the gap and the counter-rotating vortex pairs. Because of the continuous rotating energy injected into the gap zone from the shear layers in between the cylinders, the vortex pair in the current study is quite different from the vortex pairs in a free flow and is able to be maintained in Fig. 7.9A and B.

The wake flow for $Re = 200$ is maintained in 2-D mode for the entire evolution history. Although the vortex pair in the gap deforms and oscillates constantly, as seen Fig. 7.9B, leading to the periodic variations of the lift force for the front cylinder, these 3-D structures are constrained within the upper and lower shear layers and hardly impose any interference with the wake. These results were confirmed by the linear analysis in Carmo and Meneghini (2010), which reported a neutral stability at $Re \approx 250$ with the current gap distance.

Some large-scale spanwise vortical structures, represented by the **Bulge-1** in Fig. 7.10B, come into formation in $Re = 400$ flow after the linear-phase development in the upstream part of the gap vortex pairs. The gap flow thereby starts to transit into the 3-D pattern as presented in Fig. 7.9C. The detailed vortex dynamics in transition are discussed in the next section. Overall, the wake transits to the 3-D structures with the distinctive streamwise vortices and the shedding pattern are similar to Mode B instability in the fully developed flow of a single cylinder. Comparing with Mode A, Mode B instability is featured by the shorter wavelengths in spanwise and the streamwise structures are visualized as the distinctive long vortex tube located in pairs with the intermittent positive and negative signs of R_x, which are attaching to the spanwise rolling structures, as presented by the zone highlighted in dash black square in Fig. 7.9C.

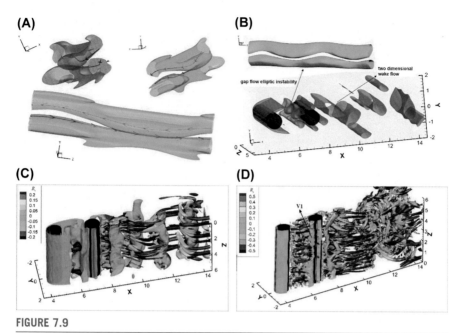

FIGURE 7.9

Vortex dynamics represented by $\|R\|_2$ iso-surfaces with R being the Liutex magnitude: (A) $\|R\|_2 = 0.4$ for $Re = 400$ flow in the elliptic instability stage in the gap and wake regions, (B) $\|R\|_2 = 0.3$ for $Re = 200$, (C) $\|R\|_2 = 0.3$ for $Re = 400$, and (D) $\|R\|_2 = 0.7$ for $Re = 800$ flows colored by R_x.

When increasing Re to 800, the gap flow gets into the more complex structures. The shear layers $Sl4$ in Fig. 7.10A strongly interact with the spanwise gap vortex and the hairpin vortex arises in the nonlinear phase, which gives rise to a variety of transition regimes as discussed in the analysis of nonlinear phase. At a higher Re, the gap vortices tend to be in a quasi-streamwise pattern. These vortices gain sufficient strengths and adhere to the spanwise vortices in the fully developed gap within which the hairpin vortex V1 can be observed in Fig. 7.9D, but in the more refined structures compared the lower Re flow.

Mode B vortex pairs arise as the secondary structures in wake, i.e., V2 in Fig. 7.9D, and in addition, the secondary structures originated from the gap exhibit more complex vortex structures when migrating downstream to the wake region in Fig. 7.9D. Generally speaking, Mode B instability is associated with the hyperbolic instability of the strongly-strained shear layer between the spanwise vortex cores as explained by Williamson (1996) and McClure et al. (2019). The wavelength in Mode B tends to decrease with Re being increased. Moreover, as exhibited by V3 in Fig. 7.9D, the spanwise vortices near wall strongly interact with the wake streamwise vortices and transform into the quasi-streamwise structures downstream, which gives rise to the bypass transition in wake turbulence of the two cylinders in tandem.

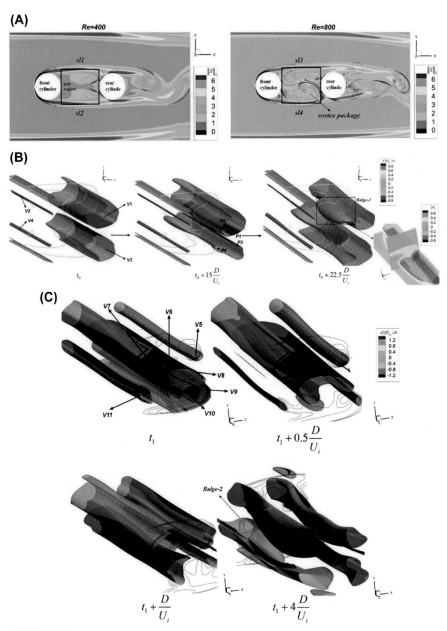

FIGURE 7.10

Evolution of gap vortices in the linear phase for: (A) shear layers denoted by $\|S\|_2$ at $Re = 400$ and $Re = 800$, and liutex iso-surface with (B) $\|R\|_2 = 0.4$ iso-surface at $Re = 400$, and (C) $\|R\|_2 = 0.8$ iso-surface at $Re = 800$. The 3-D iso-surface is colored by $\frac{\partial\|R\|_2}{\partial t}$ and the representative slides are selected to present 2-D $\|R\|_2$ contours.

7.5.4.2 Vortex dynamics of secondary instability in gap

7.5.4.2.1 Dynamics of linear instability in gap

The linear stability analysis is conducted based on the budget term study in Eq. (7.7). However, the baseflow in the stability analysis has to be considered as the corresponding periodic 2-D unsteady flow of two cylinders in tandem demonstrated in Fig. 7.7, which makes it difficult to analyze the budget terms directly. To separate the baseflow and the perturbations, the flow-field-related quantities of $u = (u_x, u_y, u_z)$ and the Liutex-Shear-related quantities of $R = (R_x, R_y, R_z)$, $S = (S_x, S_y, S_z)$ are decomposed into the summation form of the 2-D baseflow, i.e., $\widetilde{U}(x, y, t) = (\widetilde{U}_x, \widetilde{U}_y, \widetilde{U}_z)$ along with its corresponding Shear or Liutex, i.e., $\widetilde{S}(x, y, t) = (\widetilde{S}_x, \widetilde{S}_y, \widetilde{S}_z)$ or $\widetilde{R}(x, y, t) = (\widetilde{R}_x, \widetilde{R}_y, \widetilde{R}_z)$ and the 3-D infinitesimal perturbations of U', S' and R'. Based on the decomposition, the dynamics equation for Liutex, i.e., Eq. (7.7), can be written into its linearized form as

$$\frac{DR'}{Dt} = \frac{\partial R'}{\partial t} + (\widetilde{U} \cdot \nabla)R' = \nabla\widetilde{U} \cdot (R' + S') + \nabla U' \cdot (\widetilde{R} + \widetilde{S}) + \frac{1}{Re}\Delta(R' + S') - \frac{\partial S'}{\partial t}$$
$$- (\widetilde{U} \cdot \nabla)S'$$

(7.11)

Given the spanwise z being a homogeneous direction and the periodic boundary condition being imposed in z, the budgets terms in Eq. (7.11) can be decomposed into its Fourier transformation form of

$$f'(x, y, z, t) = \int_{-\infty}^{\infty} \widehat{f}(x, y, \beta, t)e^{i\beta z}d\beta$$

(7.12)

with $\widehat{f}$ denoting the Fourier mode for the terms of convection, diffusion, and Shear in Eq. (7.11). The stability analysis approaches in Carmo and Meneghini (2010) are then applied by substituting Eq. (7.12) into the linearized equation of Eq. (7.11) for the corresponding terms. The 3-D stability analysis can thereby be performed by a series of 2-D stability at the various wavenumbers of β. Since the baseflow $\overline{U}$ and the related $\overline{R}$, $\overline{S}$ are all homogeneous in the z direction, the Fourier modes $\widehat{f}(x, y, \beta, t)e^{i\beta z}$ at a given β are governed by the inverse Fourier transformation of Eq. (7.11) and each Fourier mode develops in time (t) and space of (x, y) independently at the given wavenumber of β. The following discussions are focused on obtaining the Fourier mode with a dominant instability in the gap and thereby analyzing the incremental contributions from each budget term in Eq. (7.11) to the instability.

The entire process of losing the stability is highly time-dependent and experiences two distinctive periods of the first instability with the formation in 2-D periodic von Karman vortex street and then the secondary instability with the characteristic transformation from the 2-D (spanwise) to 3-D (streamwise) vortex structures. The linear phase analysis of the secondary instability in the current study focuses on the start of the transformation from periodic 2-D to 3-D process. There-fore, it is appropriate to begin with by visualizing the vortex evolution in the early stage when the 3-D structures start to arise. Fig. 7.10B and C provide with the growing process of linear instability in the gap with $Re = 400$ and $Re = 800$, where the contours are colored by $||\boldsymbol{R}||_2$ and $\frac{\partial ||\boldsymbol{R}||_2}{\partial t}$, respectively. Since $||\boldsymbol{R}||_2$ takes its local maximum along a core line and tends to be larger in the region near the vortex core inside the iso-surfaces, the red parts of iso-surface of $\frac{\partial ||\boldsymbol{R}||_2}{\partial t}$ always protrude or bulge toward the outside and the blue parts of $\frac{\partial ||\boldsymbol{R}||_2}{\partial t}$ toward the inside.

For the lower Re, the front cylinder generates the shear layers in Fig. 7.10A and then extends and attaches to the rear cylinder downstream, which induces the vortex system restrained within the gap space in between the shear layers of $sl1$ and $sl2$ in Fig. 7.10A. The vortex system contains the four distinctive gap vortices with 2-D vortical structures, which are clearly identified by the iso-surfaces of $||\boldsymbol{R}||_2$ at the time instant of t_0. Among these vortices, $V2$ and $V4$ have the lower strengths and $V1$ and $V3$ occupy most of the gap region with the relatively higher strengths. The vortex cores of $V1$ and $V3$ are extremely off-centred and are located very close to their borders of each other where the $||\boldsymbol{R}||_2$ and $||\boldsymbol{S}||_2$ present very high derivatives as seen by the colormap in Fig. 7.10B. Due to the extremely high gradients, these parts of the iso-surfaces adjacent to $y = 0$ first lose its stability by bulging more quickly than the other parts of iso-surfaces and thereby the vortices are deformed. With the dominant unstable Fourier mode growing, the 3-D vortical structures start to arise, which are represented by the periodic bulges attaching to and growing from the main 2-D spanwise vortices as given by the **Bulge-1** in Fig. 7.10B. Toward the late-stage growth of linear instability, the stripped structures containing the local maximum of $||\boldsymbol{R}||_2$ are formed inside these bulges or vortex bubbles, i.e., the dash black zone highlighted in Fig. 7.10B. The phenomenon suggests that the sub-spanwise vortex, or vortex packet, be generated inside these 3-D structure of bulges in Fig. 7.10B.

As aforementioned, the front shear layers $sl1$ and $sl2$ extend to the surface of rear cylinder as evidently seen in Fig. 7.10A for the gap flow at $Re = 400$ and the shear layers are directly bridging to, rather than flowing over the surface of rear cylinder. Moreover, for the higher $Re = 800$ in Fig. 7.10A, more complex multi-vortex struc-tures are dynamically established. Due to the strong oscillation taking place in the vertical or cross-streamwise direction, the 2D baseflow in gap gives rise to the shear layers of $sl3$ and $sl4$ periodically detaching and impinging onto the rear-cylinder sur-face, which form a backwash into the gap zone or a separation by flowing over the surface. Based on the mechanism, the energy is injected from one side into the gap region by the backwash and meanwhile some fluid in gap zone is ejected out through the shear-layer separation on the other side. This highly dynamical process makes the gap vortex system gain and exchange remarkable momentum from the outer mainstream. Within the process, the gap vortex system in the baseflow is found con-taining three distinctive vortices with individual existence and a vortex package with two vortex cores inside the vortical compound structures as exhibited in Fig. 7.10A for $Re = 800$. The vortices behind the front cylinder denoted by $V7$ and $V11$ in Fig. 7.10C present the larger size compared to their counterparts in the $Re = 400$.

The vortex packet in Fig. 7.10C is induced by the shear layer $sl4$, which contains a system of vortices denoted as $V6, V8, V9, V10$ at the time instant t_1. However, $V9, V10$ at $t_1 + 0.5\frac{D}{U_i}$ detach from the gap vortex system and migrate downstream, leaving with $V6, V8$ staying in the gap. The blue region on the iso-surfaces in between the cores of $V6, V8$ suggests that these two vortical structures of $V6, V8$ have to be separated, with $V6$ moving backward to the front cylinder due to the shear-layer ($sl4$) backwash. Subsequently, $V11$ is squeezed by $V6$ and crushed onto the downstream side of the front-cylinder surface, which results in its vanish. Eventually at the time instant of $t_1 + \frac{D}{U_i}$ $V6$ takes over the position of $V11$.

Meanwhile, the red portion of Liutex iso-surface in Fig. 7.10C suggests that an initially new-born vortex packet starts to grow and expand in y direction, which is followed by a multi-vortex system generated antisymmetrically on the opposite side in y direction. Thereby the periodic baseflow is formed for the following linear stability analysis. Similar to the one at $Re = 400$, the late transition in gap at $Re = 800$ gives rise to the initial 3-D structures in a form of the spanwise vortices inside the bulge, i.e., **Bulge-2** in Fig. 7.10C, which later on grows from the main vortices in baseflow at the time instant of $t_1 + 4\frac{D}{U_i}$.

The 3-D structures in the linear stage mainly take the form of spanwise vortices fluctuating in z direction. The linearized transport equation of R'_z, i.e., the z component of Eq. (7.11), is applied to study the early secondary transition. The fast Fourier transformation (FFT) is utilized to calculate the 3-D unstable Fourier mode, which can be expressed in the spanwise harmonic form of $\widehat{f}_z(x, y, \beta, t)e^{i\beta z}$, $\beta = 1, 2, 3$. with β being the wavenumber.

Due to the weak instability under lower $Re = 200$ condition, the flow structures in gap are found mainly remaining in 2-D presented by Fig. 7.9B and the fluctuation only takes place in spanwise. The oscillations tend to be restricted in spanwise and not to spontaneously develop into the 3-D structures featured by the streamwise rotation of R'_x. The mechanisms were found similar to the one given by the Floquet linear analysis conducted by Carmo and Meneghini (2010). The Floquet multipliers in their study were found inclusively smaller than one so that the perturbations in the gap were not amplified and then essentially maintained in 2-D under the conditions of the unconfined gap zone with the gap distance at 2.3 and $Re = 200$. For the current study, the magnitudes of budget term in Fourier modes are peaked at their local maxima with the wavenumbers of $\beta = 2$ and $\beta = 6$.

For the gap flow at $Re = 400$, the budget terms in Eq. (7.11) need to be taken into account for the first three Fourier modes with $\beta \leq 3$ and their magnitudes reach the maximum at $\beta = 1$. Consequently, the wavelengths of the unstable Fourier modes are inclusively larger than $\frac{2\pi}{3}$, which suggests that the 3-D structures tend to bear a spanwise wavelength longer than $\frac{2\pi}{3}$. These findings provide a hint that the wavelength of the unstable Fourier mode be proportional to Re in the gap. At the time instant of $t_0 + 15\frac{D}{U_i}$ the budget terms in Eq. (7.11) were analyzed at the four representative locations of $x - y$ plane in Fig. 7.11. Specifically,

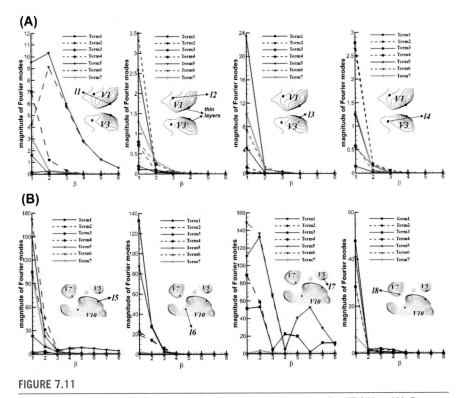

FIGURE 7.11

Magnitude variations with Fourier modes for each budget term in Eq. (7.11) at (A) $Re = 400$ and (B) $Re = 800$.

$l1$ $(x = 6.293, y = 0.171)$, $l2$ $(x = 5.865, y = 0.397)$, $l3$ $(x = 6.320, y = -0.213)$ and $l4$ $(x = 5.827, y = -0.312)$ are selected to denote the locations near the vortex cores and the outer regions of $V1$ and $V3$, respectively.

As presented in Fig. 7.11A, the convection of Liutex and Shear, namely $\left|\left(\widetilde{U} \cdot \nabla\right)\widehat{R}'_z\right|$ and $\left|\left(\widetilde{U} \cdot \nabla\right)\widehat{S}'_z\right|$, are the major terms presenting the dominant magnitudes. On the inner side of vortices, as represented by the locations of $l1$ and $l3$, the densely distributed contours in Fig. 7.11A are found with large Liutex and Shear gradients as well as their perturbations. The instability tends to be originated from these locations with the large gradients. More Fourier modes with the nonnegligible magnitudes are found in the convection terms of $\left|\left(\widetilde{U}\cdot\nabla\right)\widehat{R}'_z\right|$ and $\left|\left(\widetilde{U}\cdot\nabla\right)\widehat{S}'_z\right|$ at the location of $l1$ compared to those at the other locations. However, the short-wavelength fluctuations $\left|\frac{\partial \widehat{R}'_z}{\partial t}\right|$ with $\beta \geq 3$ are nearly balanced by $\left|\left(\widetilde{U}\cdot\nabla\right)\widehat{S}'_z\right|$ and pose very limited influence on producing the Liutex or Shear fluctuations represented by $\left|\frac{\partial \hat{R}_z'}{\partial t}\right|$ or $\left|\frac{\partial \hat{S}_z'}{\partial t}\right|$. For long-wavelength fluctuations of $\beta \leq 2$, $\left|\frac{\partial \hat{R}_z'}{\partial t}\right|$ at $\beta \leq 2$ is no longer able to balance with the corresponding $\left|(\tilde{U}\cdot\nabla)\hat{S}_z'\right|$ and therefore, the fluctuations of $\left|\frac{\partial \hat{R}_z'}{\partial t}\right|$ and $\left|\frac{\partial \hat{S}_z'}{\partial t}\right|$ both have large magnitudes in Eq. (7.11), suggesting that the 3-D structures of Shear and Liutex be developed at the wavelengths longer than π at the location of $l1$.

At the locations near $l1$ and $l3$, the transient terms of $\left|\frac{\partial R_z'}{\partial t}\right|$ and $\left|\frac{\partial S_z'}{\partial t}\right|$ are generally larger than those at the other locations, which presents the strong evidence that the perturbations are stronger at the inner side of these vortices among the cores with the large gradient zone of Liutex. It is worth mentioning that the vortex cores are always nearby and in companion with these Liutex-gradient significance spots. Thereby, with the locally strong spinning strength of vortices, these perturbations of Liutex and Shear are transferred to the inner portion of vortex system located at $l2$ and $l4$ where the spanwise sub-vortex in **Bulge-1** is generated in Fig. 7.10. The convection terms of $\left|(\tilde{U}\cdot\nabla)\hat{R}_z'\right|$ and $\left|(\tilde{U}\cdot\nabla)\hat{S}_z'\right|$ for each Fourier mode become significant in their magnitudes to be taken into account in the budget balance at these locations.

The rotating term of $\nabla\tilde{U}\cdot\hat{R}' + \nabla\hat{U}'\cdot\tilde{R}$ and shearing term of $\nabla\tilde{U}\cdot\hat{S}' + \nabla\hat{U}'\cdot\tilde{S}$ are, in general, trivial compared to the convection and hardly influential on the dominant instability modes. However, these terms become significant at large wavenumber modes due to the rapid decrease of the convection terms ($\left|(\tilde{U}\cdot\nabla)\hat{R}_z'\right|$ and $\left|(\tilde{U}\cdot\nabla)\hat{S}_z'\right|$) beyond the dominant modes, and the magnitudes of these terms even become larger than the magnitudes of the convection terms $\left|(\tilde{U}\cdot\nabla)\hat{R}_z'\right|$ and $\left|(\tilde{U}\cdot\nabla)\hat{S}_z'\right|$ for $\beta = 2$ mode at the location of $l2$.

Term1	$\frac{\partial \hat{R}_z'}{\partial t}$
Term2	$\frac{\partial \hat{S}_z'}{\partial t}$
Term3	$\left\|\nabla\tilde{U}\cdot\hat{R}' + \nabla\hat{U}'\cdot\tilde{R}'\right\|$
Term4	$\left\|\nabla\tilde{U}\cdot\hat{S}' + \nabla\hat{U}'\cdot\tilde{S}'\right\|$
Term5	$\left\|(\tilde{U}\cdot\nabla)\hat{R}'\right\|$
Term6	$\left\|(\tilde{U}\cdot\nabla)\hat{S}'\right\|$
Term7	$\frac{1}{Re}\Delta\left(\hat{R}' + \hat{S}'\right)$

Other four representative locations are chosen in $Re = 800$ flow for the budget term study, which are shown in Fig. 7.11B as $l5$ ($x = 6.348, y = -0.091$),

$l6$ $(x = 5.773, y = -0.233)$, $l7$ $(x = 6.358, y = 0.181)$ and $l8$ $(x = 5.419, y = 0.156)$ at the time instant of $t_1 + 0.5\frac{D}{U_i}$. As afore discussed in Fig. 7.10A, the shear layer $sl4$ impinges onto the surface of rear cylinder and then is backwashed into the gap, in which the transience and convection are the major events taking place. It is found that the term $\left|\frac{\partial \hat{S}_z'}{\partial t}\right|$ mainly balances with $\left|\frac{\partial \hat{R}_z'}{\partial t}\right|$ and $\left|\left(\widetilde{U}\cdot\nabla\right)\hat{R}'\right|$ balances with $\left|\left(\widetilde{U}\cdot\nabla\right)\hat{S}'\right|$ for the short-wavelength perturbations ($\beta > 3$) at $l7$. The same phenomenon occurs at the $l8$ location for $\beta = 1$ mode, where $\left|\frac{\partial \hat{R}_z'}{\partial t}\right|$ is resulted predominantly from $\left|\frac{\partial \hat{S}_z'}{\partial t}\right|$. Similar to what happens at the $l3$ location, both convection and viscous perturbations are dominant at $l5$ in the large-wavelength modes, which are the prominent feature for the balance in the Liutex-gradient significant region existing in between the vortex cores. Because of the convections induced by the vortex system of $V6, V8, V9, V10$ in Fig. 7.10C, the convection perturbation gives rise to a large $\left|\frac{\partial \hat{R}_z'}{\partial t}\right|$ at the location of $l6$ with the modes at the wavenumbers of $\beta \leq 3$.

To further study the sub-spanwise vortex, or the vortex packet, inside the bulge structure in Fig. 7.10B at the late linear instability, Fig. 7.12 provides the contour distributions of $\frac{\partial \|R\|_2}{\partial t}$ in gap at $Re = 400$. Nearby the $P3$ on the section plane of $z = 4$, the $\|R\|_2$ transient term of $\frac{\partial \|R\|_2}{\partial t}$ presents a large positive derivative and consequently, the $\|R\|_2$ contour front tends to protrude outside. On the contrary, in the vicinity of $P4$ on the same plane of $z = 4$, the negative derivative of $\frac{\partial \|R\|_2}{\partial t}$

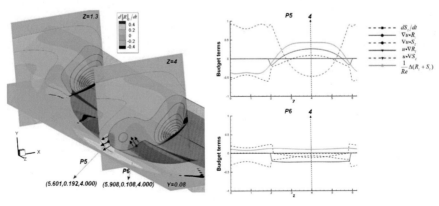

FIGURE 7.12

Generation of sub-spanwise vortex in Bulge-1 represented by the contour of $\|R\|_2 = 0.4$ at $Re = 400$.

causes the lower edge of $\|\boldsymbol{R}\|_2$ contour protruding inside. Comparing to the contours on the plane of $z = 1.3$ where no contour deformation occurs, these protrusions obviously cause the contour being deformed inside the **Bulge-1** and give rise to the sub-spanwise vortex. Moreover, based on the budget analysis, the positive $\frac{\partial \|\boldsymbol{R}\|_2}{\partial t}$ nearby $P3$ is predominantly induced by the R_z convection, i.e., $\boldsymbol{v} \cdot \nabla R_z$ and the negative $\frac{\partial \|\boldsymbol{R}\|_2}{\partial t} = -0.31$ nearby $P4$ is attributed to the summations of both the R_z convection, i.e., $\boldsymbol{v} \cdot \nabla R_z = -0.26$ and the viscous term of $\frac{1}{Re}\Delta(R_z + S_z) = -0.43$ and the Shear terms $\left(\nabla \boldsymbol{v} \cdot \vec{S}\right)_z - \frac{DS_z}{Dt} = 0.38$ in Eq. (7.7) at the point of $P4$.

7.5.4.2.2 Dynamics of nonlinear instability in gap

The nonlinear dynamics of the 3-D vortical structures become much more complicated in terms of their governing terms in dynamic equation and their spatial topologies or patterns. Eq. (7.7) can no longer be linearized into the form of Eq. (7.11) and the nonlinear governing terms in Eq. (7.7) are non-negligible.

The vortex dynamics in the gap at $Re = 400$ are characterized by the cooperative elliptic instability first explained by Leweke et al. (1998) and Carmo and Meneghini (2010). Restrained by the shear layers $sl1$ and $sl2$ emanating from the front cylinder, the core lines of spanwise vortices $V1$ and $V3$ in Fig. 7.13A are found fluctuating harmonically or cooperatively inside the gap in the nonlinear stage. The streamwise vortices are generated from the two outer ends of the bulge structures in Fig. 7.12C when the flow being fully developed. Unlike the nonlinear evolution of vortex pairs in unconfined space studied in Leweke and Williamson (1998), the shear layers of $sl1$ and $sl2$ in Fig. 7.10A constantly feed the momentum and energy into the zone occupied by $V1$ and $V3$. However, the spanwise vortices are not completely broken down into the piecewise vortices in the span direction and the quasi-streamwise vortices start to arise. Moreover, the secondary spanwise vortex in **Bulge-1** demonstrated in Fig. 7.12 is maintained fluctuating at a much longer period compared to the other 3-D structures in the fully-developed stage.

After Re is increased to 800, the nonlinear development and 3-D structures in gap are more evident with the distinctive streamwise structures in between the cylinders in Fig. 7.9D. Based on the statements in the previous section, the shear layer impinging on the rear cylinder gives rise to much different structures and interactions from the shear-layer impingement at the $Re = 400$ when comparing the flow structures in Fig. 7.10A at the two Re numbers. Consequently, the gap vortex system takes the form of flow patterns in Fig. 7.10C as the baseflow, which is characterized by the distinctive shear layer of $sl3$ and $sl4$. The cooperative elliptic instability can be similarly found in the linear development with the spanwise vortices or vortex packets oscillating in the plane normal to z, where the Liutex contours also present the **Bulge-2** structure.

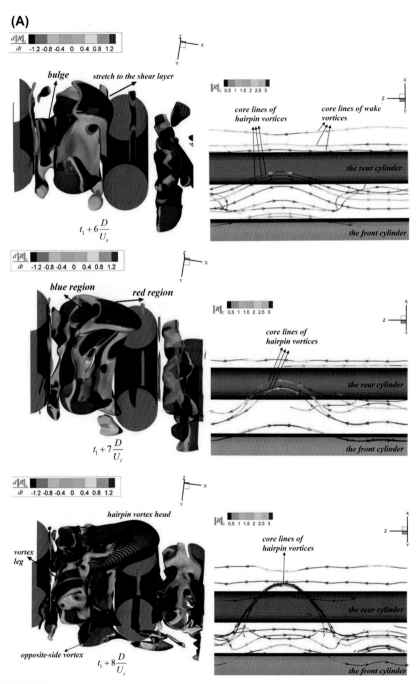

FIGURE 7.13

Hairpin-vortices generation process in gap presented by (A) the Liutex iso-surfaces and corresponding Liutex core lines; small-scale streamwise vortex generations presented by (B) iso-surfaces at the thresholds of $\|R\|_2 = 1.5$ and $\|R\|_2 = 2.0$, respectively.

(B)

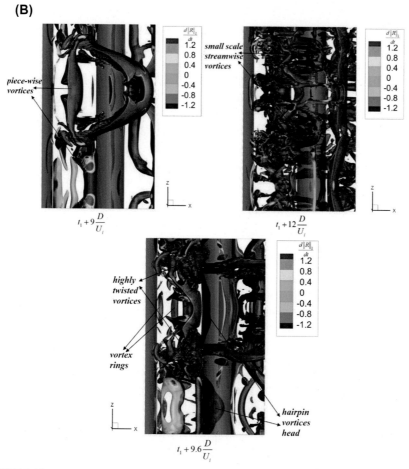

FIGURE 7.13

cont'd.

The vortex system in gap interacts more strongly with the shear layer *sl4*, which is incapable of restraining the gap vortices as occurring in the $Re = 400$ flow. When the Bulge-2 structure is growing up sufficiently in size, the spanwise vortex is ejected out of the gap by the significantly enlarged bulge structure and is stretched to the shear layer as exhibited by Fig. 7.13A at $t_1 + 6\frac{D}{U_i}$. At a subsequent time instant of $t_1 + 7\frac{D}{U_i}$ the primary head portion of a hairpin vortex originated from the spanwise vortex packets comes into formation and starts to interact with the shear layer. Meanwhile, these structures with their interactions migrate downstream by the high convection in the shear layer.

The outer red and inner blue zones on the iso-surfaces stand for the convection effects with the red and blue colors causing the remarkable rise and drop of $\frac{\partial \|R\|_2}{\partial t}$, respectively. Thereby, the stretched portion of the vortex packets starts to migrate along the surface of rear cylinder and then gives rise to the hairpin-vortex head at the next time instant of $t_1 + 8\frac{D}{U_i}$. With the gap vortices being newly generated on the opposite side in y-direction of the cylinder, the remaining portion of the vortex packets start to move back, which results in the formation of hairpin-vortex legs. Eventually, an entire hairpin vortex is generated at the next time instant of $t_1 + 8\frac{D}{U_i}$. Different from the hairpin vortex commonly found in a bypass or natural transition in a ZPGFP TBL, as reported by Li et al. (2020) and Liu et al. (2011), the current hairpin vortex is originated from the vortex packets, which is found containing multi-vortex structures in its head.

After the hairpin vortex being formed, the flow structures and patterns in gap are totally changed and are quickly transformed into the 3-D transition. The quasi-streamwise portion of the hairpin sticks out into the front shear layer and makes the shear layer be fractured into multi-pieces, which further significantly affects the gap vortex patterns and structures. During the evolution, the spanwise vortices are broken into pieces and are divided in between the streamwisely-dominated hairpin legs at $t_1 + 9\frac{D}{U_i}$ exhibited in Fig. 7.13B. Moreover, the vortex rings are found distributed in between the legs. After the legs being fractured from the hairpin vortex, the vortex structures become highly twisted in gap as shown by Fig. 7.13B at $t_1 + 9.6\frac{D}{U_i}$.

Although rotating with the orientations mainly along the spanwise of y, these highly twisted vortices can touch to and interact with the shear layer, which are gradually fractured into small-scale quasi-streamwise vortical structures migrating downstream with the strong convections in the shear layer. Eventually, these small-scale vortices mainly oriented in streamwise take the dominant role in the vortex structures and the gap turbulence transforms into the 3-D pattern of the fully-developed stage after the time instant of $t_1 + 12\frac{D}{U_i}$. Moreover, it is intriguing to notice that these highly twisted vortices divide the entire gap zone into a number of subzones where the 2D baseflow vortical structures are still maintained distinctively oriented in spanwise. Furthermore, these piecewise span-vortices are constantly migrating from the gap to the wake and thereby keep on generating the new-born hairpin vortex heads, as seen in Fig. 7.13B at $t_1 + 9.6\frac{D}{U_i}$ on the downstream side of rear cylinder. The present DNS data suggests that the hairpin vortices are the common patterns and structures in the $Re = 800$ turbulence, which can even be seen in its fully- developed stage, but at a smaller scale as demonstrated by V1 in Fig. 7.9D.

Different from the linear analysis based on Eq. (7.11) in which the terms of $\left(\nabla v \cdot \vec{R}\right)_x$ and $\left(\nabla v \cdot \vec{S}\right)_x$ are omitted, the nonlinear governing equation of

Eq. (7.7) is applied to conduct the vortex dynamics study in the nonlinear phase by focusing on investigating the balancing mechanisms in Eq. (7.7) represented by the long-wavelength modes of Liutex or Shear transient derivatives, convection, and viscous terms. In the nonlinear development, the hairpin vortices play a vital and dominant role in which the quasi-streamwise vortices are originated from the legs of these hairpin vortices and are subsequently breeding by the interactions with the gap and wake shear layers. Within the context, Eq. (7.7) provides the dynamics of $\frac{DR_x}{Dt}$ under the framework of Lagrangian coordinates, which is suitable to be applied to either mathematically analyze the nonlinear evolution of R_x or physically explained the mechanisms of streamwise vortex generation.

To investigate the detailed dynamic performance of quasi-streamwise vortices, Fig. 7.14 plotted the R_x-related budgets on a plane at $z = 2.97$ that is located by cutting cross a hairpin leg at the time instant of $t_1 + 8\frac{D}{U_i}$. For the purpose of a quantitative comparison, Table 7.1 lists each budget term at the five selected locations where $P5$ is placed on the LXC–core line and $P6-P9$ are chosen at the local extreme points of $\frac{DR_x}{Dt}$, as presented in Fig. 7.14. The hairpin legs are immersed in the shear layer from the front cylinder featured by the quite significant $\|S\|_2$.

Based on the presentation of Fig. 7.14 and Table 7.1, it can be generally found that $\frac{DR_x}{Dt}$ is primarily generated by $\frac{-DS_x}{Dt}$ and the convection terms of $\left(\nabla v \cdot \vec{R}\right)_x$ and $\left(\nabla v \cdot \vec{S}\right)_x$ pose a tangible influence on $\frac{DR_x}{Dt}$. Moreover, the viscous terms only exert their effects at the boundaries of hairpin vortex legs. Physically, the

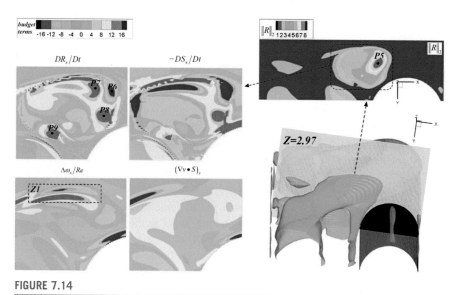

FIGURE 7.14

Budget terms at the representative points on the selected plane in gap for studying the nonlinear-phase development.

Table 7.1 Budget terms for studying the generation mechanisms of streamwise Liutex in Eq. (7.7) at the selected representative points.

Location	DR_x/Dt	$-DS_x/Dt$	$\left(\nabla v \cdot \vec{R}\right)_x$	$\left(\nabla v \cdot \vec{S}\right)_x$	$\frac{1}{Re}\Delta\omega_x$
P5	(6.61, −0.69, 2.97)	10.67	6.65	1.76	2.09	0.17
P6	(6.76, −0.78, 2.97)	−21.01	−31.61	0.86	12.09	−2.35
P7	(6.61, −0.80, 2.97)	20.82	7.31	1.69	4.34	7.47
P8	(6.76, −0.78, 2.97)	19.93	19.38	0.84	0.95	−1.24
P9	(6.76, −0.78, 2.97)	22.52	14.88	2.21	5.37	0.05

contribution from $\frac{-DS_x}{Dt}$ can be interpreted by the interactions between the shear layer and vortex leg, which produces the reallocation effects to change $\frac{-D\omega_x}{Dt}$ through the decomposited term of either $\frac{DR_x}{Dt}$ with a positive (negative) sign or $\frac{DS_x}{Dt}$ with a negative (positive), respectively. The effects of Shear ($\vec{S}$) reallocation make the directions of $\vec{S}$ and $\vec{R}$ be no longer parallel with each other. Consequently, the term of $c_c l_c$ in Eq. (7.7) arises and the term of $\left(\nabla v \cdot \vec{S}\right)_x$ starts to affect vortex development. At P5 and P9 locations where the vortices are stretched in their eigenvector direction with $\lambda_r \neq 0$, the strength of quasi-streamwise vortices is enhanced by $\left(\nabla v \cdot \vec{R}\right)_x = \lambda_r R_x$. On the other hand, the zone of Z1 is occupied by a shear dominance where the vortex leg strongly interacts with the incoming stream and meanwhile, the viscous effects are found to shape R_x due to the closeness of zone to the cylinder surface.

7.5.4.3 Vortex dynamics and wake instability
7.5.4.3.1 Dynamics of linear instability
As well known, the transition in the dual cylinder in tandem commonly starts with the first instability giving rise to the 2-D von Karman vortex streets and the periodic vortex shedding in both gap and wake, which essentially produces the baseflow for the secondary instability analysis. The similar transition stage and regime were also reported in the unconfined wake of a single cylinder investigated by Williamson (1988, 1996).

Subsequent to the first instability, the von Karman vortices with the 2-D structures evolve into secondary instability development, which include a series of interactions between the Liutex rotation of $\vec{R}$ and the Shear motions of $\vec{S}$ in Fig. 7.15. However, the confined walls are located sufficiently close to the von Karman streets, as measured by a blocking ratio of $B = 1/4$. As a result, the wall shear layers significantly interact with the wake vortices from the cylinders in both $Re = 400$ and $Re = 800$ transitions.

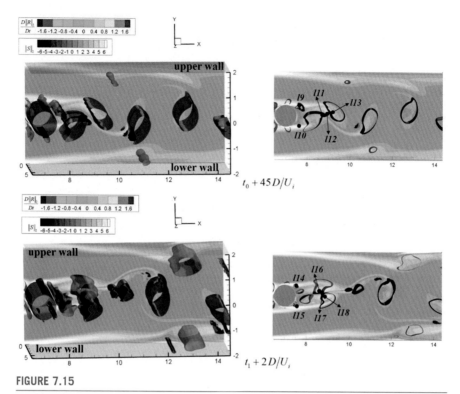

FIGURE 7.15

von Karman vortex streets presented by Liutex magnitude of $\|R\|_2 = 0.4$ and shearing-magnitude $\|S\|_2$ contours on a cross-section plane of $x - y$ in the stage of linear instability for the $Re = 400$ and $Re = 800$ wakes at two time instants of $t_0 + 45\frac{D}{U_i}$ and $t_1 + 2\frac{D}{U_i}$, respectively.

    The increase of Re tangibly enhances the rotating strengths of the channel-wall vortices and on the other hand, the cylinders inside the channel significantly affect the flow structures and behaviors near the channel walls. These interactions were applied to induce a bypass transition and to enhance the near-wall heat transfers by Yoon et al. (2009) and Pandit et al. (2014). In addition, the secondary linear instability in the wake gives rise to the perturbations in both wake and on the channel walls, which initially induces the oscillations along the spanwise direction and eventually leads to the transition to a fully developed turbulence.

    A total of 10 probing points equally divided for the wakes of $Re = 400$ and $Re = 800$ are chosen to investigate the linear instabilities based on the perturbation budget study using Eq. (7.11). As shown in Fig. 7.16, these probing points are selected at the representative locations of (1) $l9, l10$ and $l14, l15$ close to the separation points of rear cylinder; (2) $l11, l12, l13$ and $l16, l17, l18$ in the wakes of shear layers and intensive Liutex rotations for $Re = 400$ and $Re = 800$, respectively. With $\widehat{S}'_z$ standing for

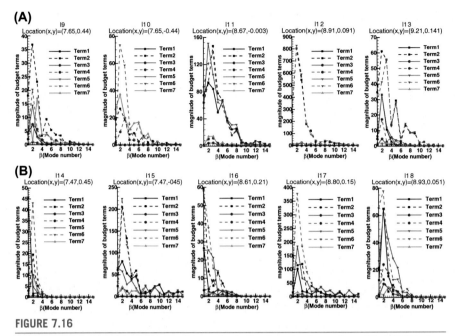

FIGURE 7.16

Budget term variations with the Fourier modes in the linearized equation of Eq. (7.11) at the probing point locations presented in Fig. 7.15.

the perturbation of spanwise shear, the term of $\frac{\partial \bar{S}_z'}{\partial t}$ presents, in Fig. 7.16, its local maxima at the modes with the equivalent wavelengths of πD, $\frac{\pi D}{2.5}$ corresponding to the instability of Mode A, i.e., a wavelength of $3D - 4D$, and Mode B, i.e., a wavelength of about $1D$. The dominant unstable modes are found independent on Re for the wake transitions at both $Re = 400$ and $Re = 800$ as presented in Fig. 7.16. Moreover, the comparison between Figs. 7.11 and 7.16 confirms the conclusions made in Carmo and Meneghini (2010) that the instability modes in wake are generally wider than those in gap, which suggests that the gap vortex dynamics, as presented in previous sections, impose strongly inferences on the wake through migrating a variety of gap vortex structures into the wake zone and meanwhile the wake dynamics generates its own instability modes.

Within the context, the $\beta = 1$ Fourier mode is obviously induced by the gap instability and is transferred into the wake of rear cylinder through the shear layers. In particular, the $\beta = 1$ mode exhibits the larger magnitudes at $Re = 800$, for instance, at the location of $l = 14$, since the gap vortices are no longer restrained by the shear layers at the Re presented in Fig. 7.10A.

In fact, most of the existing studies were mainly focused on the instability of the von Karman street or the wake flow. The current chapter argues that the instability modes with the wavelengths of Mode A and B already exist in the shear layers from rear cylinder. With the first spanwise vortex appearing in the wake of rear cylinder, the shear-layer instability starts subjecting to the influences from the wake vortices by the convection. Based on the argument, the locations of $l11, l12, l13$ for $Re = 400$ and $l16, l17, l18$ for $Re = 800$ are chosen for the representative budget analysis in Fig. 7.15 where the locations of $l11, l13, l16, l18$ are placed in the wake vortices and the points of $l12, l17$ are set in the shear-layer zone between the von Karman vortices.

Fig. 7.16 presents that the larger magnitudes are generally found at the modes of $\beta = 2.5$. The mode at $\beta = 2$ corresponds to the secondary instability of Mode A, which gives rise to the streamwise vortex pairs with a wavelength of π in the shear-layer zone meandering between the Karman vortices. From the existing studies of a single-cylinder wake, Mode A's secondary instability was found particularly in the lower Re transition. However, when making the comparison of $\frac{\partial R_z}{\partial t}$ at the locations of $l12$ and $l17$, the term tends to be smaller in the $Re = 400$ wake, but presents the larger magnitudes in the $Re = 800$ wake, suggesting that the secondary instability of Mode A takes place in a higher Re wake.

A Re-delay in the occurrence of Mode A secondary instability is confirmed by Fig. 7.18, which is likely attributed to the interactions between the gap vortices and the shear layers by magnifying the perturbations with the larger wavelengths. The mode at $\beta = 5$ stands for the Mode B's secondary instability, which induces the streamwise vortex pairs at the wavelength of about $1D$ in spanwise. These vortices commonly have their presences in the wakes of both $Re = 400$ and $Re = 800$.

7.5.4.3.2 Dynamics of nonlinear instability

The nonlinear development of quasi-streamwise vortices are found in three characteristic types. The first type is similar to the wake in an unconfined single-cylinder at the lower $Re = 400$, which initially presents Mode B's instability in the shear-layer zones between the von Karman vortices and eventually results in the vortex pairs with the opposite sign of R_x.

The other two types are found in the wake of higher $Re = 800$. As discussed before, for the wake at $Re = 800$, the gap vortices and shear layers migrate and transfer the 3-D instability from the gap to wake zone, which induces a much quicker transition in the current wake than the one in a single cylinder. In addition to the vortex pairs due to Mode B's instability, the vortices from Mode A's instability are also found with the characteric larger wavelengths around $3D$. The rupture and bend of spanwise vortices are the major causes of these vortical structures as explained by McClure et al. (2019). Moreover, the hairpin vortices originated from the gap can also lead to the quasi-streamwise vortices in wake. To further detect the nonlinear dynamics and the 3-D structures' mechanism in wake, the following discussions are concentrated on the three characteristic types of quasi-streamwise vortices from Mode A, Mode B, and Hairpin.

The streamwise vortex pairs induced by Mode B instability are mainly found in the shear layers of S defined by the **R-S** decomposition and represented by the

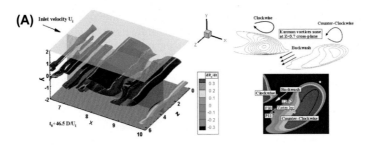

Budget term for R_x generation mechanisms at P10 and P11

Budget	Location	$\dfrac{DR_x}{Dt}$	$-\dfrac{DS_x}{Dt}$	$(\nabla v \bullet R)_x$	$(\nabla v \bullet S)_x$	$\dfrac{\Delta \omega_x}{Re}$	$-\dfrac{DS_x}{Dt} + (\nabla v \bullet S)_x$
P10	(8.70, -0216)	10.67	6.65	1.76	2.09	0.17	4.241
P11	(8.71, -0289)	-21.01	-31.61	0.86	12.09	-2.35	-1.584

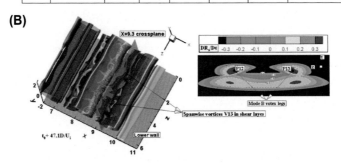

Budget term for R_x generation mechanisms at P12 and P13

Budget	Location	$\dfrac{DR_x}{Dt}$	$-\dfrac{DS_x}{Dt}$	$(\nabla v \bullet R)_x$	$(\nabla v \bullet S)_x$	$\dfrac{\Delta \omega_x}{Re}$	$-\dfrac{DS_x}{Dt} + (\nabla v \bullet S)_x$
P12	(-0.425, 4.393)	20.82	7.31	1.69	4.34	7.47	-0.516
P13	(-0.428, 3.502)	19.93	19.38	0.84	0.95	-1.24	0.866

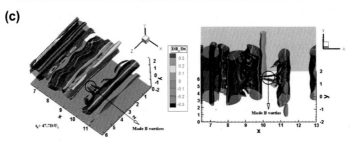

FIGURE 7.17

First and second developing stages of the vortex pairs due to Mode B instability in the $Re = 400$ wake with the iso-surface at $\|R\|_2 = 0.4$.

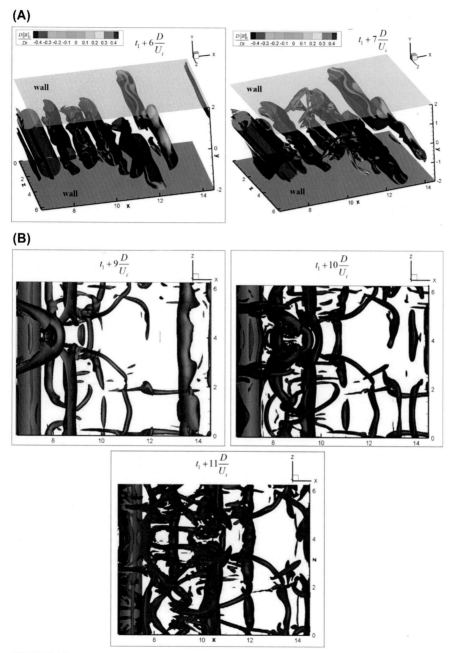

FIGURE 7.18

Vortex pairs from Mode B instability in the early nonlinear transition of wake represented by (A) the iso-surfaces of $\|R\|_2 = 0.4$ and the hairpin vortex migration with its effects on the wake vortical structures exhibited by (B) the iso-surfaces of $\|R\|_2 = 0.6$.

streak structures wrapping around the von Karman vortex streets, which are reported in relation to the hyperbolic instability in Carmo and Meneghini (2010). Based on the recent vortex identification, particularly the Liutex and Shear quantities in the 3rd-G VI, the current study presents more accurate and detailed dynamics of these vortex pairing structures by discovering the two developing stages with their distinctive features and mechanisms.

In the first stage, the initial vortex pairs due to Mode B's instability are formed in the shear layer. The instability in linear phase causes the deformation of wake vortices non-uniformly distributed along the spanwise direction and a subsequent growth of Liutex bulges from the spanwise vortex structures at the time instant of $t_0 + 46.5 \frac{D}{U_i}$ in Fig. 7.17A. As conducted under Langrangian framework and presented in Fig. 7.17A, the budget analysis of $\widetilde{R}_x$ provides the formation mechanisms of these vortex pairs on the plane of $z = 3.7$ cutting across one of Mode B's vortices with $\widetilde{R}_x > 0$. The stripped zone of Z2 is found located on the bulge where $\widetilde{R}_x$ exhibits a significant positive material derivative of time. The balancing mechanism of Eq. (7.7) on P10 in Fig. 7.17A suggests that the positive $\frac{D\widetilde{R}_x}{Dt}$ is mainly attributed to the shear reallocating term of $-\frac{D\widetilde{S}_x}{Dt}$, which provides the evidence that the shear-layer instability, indeed, generates the Mode B vortex pairs.

Given $\widetilde{R}_x$ being generated by the shear-layer instability, the initial vortex pair in streamwise comes into existence attached to the bulge structure in Fig. 7.17A at time instant of $t_0 + 46.5 \frac{D}{U_i}$. With $\widetilde{R}_x$ being further enhanced, $\widetilde{R}_2$ begins to grow out from the zone of Z2, which is found being extended in the Z2 demonstrated by the contours or iso-surfaces of $||\widetilde{R}||_2$ in Fig. 7.17A. Therefore, these distinctive streamwise vortex structures, reflected by the bulge of $||\widetilde{R}||_2$ contours or iso-surface, are attached to the main spanwise vortex as a portion of the von Karman vortex streets. However, at the point of P11 in Fig. 7.17A adjacent to the contour of $||\widetilde{R}||_2 = 0.4$, $\frac{D\widetilde{R}_x}{dt} < 0$ and the associated decrease of $\widetilde{R}_x$ indicate that the streamwise vortices have to be separated from the main von Karman vortex in spanwise. As a consequence, the streamwise vortex legs come into view and stick out into the shear layer, as seen in Fig. 7.17B and C.

The second stage stands for the development of the elongated vortex legs. Due to the spanwise rotation of the von Karman vortices in Fig. 7.17B, the quasi-streamwise vortex legs are induced and approach to the wake backwash region in Fig. 7.17A as mentioned before, where the vortex legs tend to be stretched with $\widetilde{\lambda}_r > 0$. In the process, the term of $\left(\nabla \widetilde{v} \cdot \widetilde{R}\right)_x = \widetilde{\lambda}_{cr} \widetilde{R}_x$ starts to play a dominate role in the budget balance of $\frac{D\widetilde{R}_x}{Dt}$ in Eq. (7.7). With the budget terms in the vortex legs being examined at P12 and P13 in Fig. 7.17B, it is found that the shear ($\frac{D\widetilde{S}_x}{Dt}$) and viscous ($\frac{1}{Re} \Delta(\widetilde{R}_x + \widetilde{S}_x)$) terms are, in general, tangible. However, the approximate balancing relation, i.e., $\left(\nabla \widetilde{v} \cdot \widetilde{R}\right)_x + \frac{1}{Re} \Delta(\widetilde{R}_x + \widetilde{S}_x) - \frac{D\widetilde{S}_x}{Dt} \approx 0$, found in this specific region gives rise to the mechanism of $\frac{D\widetilde{R}_x}{Dt} \approx \left(\nabla \widetilde{v} \cdot \widetilde{R}\right)_x = \widetilde{\lambda}_{cr} \widetilde{R}_x$ which leads to the $\widetilde{R}_x$ generation of

$\widetilde{R}(P12 \text{ or } P13, t) \approx \widetilde{R}(P12 \text{ or } P13, t_0) e^{\int_{t_0}^{t} \widetilde{\lambda}_r (X_0, \tau) d\tau}$. Thereby, the amplification factor $e^{\int_{t_0}^{t} \widetilde{\lambda}_r (X_0, \tau) d\tau} > 1$ causes $\widetilde{R}_x$ being enhanced, which makes the streamwise vortex legs extend upstream touching and interacting with the von Karman vortex as exhibited in Fig. 7.17C. Within the context, the second stage is distinguished by the formation of vortex legs as a result of the stretching effects due to the term of $(\nabla \widetilde{v} \cdot \widetilde{R})_x = \widetilde{\lambda}_{cr} \widetilde{R}_x$ and the distinctive vortex pair structure can be found at the time instant of $t_0 + 47.7 \frac{D}{U_i}$ in Fig. 7.17C. Moreover, the streamwise vortices at *V12* denoted in Fig. 7.17C can be noticed in the shear layer among the von Karman vortices at $t_0 + 48.3 \frac{D}{U_i}$, which are found to develop into Mode B vortex pairs with a larger spanwise structure. Generally speaking, in the $Re = 400$ wake, the head structures of these streamwise vortices keep on constantly attaching to the spanwise vortices and the pattern of von Karman vortex street is preserved in the near field of wake for the entire fully-developed stage as demonstrated by the previous Fig. 7.9C.

Before fully developing into Mode B quasi-streamwise vortices in the shear layers between the von Karman vortex streets, Fig. 7.17B shows that the vortices of *V13* with long size in spanwise can also be noticed in the shear layers at the time instant of $t_0 + 47.1 \frac{D}{U_i}$. These spanwise vortical structures are discontinuously distributed near the bulges of main von Karman vortices. In addition, the bending effect of these spanwise vortices can generate Mode A vortex structures in the $Re = 800$ wake, but the similar structures are not found in the corresponding $Re = 400$ wake, which is referred as the Re-delay mentioned early.

Regarding the wake at a higher $Re = 800$, the vortex pairs induced by Mode B instability can be further found in the early stage of nonlinear development as given in Fig. 7.18A at $t_1 + 7 \frac{D}{U_i}$. Under the circumstance, the vortex legs become more elongated when approaching the confined walls of the channel plate. Mode A vortex pairs, as distinctively characterized by their π wavelength in spanwise, can also be typically captured in the late transition stage, as exhibited in Fig. 7.18B. From the physics perspective, the spanwise vortices in wake, originated from the shear-layers streaks wrapping around the von Karman vortices, can be fractured and meanderingly twisted to give rise to Mode A vortices, as explained by McClure et al. (2019). In addition, the various instabilities in gap, after being transferred into the wake, can also lead to the distinctive 3-D wake vortices in the late stage of development. Fig. 7.18B demonstrates the quasi-streamwise vortices in wake as the result of prominent interactions with the hairpin vortex from gap.

When the hairpin vortical structures migrating from gap into the wake, the vortex-head structure is first found not being able to survive for long and die out quickly. However, the vortex leg successively reaching the wake can survive but be transformed into quasi-streamwise structures with the smaller scales, which is a quite intriguing phenomenon deserve a further investigation. At the time of $t_1 + 10 \frac{D}{U_i}$ in Fig. 7.18B, the first complete hairpin vortex gets into the wake and thereafter, the core structure of the 2-D von Karman vortex is quickly demolished and

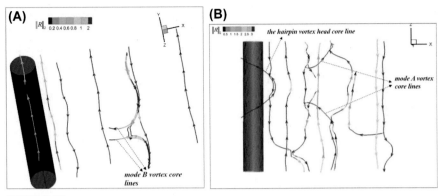

FIGURE 7.19

(A) Mode B vortex core lines at $t_0 + 47.7\frac{D}{U_i}$ in $Re = 400$ transition; (B) Mode A and typical hairpin core lines at $t_1 + 9\frac{D}{U_i}$ in $Re = 800$ transition.

is distinctively replaced by the elongated quasi-streamwise vortices at $t_1 + 11\frac{D}{U_i}$. Meanwhile, the quasi-streamwise vortices from gap in Fig. 7.13 can also stream into the wake and promote the effects to transit the wake spanwise vortices into more streamwise-oriented vortical structures in the downstream wake. Hence, it is evident in Fig. 7.9D that the $Re = 800$ wake gives rise to the featured presence of highly mixed and densely populated streamwise and twisted spanwise vortices in the fully-developed wake turbulence.

To more limpidly demonstrate the vortex structure evolution and their interactions, Fig. 7.19 presents the corresponding LXC-core-lines structures induced by Mode A and Mode B instabilities. These core-line patterns along with the iso-surface structures evidently confirm the formation mechanisms and their corresponding vortical structures generated from these two instability modes. The original core lines, representing the spanwise vortices in the original wake, experience the processes of rupture, bend, and twist, eventually leading to Mode A core lines. As a result, the core lines are separated with the structures identified as an original vortex from Mode A instability. However, Mode B core lines tend to be connected or attached to those spanwise vortex, reflecting the fact that Mode B vortices are actually grown from the spanwise vortex in the two distinctive developing stages as discussed early.

7.5.5 Summary

The secondary instabilities in gap and wake are investigated by the 3rd-G VI in terms of the transformation from the spanwise to the quasi-streamwise vortices, the vortex structure characteristics, and the dynamics of these vortical structures by the well-defined Liutex and Shear quantities and the $R - S$ decomposition of vorticity. The summary is provided as follows:

(1) The Liutex transport equation obtained by the Helmholtz equation and the $R - S$ decomposition is applied to study the vortex or Liutex dynamics, instead of vorticity dynamics. The budget in the equation is interrogated for the instability mechanisms attributed to the convection, viscous, and shear terms, which can well explain the Liutex spontaneous behaviors.

The viscous and shear terms under Langrangian framework are found responsible for the spontaneous Liutex behaviors of enhancement caused by the vortex stretching due to $\lambda_r > 0$ with the real eigenvalue λ_r being the velocity gradient tensor. When omitting viscous terms, the general solution of Liutex transport equation suggests that the spontaneous behaviors of Liutex lead to the 2-D non-uniform deformation of vortices when λ_r being unevenly distributed, but not responsible for the 3-D transition by generating, twisting, and bending the vortices. The initial streamwise vortex is generated mainly due to the shearing instability, which can be splitted into the shearing and rotating reallocations when the Liutex(R) and Shear(S) orientations are not in-line with each other. Based on these findings, the formation mechanisms of quasi-streamwise vortices are investigated via the budget analysis of Liutex transport equation.

(2) For the transition in gap, the flow was found firstly losing the secondary stability in the $Re = 400$ and 800 flows with the current gap distance. The dominant unstable Fourier modes were obtained in the linear analysis, which found the transition being induced mainly by a convective instability. In the subsequent elliptic instability as reflected by the vortex-pair characteristics, the gap vortices initially presented cooperative fluctuation for all the Re cases. The gap vortices at $Re = 400$ are completely restrained in between the shear layers shedding from the front cylinder and thereafter, connecting to the rear-cylinder downstream. The transition of 3-D structures grows out as a bulge from the spanwise vortices, which are characterized by the R_z periodically distributed in the spanwise and are in the form of the piecewise span vortex.

However, for $Re = 800$ transition, the shear layers from front cylinder are firstly attached to the front surface of rear cylinder and the shear layers start to interact with the vortex system in gap. Thereafter, the vortex evolution experiences a number of stages. The DNS data suggests that the vortex packets are generated in the gap at $Re = 800$. When the linear instabilities are grown up into a sufficiently strong state, a portion of the gap vortices starts to touch the shear layers and thereby to produce a head of hairpin-patterned vortex by the shear-layer convection effects. After the hairpin in the gap is formed and swept downstream into the wake, the initial quasi-streamwise vortices come into view as the leg of hairpin. Consequently, the 3-D gap transition distinctively takes place and transforms from the main vortex system in spanwise into the fully-developed state with the prominent quasi-streamwise vortices distributed densely in gap. Within this regard, the current investigations interrogate the Re effects on the structures of gap vortices and their related dynamics in the transition enabled by the 3rd-G VI methods.

(3) The various secondary instabilities in the gap induce the distinct quasi-streamwise vortices in the downstream wake. The gap instability at $Re = 400$, due to the vortices being highly restrained in gap, is found not being able to directly impose their influence onto the wake. Consequently, Mode B vortex pairs are found in the wake, similar to the single-cylinder wake. The budget analysis of Liutex dynamics affirms that the initial streamwise Liutex is generated by the shear instability. Moreover, for the first time, the DNS data authenticate that the two prominent developing stages of Mode B vortex pairs are corresponding to the shear instability and vortex stretching with a variety of balancing mechanisms governed by the Liutex dynamics equation.

For the transition at $Re = 800$, the DNS data demonstrated both vortex structures of Mode A and B secondary instabilities. The delayed Mode A instability to $Re = 800$ in wake is induced by the gap interaction different from the one at $Re = 400$. Under this circumstance, the hairpin vortex formed in gap migrates downstream into the wake and generates the quasi-streamwise vortices directly, rather than indirectly through the perturbation propagation in $Re = 400$. The migrations of hairpin vortex and other streamwise vortices from the gap to the wake significantly accelerate the wake transiting into its 3-D and fully-developed state.

(4) Since vorticity is proved far away from accurately representing vortex, Liutex must be applied for vortex dynamics and the R-S decomposition of vorticity becomes the key in establishing the vortex dynamic model. The secondary instability is investigated based on Liutex and Shear in both the linear and nonlinear stages, which is found dominant in the transition to turbulence. Given the fact that any existing VI methods based on iso-surface type can only present a portion of vortex structure while only the Liutex core line but nothing else can display the complete vortex skeletons and structures, the Liutex core line as a unique tool must be applied to visualize and study the vortex/Liutex dynamics.

Liutex similarity, structure, and asymmetry in turbulent boundary layer

As a new physical quantity exactly representing fluid rotation or vortex, Liutex would certainly play an important role in turbulence generation and sustenance as vortices are the building blocks and muscles of turbulence. The introduction of Liutex may open a new door for quantified turbulence research in comparison with qualitatively turbulence research based on observations, graphics, movies, simplifications, assumptions, guesses, or hypothesis. This chapter will present several applications of Liutex in turbulence research, including Liutex similarity, structures, and asymmetry in turbulent boundary layer.

8.1 Liutex similarity in turbulent boundary layer

8.1.1 Short review

It is generally accepted that Kolmogorov's 1941 theory (K41 for short, Kolmogorov, 1941a,b,c) of similarity hypotheses and the $-5/3$ law for energy spectrum are the most important theoretical achievement and landmark in turbulence research. However, the assumptions of sufficiently high Reynolds number and isotropy of turbulence that K41 is based upon are seldom satisfied in boundary layer flows, and thus discrepancy usually exists between the $-5/3$ law and the results from direct numerical simulation (DNS) of wall-bounded turbulence, especially for low-to-moderate Reynolds number flows. In contrast, the Liutex vector is a kinematic quantity used to represent the rigid rotation part of fluid motion. Liutex is free from viscous dissipation and thus independent of Reynolds number. It has been found that in a moderate Reynolds number turbulent boundary layer ($Re_\theta \approx 1000$), both the frequency and wavenumber spectrum of Liutex accurately match a $-5/3$ law in the high frequency subrange (Xu et al., 2019), much better than the turbulence energy spectrum, while vorticity and other popular vortex identification methods, the Q criterion, for example, fail to offer such a distinguished feature due to stretching and shearing contamination. This unique feature of Liutex is called Liutex similarity (Xu et al., 2019b) and would be a significant contribution to turbulence research and turbulent modeling.

Before diving into the Liutex similarity, K41 is briefly discussed in this section. Richardson (1920, 1922) first introduced the concept of energy cascade via his famous rhyme: "Big whorls have little whorls, which feed on their velocity, and little whorls have lesser whorls, and so on to viscosity." According to Richardson's

Liutex and Its Applications in Turbulence Research. https://doi.org/10.1016/B978-0-12-819023-4.00003-3
Copyright © 2021 Elsevier Inc. All rights reserved.

description, large-scale eddies are driven by external forces that feed energy to the fluid, which in turn transfer energy to small-scale eddies. And small-scale eddies will transfer energy to eddies with smaller scale. This process continues until the smallest scale, called the Kolmogorov scale η, is achieved such that eddies will be dissipated by viscosity and the kinetic energy converts to thermal energy.

Richardson's concept of energy cascade prompts Kolmogorov to propose the famous K41 for homogeneous, isotropic, and incompressible turbulence in the form of three hypotheses, namely the local isotropy hypothesis, the first similarity hypothesis, and the second similarity hypothesis. The local isotropy hypothesis postulates that the small scales in turbulence $(l \ll l_0)$ are statistically isotropic for sufficiently high Reynolds number with the largest eddy size of l_0. Kolmogorov's first similarity hypothesis further assumes that the local isotropic statistics of the small scales have a universal form uniquely determined by v (viscosity) and ε (energy dissipation), which leads to the definition of Kolmogorov scale $\eta = \left(v^3 / \varepsilon\right)$ where dissipation becomes effective. The second similarity hypothesis states that there exists an inertial subrange $(\eta \ll l \ll l_0)$ where statistics solely depends on ε, if the Reynolds number is sufficiently high. Based on dimensional analysis, the energy spectrum $E(\kappa)$ in this range must be of the following form

$$E(\kappa) = C_1 \varepsilon^{2/3} \kappa^{-5/3} \tag{8.1}$$

where C_1 represents a universal constant. The energy spectrum $E(\kappa)$ is interpreted as the energy associated with the wavenumber κ

$$k = \int_0^\infty E(\kappa) d\kappa \tag{8.2}$$

where k is the mean turbulent kinetic energy. Eq. (8.1) is referred as the $-5/3$ law and considered as one of the most famous results of K41 to provide real quantified predictions of turbulence, which was later confirmed by considerable experiments and numerical simulations.

K41 envisaged a cascade of kinetic energy from large scales to small scales and identified an inertial subrange where viscous effects are negligible and information of large eddies is lost or "forgotten," given that the Reynolds number is sufficiently large. This vivid energy cascade description, however, has never been directly observed even with nowadays extraordinarily advanced experimental and numerical simulation techniques. The vortex breakdown process, which transfers energy from large to small scales, is never presented from an instantaneous perspective neither experimentally nor numerically. Furthermore, the $-5/3$ law only matches well for isotropic turbulence, which is an ideal model and thus is not of great importance in practice, thus cannot be confirmed in boundary layers. For low-to-moderate Reynolds number and wall-bounded turbulence, since the assumptions of high Reynolds number and isotropy are no longer valid, the $-5/3$ law is questionable. Critiques on the energy cascade and K41 began with Landau's objection of universality (Landau and Lifshitz, 1987), and revisit on the classical turbulence generation theory has been appealed by many researchers, including Adrian (2007) and Liu et al. (2014).

The Liutex vector is a newly defined physical quantity to extract the rigid rotation from fluid motion, with its direction representing the local rotational axis and magnitude representing twice the local angular speed of rigid rotation (see Chapter 3). Liutex extracts the rigid rotation part out from the fluid motion, which is free from viscous dissipation and thus independent of Reynolds number. In the following, two Liutex hypotheses are introduced.

8.1.2 Liutex local isotropy hypothesis

Liutex local isotropy hypothesis means that the small-scale Liutex ($l \ll l_0$) in turbulence behaves statistically isotropic. This hypothesis differs from Kolmogorov's local isotropy hypothesis in that: (1) it is independent of Reynolds number and thus is applicable to low and moderate Reynolds number flows that can sustain turbulence, (2) the isotropy is now just enforced on the physical quantity of Liutex, which only represents the rigid rotation part of the fluid motion. All viscous effects are excluded. However, this hypothesis should be used with caution. Note that the Liutex similarity only stands for small length scales (high frequences). The main purpose of this hypothesis is to remove or relax some strict assumptions of K41 and thus obtain more practical results that could be used for turbulence subgrid modeling. Although no Reynolds number constraint is imposed in the hypothesis, the Liutex statistics would be anisotropic for large-scale structure or transitional flow as the large vortex plays a central role in this case. In the current study, only fully developed turbulent flow with mediate Reynolds numbers is considered and only streamwise 1D spectrum analysis is conducted.

8.1.3 Liutex similarity hypothesis

Liutex similarity hypothesis means that the statistics of small-scale Liutex ($l \ll l_0$) have a universal form. As discussed in the previous chapters, the Liutex vector represents the rigid rotation part of the flow motion, which is free from stretching and shearing contamination. Therefore, it is reasonable to assume that the small-scale, no matter how small (under the continuum hypothesis), Liutex would not be affected by viscosity. But the determining factor of the universal form will not be explicitly given here. Instead, the $-5/3$ law of Liutex spectrum is presented from the numerical perspective, by investigating the DNS data of a turbulent boundary layer.

8.1.4 $-5/3$ law of Liutex spectrum

The Fourier transform of the Liutex magnitude can be written as

$$R(x) = \int_{-\infty}^{\infty} \widehat{R}(\kappa)\exp(ikx)d\kappa \qquad (8.3)$$

where $\left|\widehat{R}(\kappa)\right|$ represents the contribution of corresponding wave number κ to the Liutex magnitude $R(x)$. $E_L(\kappa) = \left|\widehat{R}(\kappa)\right|$ is used to denote the Liutex spectrum. In the following, it is discovered from the spectrum analysis of the DNS data that $E_L(\kappa)$ has the form

$$E_L(\kappa) = C_2 \phi k^{-5/3} \tag{8.4}$$

where C_2 is a constant and ϕ has the dimension of $\mathrm{m}^{-2/3}\mathrm{s}^{-1}$ to compensate the dimensions in Eq. (8.4).

The DNS data of a zero-pressure-gradient boundary layer transition is obtained by the code DNSUTA. DNSUTA applies a sixth-order compact scheme in the streamwise and normal directions and the pseudo-spectral method with periodic boundary condition in the spanwise direction. In addition, an implicit sixth-order compact filter is adopted to eliminate spurious numerical oscillations caused by central difference scheme. A third-order TVD (total variation diminishing) Runge−Kutta method is employed for time integration. The thorough introduction of DNSUTA and the simulation details are provided in Liu et al. (2014) and Liu and Chen (2011).

Taylor's frozen-turbulence hypothesis (Taylor, 1938) is invoked to approximate the spatial correlations by temporal correlations. A probe is set in the log-law region of the fully developed turbulent flow where the transition is completed and the Reynolds number based on the displacement thickness Re_δ is approximately equal to 1000. Fourier analysis is then applied to the time-series information of the probe. The frequency spectrums of several quantities including Liutex, vorticity, and the Q criterion are shown in Fig. 8.1. It can be clearly found that only the Liutex

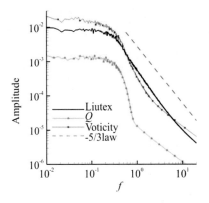

FIGURE 8.1

Frequency spectrums of Liutex, vorticity, and Q.

Reproduced from W. Xu, Y. Wang, Y. Gao, J. Liu, H. Dou, C. Liu, Liutex similarity in turbulent boundary layer, Journal of Hydrodynamics 31 (6), 2019, 1259–1262, with the permission of Journal of Hydrodynamics.

spectrum matches the −5/3 law almost exactly, while both of the spectrums of Q and vorticity significantly deviate from any power law spectrum. One possible reason of this excellent Liutex spectrum could be that Liutex is the rigid rotation part of the flow motion, which is free from stretching and shearing contaminations and any viscous effect. On the order hand, vorticity, Q and other vortex identification methods will certainly suffer from stretching and shearing contaminations and thus no power law spectrum could be observed (Gao et al., 2018).

Fig. 8.2 shows the wavenumber spectrum of Liutex in the log-law sublayer along the streamwise direction. Despite the turbulent boundary layer being only approximately homogeneous in the longitudinal direction and the number of points is limited by the length of the computational domain, a clear −5/3 law can be observed. On the other hand, the energy spectrum, as shown in Fig. 8.3, marginally follows the −5/3 law in a much smaller frequency range due to the limited mesh size of DNS. The stronger universality of Liutex −5/3 law over K41 −5/3 law comes from the fact that Liutex represents the rigid rotation part, which is free from viscous effect while the inertial subrange can only be asymptotically identified in K41 as Reynolds number approaches to a sufficiently high value and $l \gg \eta$. Therefore, the Liutex similarity of −5/3 law removes the strict requirement of the high Reynolds number of K41 and can be applied more generally, which could serve as a better foundation for turbulence subgrid modeling.

The discovery of the astonishing Liutex similarity provides an evidence of the power of Liutex in turbulence research. But currently, the physical meaning of ϕ with dimension $m^{-2/3}s^{-1}$ in Eq. (8.4) is still not clear. Accordingly, it could be a topic for future study. Another important direction is the turbulence modeling based on the Liutex similarity for large eddy simulation (LES). It is believed that the universality of the Liutex similarity will help construct a more accurate model to mimic the physics of the unresolved turbulent motion.

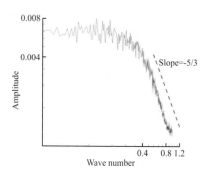

FIGURE 8.2

Wavenumber spectrum of Liutex.

Reproduced from W. Xu, Y. Wang, Y. Gao, J. Liu, H. Dou, C. Liu, Liutex similarity in turbulent boundary layer,
Journal of Hydrodynamics 31 (6), 2019, 1259–1262, with the permission of Journal of Hydrodynamics.

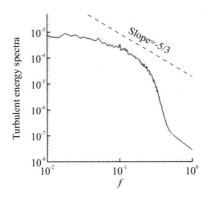

FIGURE 8.3

Energy spectrum.

Reproduced from W. Xu, Y. Wang, Y. Gao, J. Liu, H. Dou, C. Liu, Liutex similarity in turbulent boundary layer, Journal of Hydrodynamics 31 (6), 2019, 1259–1262, with the permission of Journal of Hydrodynamics.

8.2 Coherent vortex structures in late boundary layer transition

8.2.1 Short review

Wall-bounded turbulent flow is considered to be one of the most challenging problems in fluid mechanics and has attracted lots of attentions for a long time due to its paramount importance to both theoretical research and engineering applications such as flow control and drag reduction (Kurz et al., 2013; Li and Liu, 2017). It has been found that there exist many spatially coherent, temporally evolving vortical motions, also referred to as coherent structures in wall-bounded turbulent and transitional flows. Theodorsen (1952) first proposed the horseshoe vortex model to describe the experimental observation in wall-bounded turbulent flow. Kline et al. (1967) found that the long streamwise streaks of hydrogen bubbles in the near-wall region by experiment and that the spanwise spacing of these streaks were about 100 wall units. They believed that the instability of these streaks plays a vital role in turbulence generation. Falco (1977) gave a well-known visualization of a low Reynolds number turbulent boundary layer that illustrates several known types of coherent structures. Robinson (1991) believed that the coherent structure is responsible for the production and dissipation of turbulence in a boundary layer, and the study of turbulence structures is extremely important to understand and control the turbulent boundary layer. Working with Kline, Robinson (1989) gave a summary of the structures; he observed that quasi-streamwise vortices are located close to the wall, arches, or horseshoe vortices in the wake region and a mixture of quasi-streamwise vortices and arches in the logarithmic layer. Liu and Liu (1995, 1997), Bake et al. (2002), and Wu and Moin (2009) obtained fully developed turbulent flow with a forest of hairpin vortices through DNS. Scaling theories coupled with the notion of coherently organized motions were first proposed by Townsend (1956, 1976) and advanced by Perry and Chong (1982). They gave a model of individual hairpin vortices scattered randomly in the flow and found that the vortices are statistically independent of each other. They concluded the wall layer as a forest of single layer hairpin vortices, which can be modeled with simplified shapes in a hierarchy of scales above the wall. Yan et al. (2014a, b) and Yan and Liu (2014) found that the hairpin vortex is a combination of the Λ-vortex roots and vortex rings. Λ-vortex roots and vortex rings are formed separately, and the mechanism of the Λ-vortex self-deformation to hairpin vortex does not exist. Liu et al. (2014) believed that turbulence is an inherent property of fluid flow, although the external disturbance is needed. The nature of turbulence generation is that fluids, away from the wall, cannot tolerate the shear, and shear must transfer to rotation. Therefore, shear layer instability is the mother of turbulence. Adrian (2007) believed hairpins could be autogenerated to form packets that populate a significant fraction of the boundary layer, and he addressed the important role of hairpin vortex ejections and sweeps. Marusic and McKeon (2010) mentioned that there is still a dichotomy on whether the hairpin vortex exists or not. Schoppa and Hussain (2002) thought complete hairpin vortices do not exist in wall-bounded turbulence. Although considerable efforts have been devoted to such a problem, the mechanism of wall-bounded turbulence generation and sustenance remains a mystery. Therefore, more work must be conducted to get a deep understanding about the vortex structure of the low Reynolds number turbulent boundary layer.

In this section, a high-order DNS with nearly 60 million grid points and 400,000 time steps is performed to investigate the coherent structure for the low Reynolds number turbulent boundary layer flow. Meanwhile, a new Lagrangian property experiment technique, which is named as the moving single frame and long exposure imaging (MSFLE) method, is carried out for measuring the coherent structure in a turbulent boundary layer. The qualitative comparison of numerical simulation and experiment will provide a deeper understanding on the coherent structures of a transitional and turbulent boundary layer with low Reynolds number, especially for multilevel vortex structures.

8.2.2 Case setup

8.2.2.1 DNS case setup

Fig. 8.4 shows the computational domain where x, y, z represent the streamwise, spanwise, and wall-normal directions, respectively. There are $1920 \times 128 \times 241$ grid points in the computation domain. The points are distributed uniformly in the streamwise and spanwise directions and stretched in the wall-normal direction to ensure the grid has enough resolution to capture all small length scales. The first grid interval is set to 0.43 in wall units ($z^+ = 0.43$) in the normal direction. Table 8.1 gives the details of the flow conditions including Mach number, Reynolds number, and so on. Here δ_{in} is the inflow displacement thickness, and other parameters are nondimensionalized by δ_{in} as reference length. L_x and L_y are the length of

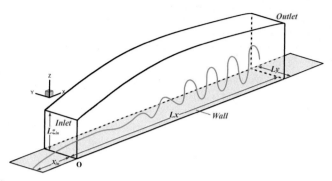

FIGURE 8.4

Computational domain sketch.

Reproduced from Chapter 7 of Bentham books, Liutex-based and other mathematical, computational and experimental methods for turbulence structure, Volume 2, ISSN: 2589-2711, ISBN: 978-981-14-3758-8, April 2020 with the permission of Bentham Books.

computational domain in the x (streamwise) and y (spanwise) directions, and $L_{z_{in}}$ is the height of the inlet. x_{in} is the distance between the leading edge and inlet. T_w and T_∞ represent the wall temperature and freestream temperature. The Reynolds number of the inlet is defined as $Re = \rho_\infty U_\infty \delta_{in}/\mu_\infty$, where ρ_∞ is the density, U_∞ the velocity at the freestream, and μ_∞ the dynamic viscosity.

The inflow boundary condition is given by the following equation

$$Q = Q_{\text{Blasius}} + A_{2d}q'_{2d}e^{i(\alpha x - \omega t)} + A_{3d}q'_{3d}e^{i(\alpha x \pm \beta y - \omega t)} \tag{8.5}$$

where Q contains three velocity components, pressure, and temperature, and Q_{Blasius} represents the Blasius solution. A_{2d} and A_{3d} are amplitudes of the 2D and 3D inlet disturbance, ω is the frequency of inlet disturbance, α and β are streamwise and spanwise wave number of the inlet disturbance. The second and third terms denote the 2D and 3D Tollmien−Schlichting (T-S) waves. The nonslipping and adiabatic condition is applied to the wall and the nonreflecting condition is applied for the outflow boundaries. Periodic conditions are enforced at the spanwise boundaries.

In the streamwise and normal directions, a six-order compact scheme (Lele, 1992) is applied for the spatial discretization. A pseudo-spectral method is used in the spanwise direction. In addition, to eliminate the spurious numerical oscillations caused by the central difference scheme, a high-order spatial scheme filtering is adopted. A third order total variation diminishing (TVD) Runge−Kutta scheme (Shu, 1988) is employed.

Table 8.1 Flow parameters.

M_∞	Re	x_{in}	L_x	L_y	$L_{z_{in}}$	T_w	T_∞
0.5	1000	$300.79\delta_{in}$	$780.03\delta_{in}$	$22\delta_{in}$	$40\delta_{in}$	273.15K	273.15K

8.2.2.2 Experiment setup

Although the DNS results for the late boundary layer transition are well validated, it is still needed to compare the vortex structure qualitatively with experiment for early turbulent flow with low Reynolds number. According to the velocity profile, flow in the last section of the DNS domain after $x = 900$ has become turbulent (see Fig. 8.5) and is comparable with the experiment work for turbulent flow with the low Reynolds number.

The MSFLE method is based on the single frame and single exposure imaging (SFSE) method (Chen and Cai, 2015). SFSE can record the trajectory of tracer particles clearly by setting proper exposure time.

The experiment is conducted in a low-speed circulating water tunnel. The size of the water tunnel is 2500 mm × 104 mm × 90 mm. The definition of the coordinate system and measurement system is shown in Fig. 8.5. Coordinates x, y, z are defined as the streamwise, spanwise, and wall-normal directions, respectively, and the origin point is located at the front of the water tunnel. A plexiglass plate is placed in the middle of the water tunnel, and the size of the plate is 1500 mm × 78 mm × 4.7 mm, which is located from $x = 700$ mm to $x = 2200$ mm. The leading edge of the plate is shaped as an ellipse with a ratio of

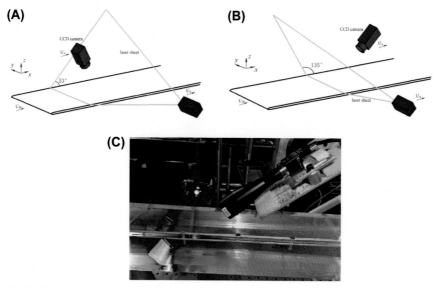

FIGURE 8.5

Definition of the coordinate system and measurement system. (A) Laser sheet placed with 53 degrees inclination to the flow direction. (B) Laser sheet placed with 135 degrees inclination to the flow direction. (C) Picture of experimental equipment with laser sheet placed at 135 degrees.

Reproduced from Chapter 7 of Bentham books, Liutex-based and other mathematical, computational and experimental methods for turbulence structure, Volume 2, ISSN: 2589-2711, ISBN: 978-981-14-3758-8, April 2020 with the permission of Bentham Books.

4:1 for the major axis to the minor axis as shown in Fig. 8.5. Tracer particles are hollow glass microspheres with an average diameter of 20 μm, and the density of tracer particles is 0.6 g/mL. The light source is 450 nm wavelength continuous laser diode, which produces a 1 mm thick laser sheet. Images are captured by a CMOS camera, which has a resolution of 1280 × 1024 pixels. The entire measurement system is installed on a horizontal guide rail, which can move along the streamwise direction.

The laser sheet is placed in two different positions as shown in Fig. 8.5. In Fig. 8.5A, the angle between the laser sheet and flow direction is 53 degrees; the camera is perpendicular to the laser sheet. In Fig. 8.5B, the angle is increased to 135 degrees, but the camera stays perpendicular to the laser sheet. Fig. 8.5C shows the picture of the experimental equipment with laser sheet placed at 135 degrees to the streamwise direction.

The test conditions can be described as follows. Inflow speed is $U_\infty = 85$ mm·s^{-1}. Because the velocity inside the boundary layer is slower than the main flow, the speed of the camera is set as $U_c = 0.94U_\infty$. Measuring range is from $x = 1440$ mm to $x = 2190$ mm, and the Reynolds numbers of these two positions are $Re_\theta = \theta U_\infty / \nu = 429$ and 750, respectively, where θ is the momentum thickness of the boundary layer and ν is the coefficient of the kinematic viscosity.

During these experiments, the camera captured the coherent structure of flow continuously with the MSFLE method. The single frame exposure time is 200 ms, and the time interval between two frames is less than 70 μs The magnification factor of the camera is 0.14, the sensor resolution is 1280 × 1024 pixels, and the size of the pixel is 4.8 × 4.8 μm^2. Therefore, the active area is 6.144 × 4.915 mm^2, and the field of view is 43.89 × 35.11 mm^2.

8.2.3 Results and discussions

8.2.3.1 Comparison between DNS and experiment results

Note that the experimental setup is not exactly the same as DNS's setup especially in the inflow conditions, but both can be classified as a "low Reynolds number turbulent flow." The side wall influence in the experiment can be ignored in the middle of the test section where the flow can still be considered as a turbulent boundary layer while DNS assumes a periodic boundary condition in the spanwise direction. Since turbulent flow is fluctuated, it is difficult to ensure the flow initial condition and inflow boundary condition are the same for both experiment and DNS. Quantitatively instantaneous comparison is not realistic, but qualitative comparison is possible.

Fig. 8.6A shows the coherent structures in a low Reynolds number but fully developed turbulent boundary layer obtained by the MSFLE experimental method; the laser sheet is placed with 53 degrees inclination to the flow direction. Fig. 8.6B gives the particle trace lines calculated by the DNS data. Nearly 30,000 seed points were placed, and the points were picked from one plane, which has the same direction as the laser sheet. In Fig. 8.6A, the bright dots indicate the tracer particles moving mainly in the streamwise direction, and the long bright lines denote the tracer particles moving in the spanwise or wall-normal direction since a dot means the

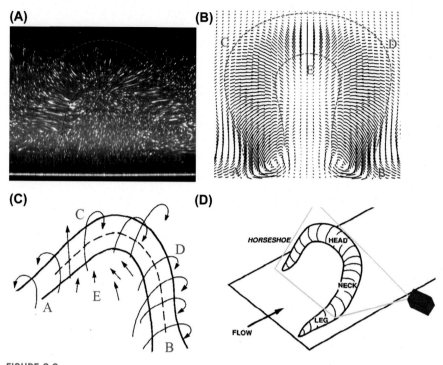

FIGURE 8.6

(A) Horseshoe vortex structure obtained by MSFLE method. (B) Horseshoe vortex structure obtained by DNS data. (C) Sketch of five different sections of the horseshoe vortex and velocity directions. (D) Sketch of the relative position of the laser sheet and horseshoe vortex.

(A) and (B) reproduced from Chapter 7 of Bentham books, Liutex-based and other mathematical, computational and experimental methods for turbulence structure, Volume 2, ISSN: 2589-2711, ISBN: 978-981-14-3758-8, April 2020 with the permission of Bentham Books. (C) modified from Theodorson (1952); (D) modified from Robinson (1991).

particle moving direction is orthogonal to the laser sheet. The structure obtained in Fig. 8.6A is a horseshoe structure. To demonstrate this, we divided the horseshoe vortex into five different sections: sections A and B representing left and right legs, sections C and D for left and right necks, and section E for the center of the horseshoe vortex as shown in Fig. 8.6C. Fig. 8.6D shows the relative position between the laser sheet and the horseshoe vortex. The laser sheet intersects through the upper part of the horseshoe vortex. If a horseshoe vortex shown in Fig. 8.6C is cut by a plane shown in Fig. 8.6D, then the flow directions of these five different sections are determined. It is easy to see that the local flow direction in section A is toward to the left, to the right in section B, to the upper left in section C, to the upper right in section D, and opposite to the streamwise direction in section E Two vortex cores at the bottom of sections A and B are clearly found in Fig. 8.6A. This is because the laser sheet cut through the left and right legs of the horseshoe. The horseshoe-like structure is also obtained by the DNS result as shown in Fig. 8.6B. The result is similar to the one obtained by the MSFLE experiment.

Fig. 8.7A gives the MSFLE experimental result with the laser sheet placed at 135 degrees inclination to the flow direction. Fig. 8.7B gives the particle trace calculated by the DNS data with seed points picked from a plane that has the same direction as the laser sheet. Fig. 8.7B is colored by the Omega method and the red area represents a large value of Omega, which most likely captures the vortex cores. The vortices denoted by the particle trace lines are in accordance with the vortex cores represented by the Omega contour. As shown in Fig. 8.7A, seven streamwise vortices are captured by the MSFLE experiment, and these vortices are believed to be the legs of the hairpin vortices. It is found that the vortices are located in different layers. A total of three layers of vortices can be observed: vortices 1, 4, and 7 lie in the first layer near to the wall; vortices 2 and 5 are observed in the middle layer; and vortices 3 and 6 are found on the top layer. This phenomenon is also observed by the DNS result as shown in Fig. 8.7B. Similar multilevel vortex structures are obtained by calculating particle trace lines. Totally 8 vortices are captured. Vortices 1 and 8 lie in the first layer near the wall; vortices 2, 4, 5, and 7 appear in the middle layer; and vortices 3 and 6 are in the top layer.

8.2.3.2 Multilevel vortex structures
Multilevel vortex structures are observed in the low Reynolds number turbulent boundary layer flow.

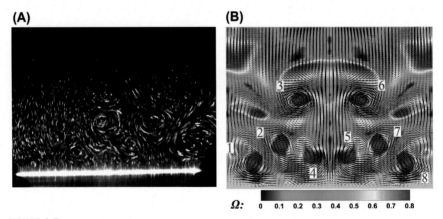

FIGURE 8.7

(A) Multilevel hairpin vortex legs obtained by the MSFLE method with laser sheet placed at 135 degrees to intersect the streamwise direction. (B) Multilevel hairpin vortex legs obtained by the DNS data colored by the Ω contour.

Reproduced from Chapter 7 of Bentham books, Liutex-based and other mathematical, computational and experimental methods for turbulence structure, Volume 2, ISSN: 2589-2711, ISBN: 978-981-14-3758-8, April 2020 with the permission of Bentham Books.

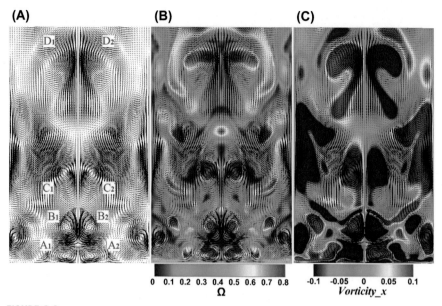

FIGURE 8.8

(A) Multilevel vortex structures identified by trace lines for $x = 913$ in a wall-normal and spanwise plane. (B) Multilevel vortex structures colored by Omega. (C) Multilevel vortex structures colored by the x-direction component of vorticity.

Reproduced from Chapter 7 of Bentham books, Liutex-based and other mathematical, computational and experimental methods for turbulence structure, Volume 2, ISSN: 2589-2711, ISBN: 978-981-14-3758-8, April 2020 with the permission of Bentham Books.

Fig. 8.8A shows the particle trace lines calculated by the DNS data with the seed points picked from a vertical plane located at $x = 913$. As seen, a total of four levels of vortices can be observed. In the middle of the plane, there are four pairs of hairpin vortex legs denoted by $A_1 - D_1$ and $A_2 - D_2$. In the following, it is here only focused on the left legs of these vortices denoted by $A_1 - D_1$. Fig. 8.8B is the trace line colored by the Omega method. The particle trace lines coincide with the vortex cores indicated by the Omega method (the red areas). It is easy to find that the rotating directions of the two vortex legs A_1 and A_2 are opposite. The other three pairs of vortices show the same trend. However, it is difficult to distinguish the vortex rotating direction by the trace line. In such a case, vorticity is used to distinguish the clockwise vortices and counterclockwise vortices as $\vec{\omega} \cdot \vec{R} > 0$ is part of Liutex definition. Fig. 8.8C is the trace line colored by the x-direction component of the vorticity. The red color denotes that the vortex rotates counterclockwise, and the blue color means that the vortex rotates clockwise. It is evident that the same pair of vortex legs, A_1 and A_2 for example, has different colors, which means they are counter-rotating pair. Apparently, the results coincide with the analysis earlier. So, it is appropriate to identify the hairpin or Lambda vortex rotational direction (clockwise or counterclockwise) by the x-direction component of the vorticity.

Fig. 8.8C shows that the rotating directions of the first level vortex A_1 and second level vortex B_1 are opposite in the rotation direction. However, the upper three levels of the vortices B_1, C_1, and D_1 have the same rotating directions, and they are all counterclockwise vortices. It means that the lower vortices are not simply induced by the upper-level vortices, contradicting to what proposed by Adrian (2007). If the lower-level vortices are simply induced by upper-level vortices, the rotating direction should be opposite between two levels, which is clearly different from the observation.

From both Figs. 8.7 and 8.8, it is observed that in the low Reynolds number turbulent boundary layer flow, the size of the upper-level vortex is not twice the lower level vortex, and sometimes they have almost the same size. The eddy cascade phenomenon indicated by Richardson (1922) is not observed in the low Reynolds number turbulent boundary layer flow. Therefore, it is hard to conclude that the energy is transported from the large vortex to small vortices through vortex breakdown process that has not been found by either experiment or DNS.

8.2.3.3 Multilevel vortices ejections and sweeps

The observation here shows that the energy of vortices inside the boundary layer comes from the main flow. The question is how the energy is transported from the main flow to the boundary layer and how the new vortices are generated. Kolmogorov (1941a) proposed that the large eddies pass energy to small eddies through "vortex breakdown." However, it has never been observed in the transitional and turbulent boundary layer yet. In the earlier experiment and DNS results, there is no vortex breakdown phenomenon observed in the low Reynolds number turbulent boundary layer flow. Adrian (2007) demonstrated that there are ejections and sweep motions inside the boundary layer, which are induced by the hairpin vortices. Ejection represents the negative streamwise fluctuations, which lifts flow away from the wall. Thus, the low-speed flow near the wall is ejected into the high-speed upper zone. Sweep represents the positive streamwise fluctuations, which brings the high-speed fluid toward the wall. Thus, the high-speed flow sweeps into the low-speed area near the wall. This phenomenon has been observed by the DNS data as shown in Fig. 8.9.

Fig. 8.9A displays the sketch of a single hairpin vortex structure (identified by the Omega iso-surface with $\Omega = 0.52$) colored by wall-normal velocity, and the vector field denotes the sweeps and ejections around the vortex legs. Fig. 8.9B shows the sweeps and ejections around the vortex neck. The red areas mean that the fluid is going up to the main flow corresponding to the ejection event. The red areas mainly locate at the inboard region of the vortex legs. The velocity vector clearly shows the hairpin vortex ejection event exists near the inboard region of the vortex legs and necks. The blue areas indicate that the fluid is going down to the wall corresponding to the sweep events. The sweep events mainly happen in the outboard region of the hairpin vortex legs.

Five monitoring points from the hairpin vortex are picked as shown in Fig. 8.9C, and the wall-normal velocities of these monitoring points are listed in Table 8.2: P_1 is

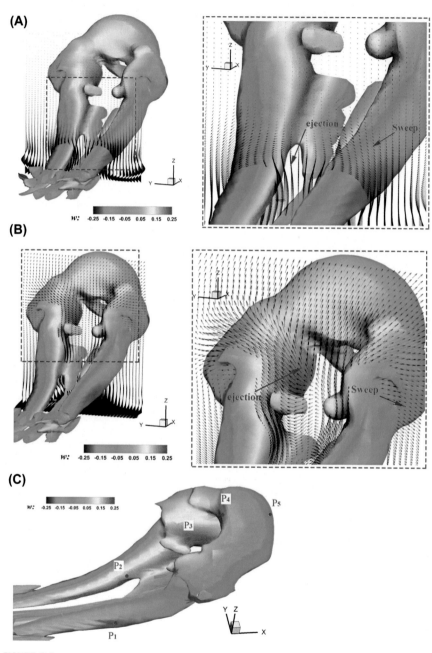

FIGURE 8.9

placed in the outboard region of the vortex legs, showing the sweep strength of vortex leg; P_2 and P_3 are placed in the inboard region of the hairpin vortex, representing the ejection strength of vortex leg and neck; P_4 and P_5 represent the ejection and sweep strength of the hairpin vortex rings. As seen, the wall-normal velocities of vortex legs and necks are $-0.21U_\infty, 0.19U_\infty$, and $0.22U_\infty$, respectively. These magnitudes are about twice as large as the wall-normal velocities in the area of the vortex ring, which are $0.1U_\infty$ and $-0.06U_\infty$. The ejection or sweep strength is much stronger in the hairpin legs and necks than in the ring area.

Fig. 8.10A shows three vortices that are selected in two levels. Vortices B and C are in the first level and vortex A is in the second level. These three vortices are colored by wall-normal velocity. Fig. 8.10B is half of the same vortices shown in Fig. 8.10A, which is cut through at the central plane.

A total of 7 monitoring points are placed as shown in Fig. 8.10B: monitoring points P_1–P_3 are placed on the inboard region of the three vortex legs and monitoring points P_4–P_7 are placed in the rings of vortex A and B, respectively. The wall-normal velocities of these monitoring points are listed in Table 8.3. It can be seen that the wall-normal velocities of P_2 and P_3 are positive and wall-normal velocity of P_1 is negative, which means the rotational directions of the left legs of vortex A and B are counterclockwise, and the rotating direction of the left leg of vortex C is clockwise as denoted by the black arrow in Fig. 8.10B. The rotational directions of the first level vortex C and the second level vortex A are opposite, which is coincident with the result shown in Fig. 8.13. From Table 8.3, it can also be found that the wall-normal velocity strengths of vortex legs are $-0.33U_\infty, 0.22U_\infty$, and $0.12U_\infty$, respectively. These are about three to five times stronger than the rings, which are $0.04U_\infty, -0.06U_\infty, 0.1U_\infty$, and $-0.08U_\infty$. The observation clearly shows the ejection or sweep strength is much stronger in the hairpin leg and neck areas than in the ring area.

Vortex C does not have a vortex ring because vortex C is a clockwise rotational vortex, which cannot form the ring head since the ring must be a counterclockwise vortex due to the profile of the Blasius velocity. The rotating directions of the rings of vortices A and B are denoted by the black arrow in Fig. 8.10A. The upper ring head is higher and its velocity must be larger than the lower part, and the vortex ring can only rotate in the direction shown in Fig. 8.10A. If the legs rotate in the opposite direction just like vortex C, the vortex ring cannot rotate in an opposite direction as a part of the hairpin vortex, which would be impossible for vortex C to

Hairpin vortex is identified by iso-surface with $\Omega = 0.52$ and colored by wall–normal velocity. (A) Ejection and sweeps near the hairpin vortex legs. (B) Ejection and sweeps near the hairpin vortex neck. (C) Sketch of five monitor points.

Reproduced from Chapter 7 of Bentham books, Liutex-based and other mathematical, computational and experimental methods for turbulence structure, Volume 2, ISSN: 2589-2711, ISBN: 978-981-14-3758-8, April 2020 with the permission of Bentham Books.

Table 8.2 Wall-normal velocity of five monitoring points.

	P_1	P_2	P_3	P_4	P_5
V_z/U_∞	−0.21	0.19	0.22	0.1	−0.06

Reproduced from Chapter 7 of Bentham books, Liutex-based and other mathematical, computational and experimental methods for turbulence structure, Volume 2, ISSN: 2589-2711, ISBN: 978-981-14-3758-8, April 2020 with the permission of Bentham Books.

have a vortex ring. This would lead to a velocity of the upper ring area smaller than the one in the lower ring area. Therefore, clockwise rotation vortex legs cannot form a vortex ring. It also proves that the hairpin vortex is not a single vortex, but composed of two different parts: the vortex legs and the vortex ring.

Fig. 8.11A shows the same vortex structures as shown in Fig. 8.8, which are colored by the wall-normal velocity. The red area means the fluid is going up to the main flow, and the blue area indicates the fluid is going down to the wall. Fig. 8.11B is colored by the streamwise velocity and Fig. 8.11C shows the positions of 8 monitoring points. The wall-normal velocity and streamwise velocity of the eight monitoring points are listed in Table 8.4.

As seen in Fig. 8.11A, the vortex center is located between the red and blue area, and from Fig. 8.8 it is easy to find that the vortices in the upper three level B_1, C_1, and D_1 are counterclockwise vortices, but the first level vortex A_1 is a clockwise vortex. The monitoring points P_1, P_2, P_3, and P_4 are located in the inboard region of vortex legs. From Table 8.4, it can be seen that the wall-normal velocities of P_2, P_3, and P_4 are upward. Thus, the low-speed flow in the inboard region of second-level vortex

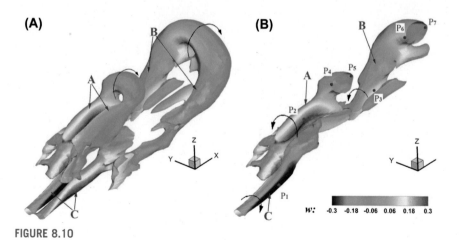

FIGURE 8.10

(A) Two-level hairpin vortex structures colored by wall-normal velocity (B) half of the vortex structure cut by the central plane.

Reproduced from Chapter 7 of Bentham books, Liutex-based and other mathematical, computational and experimental methods for turbulence structure, Volume 2, ISSN: 2589-2711, ISBN: 978-981-14-3758-8, April 2020 with the permission of Bentham Books.

Table 8.3 Wall-normal velocity of seven monitor points.

	P_1	P_2	P_3	P_4	P_5	P_6	P_7
V_z/U_∞	−0.33	0.22	0.12	0.04	−0.06	0.1	−0.08

Reproduced from Chapter 7 of Bentham books, Liutex-based and other mathematical, computational and experimental methods for turbulence structure, Volume 2, ISSN: 2589-2711, ISBN: 978-981-14-3758-8, April 2020 with the permission of Bentham Books.

legs (B_1 and B_2) is ejected upward. Therefore, because the rotation direction of the third level vortices (C_1 and C_2) is the same as the second level, and the low-speed flow lifted by the second-level vortex will continue to be lifted up by the ejection of the third-level vortices and then be lifted up a third time by the fourth-level vortices (D_1 and D_2). Therefore, the low-speed flow between the vortex legs near the wall is lifted up to the main flow by the multilevel vortices, which are corotating with the clockwise rotation. In Fig. 8.11B, the low-speed zone inboard of the two vortex legs can be observed. The streamwise velocities of points P_2, P_3, and P_4 are $0.43U_\infty$, $0.64U_\infty$, and $0.77U_\infty$, respectively, which are obviously smaller than the neighboring regions. This is a key point to understand why the flow streaks or

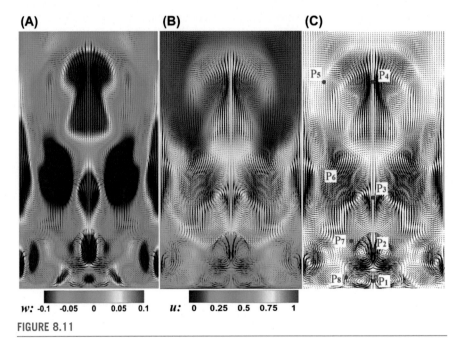

(A) **(B)** **(C)**

w: -0.1 -0.05 0 0.05 0.1 u: 0 0.25 0.5 0.75 1

FIGURE 8.11

Multilevel vortex structures colored by (A) wall-normal velocity. (B) Streamwise velocity. (C) Sketch of eight monitoring points.

Reproduced from Chapter 7 of Bentham books, Liutex-based and other mathematical, computational and experimental methods for turbulence structure, Volume 2, ISSN: 2589-2711, ISBN: 978-981-14-3758-8, April 2020 with the permission of Bentham Books.

Table 8.4 Wall-normal velocity and streamwise velocity of monitoring points.

	P₁	P₂	P₃	P₄	P₅	P₆	P₇	P₈
V_z/U_∞	−0.11	0.27	0.18	0.23	−0.02	−0.15	−0.06	0.17
V_x/U_∞	0.39	0.43	0.64	0.77	0.995	0.85	0.8	0.22

Reproduced from Chapter 7 of Bentham books, Liutex-based and other mathematical, computational and experimental methods for turbulence structure, Volume 2, ISSN: 2589-2711, ISBN: 978-981-14-3758-8, April 2020 with the permission of Bentham Books.

white and black stripes can be formed, which is caused by the multilevel ejections. The black/white strips strictly require multilevel vortices to have the same rotation direction (clockwise, for example), challenging the simple inducement theory (Adrian, 2007) because the simple inducement theory must lead to vortices with opposite rotation directions.

Now let us consider outboard regions of the hairpin vortex legs. Points P_5, P_6, P_7, and P_8 are located in the outboard region of vortex legs. From Table 8.4, it can be seen that the fluids at points P_2, P_3, and P_4 are going down to the wall and represent the sweep event. The high-speed main flow is brought toward the wall by the top-level vortex D_1; then the third-level vortex C_1 continues to drive the high-speed flow toward the wall and so does the second-level vortex B_1. In this way, the high-speed flow near the outboard hairpin vortex legs is brought down three times to the near-wall area by the multilevel vortices (see Fig. 8.8). In Fig. 8.11B, the high-speed zone outboard of the second and third-level vortices can be observed. The streamwise velocities of point P_5, P_6, and P_7 are $0.995U_\infty$, $0.85U_\infty$, and $0.8U_\infty$, respectively, which are obviously greater than their neighboring regions. In this way, the energy is transported by the multiple sweeps from the main flow into the different lower levels of the turbulent boundary layer. Again, these new observations about the sweeps also challenge the simple vortex inducement theory.

The multilevel ejections and sweeps mentioned earlier not only transport the energy from the main flow to the boundary layer but also form a lot of strong shear layers. The fluid cannot tolerate high shear and the shear layer must transfer into the vortices. So, more small vortices will be developed due to multilevel hairpin vortices ejections and sweeps. It is confirmed that small-scale vortices are not created by large vortex breakdown and the energy is not transported by the eddy cascade. Both eddy cascade and vortex breakdown are not observed by either experiment or DNS results in the turbulent boundary layer flow with low Reynolds number. Therefore, none of them can be confirmed.

One thing must be explained here that although the last section of the DNS domain can be considered as near the fully developed turbulent flow based on the velocity profile, the DNS results provided here are mostly in the late flow transition stage and, therefore, are still almost symmetric. Note that there is no symmetric turbulence unless in a statistic sense.

8.2.4 Summary

An elaborate DNS for late boundary layer transition has been performed. The DNS results are qualitatively compared with a new Lagrangian property experimental technique named the moving single frame and long exposure imaging (MSFLE) method to obtain a deeper understanding on the coherent structures of a transitional and turbulent boundary layer with low Reynolds number. Multilevel vortex structures are clearly observed by both experiment and DNS. It is found that there are multilevel corotating vortices, showing how energy is transported from the main flow to the bottom of the boundary layer. The results also show that the lower-level vortices cannot simply be produced by the upper-level vortex inducement. There are multilevel hairpin vortex ejections and sweeps inside the boundary layer. The ejections and sweeps are much stronger around the hairpin legs and necks than those in the ring area. This clearly shows that the ring-like vortices are the production of the strong vortex neck rotation. In conjunction with ejections and sweeps, a lot of strong shear layers are produced. Because fluid cannot tolerate the strong shear, the shear layer must turn to rotation, or S transforms to Liutex and forms many vortices. This would help to reveal the mechanism of multilevel vortices and turbulence generation. Although the upper-level vortices could be larger than the lower ones, the lower-level vortices sometimes have the same size as the neighboring upper-level vortices. The vortex cascade and large vortex breakdown are not observed by either DNS or experiment.

It is very encouraging that the experimental observations match the DNS visualization closely in the coherent vortex structure of a fully developed turbulent boundary with low Reynolds number. The DNS visualization matching the experimental observation demonstrates that our results are reliable qualitatively in the transient sense although the case of DNS is pretty much a late flow transition and the experiment should be low Reynolds number turbulence. They are still matching qualitatively.

Based on the DNS simulation and MSFLE experiment, the following conclusions can be made:

1. Multilevel hairpin vortex structures are observed by both the experiment and DNS results. The upper three-layer vortices rotate in the same direction. These new findings on corotating vortices provide evidence that the lower-level vortices cannot be simply induced by the higher-level vortices.
2. In the low Reynolds number turbulent boundary layer flow, the size of the upper layer vortices is not twice as big as the lower layer vortices; sometimes the upper- and lower-level vortices are nearly equal in size. The Kolmogorov's vortex breakdown process and Richardson's eddy cascade phenomena are not observed.
3. There are multilevel hairpin vortex ejections and sweeps inside the turbulent boundary layer. The strength of ejections or sweeps is much higher near the hairpin legs and necks than in the ring area. The low-speed flow near the wall is lifted up by multiejection events, and the high-speed flow is brought toward the wall by multilevel sweeps. The energy is transported from the main flow to the boundary layer by the multilevel sweep process.

4. Multilevel hairpin vortex ejections and sweeps are responsible for the formation of streamwise streaks; these low-speed streaks are the result of the multilevel vortex ejection process rather than the driving force of the formation of hairpin vortices.

5. Multilevel vortex ejections and sweeps will produce many strong shear layers. However, fluid cannot tolerate high shear and the shear layers will turn into vortices, which could reveal the mechanism of multilevel vortices generation.

8.3 Mathematical foundation of turbulence generation

8.3.1 Short review

Turbulence is generally acknowledged as one of the most complex phenomena in nature (Batchelor, 2000; Tennekes and Lumley, 1972). In 1883, Reynolds revealed the complex flow pattern through his famous round tube experiment. After that, a large number of scientists engaged in turbulence research have solved a large number of engineering problems. However, due to the complexity of turbulence, the universal mechanism of turbulence generation is yet to be fully understood. This leads to a variety of turbulence generation theories. Richardson described vortex cascade generated by large vortex breakdown (1922). But from many DNSs, the vortex cascade has never been observed. Kolmogorov accepted the idea of the Richardson energy cascade and vortex breakdown and thought that the large eddies passed energy to small eddies through vortex breakdown, then continue to smaller scale, until the Kolmogorov's smallest scale (1941a). However, nowadays according to the most accurate experimental equipment, no one can confirm that turbulence is caused by vortex breakdown. Currently, no mathematical principle has been found to explain the generation of turbulence, especially the occurrence and evolution of asymmetric structures in the boundary layer flow transition process. This section attempts to provide a mathematical foundation of turbulence generation (Liu et al., 2019a) based on the bias definition of the Liutex vector.

Since turbulence is closely related to the vortex, it can be concluded that there exists no turbulence without Liutex. According to DNSs and experiments, forest of hairpin vortices has been found in transitional and low Reynolds number turbulent flows, while one-leg vortices are predominant in fully developed turbulent flows. This section demonstrates that the symmetric hairpin vortex is unstable. The hairpin vortex will be weakened or lose one leg by the shear and Liutex interaction, based on the Liutex definition and mathematical analysis without any physical assumptions. The asymmetry of the vortex is caused by the interaction of symmetric shear and symmetric Liutex since the smaller element of a pair of vorticity elements determines the rotational strength. For a 2D fluid rotation, according to the Liutex strength definition, if a shear disturbance affects the larger element of the nondiagonal elements, the rotation strength will not be changed, but if the shear disturbance affects the smaller element, the rotation strength will be immediately changed due to the bias definition of the Liutex strength. For a rigid rotation, if the shearing part of the vorticity and Liutex present the same directions, e.g. clockwise, the Liutex strength will not be changed. If the shearing part of the vorticity and Liutex present different directions, e.g., one clockwise and another counterclockwise, the Liutex strength will be weakened. Consequently, the hairpin vortex could lose the symmetry and even deform to a one-leg vortex. The one-leg vortex cannot keep balance, and the chaotic motion and flow fluctuation are doomed. This is considered as the mathematical foundation of turbulence formation. The DNS results of boundary layer transition are used to justify this theory (Liu et al., 2019a).

8.3.2 Necessary condition for turbulence

Turbulence or turbulent flow is a common type of fluid motion characterized by chaotic changes in pressure and flow velocity. Turbulence is commonly observed in everyday phenomena and most realistic engineering flows. Feynman (1955) has described turbulence as the most important unsolved problem in classical physics. Turbulent flow is very irregular, diffusive, dissipative, and chaotic. Vortex is the building block and muscle of turbulence with variety of vortices of different sizes and rotational strengths (Liu et al., 2014). Without vortex, there would be no turbulence. Without asymmetric vortex, there would be no turbulence.

8.3.3 Interaction of 2D shear and Liutex

A simple shear tensor can be described as $\begin{bmatrix} 0 & s \\ 0 & 0 \end{bmatrix}$ or $\begin{bmatrix} 0 & 0 \\ s & 0 \end{bmatrix}$ for 2D. Let us add the shear disturbance to the channel flow, we will get $\begin{bmatrix} 0 & -2y+s \\ 0 & 0 \end{bmatrix}$ or $\begin{bmatrix} 0 & -2y \\ s & 0 \end{bmatrix}$ Although R is zero at the beginning, the first shear does not generate Liutex, but the second one does. This clearly shows shear may or may not generate rotation depending on the shear direction. If the shear disturbance increases the larger element of the base shear/Liutex, there is no change in the rotation strength. If the shear disturbance decreases the smaller element of the base shear/Liutex, the rotation strength will be reduced. Consider the interaction of a shear and a rigid rotation in which the tensor of the rigid rotation is $\begin{bmatrix} 0 & \omega \\ -\omega & 0 \end{bmatrix}$ and the shear is $\begin{bmatrix} 0 & s \\ 0 & 0 \end{bmatrix}$ By adding them together, it will be $\begin{bmatrix} 0 & \omega+s \\ -\omega & 0 \end{bmatrix}$. Assume ω is positive, there are two totally different cases: (A) If s is positive, R will have no change no matter how big s is, according to the definition of $R = 2\min\left\{\left|\frac{\partial u}{\partial y}\right|, \left|\frac{\partial v}{\partial x}\right|\right\}$; (B) If s is negative, there are several possibilities: (1) If $|s| < \omega$, then $R = 2\{\omega - |s|\}$; (2) if $|s| = \omega$, then $R = 2\{\omega - |s|\} = 0$; (3) if $|s| > \omega$, then $R = 0$ since $(-\omega)(\omega+s) > 0$. From this example, it can be clearly found that the interaction of shear and Liutex could make no changes on R if the shear disturbance has the same direction as the rotation; however, R will be reduced or even the fluid rotation is stopped if the shear has the opposite direction from the rotation. Now, consider another example with $\begin{bmatrix} 0 & \omega + s_1 \\ -\omega & 0 \end{bmatrix} + \begin{bmatrix} 0 & s_2 \\ 0 & 0 \end{bmatrix}$. Assume ω and s_1 are positive: (1) If s_2 is positive, $R = 2\omega$ will have no change; (2) If $s_2 < 0$ and $|s_2| < |s_1|$, then $R = 2\omega$ has no change either; (3) If $s_2 < 0, |s_2| > |s_1|$ and $\omega + s_2 + s_1 > 0$, then $R = 2(\omega + s_2 + s_1)$, otherwise $R = 0$. From these examples, several questions and answers arise: (1) does shear change Liutex? Not sure and it depends; (2) if shear has the same sign as the rotation, the rotation strength will not change; (3) if the shear and rotation have opposite signs with $|s_2| < |s_1|$, the rotation strength will still not change; (4) if the shear and rotation have opposite signs with $|s_2| > |s_1|$, the rotation strength will change. This gives a clue that the interaction of shear and Liutex is not reversible: the increase of shear does not increase the strength of rotation if the larger element increases, but the decrease of shear may reduce the strength of rotation. In any cases, the change of smaller element by shear will cause the change of the rotation strength. Although these are analyses for 2D cases, this basic physics of shear and Liutex interaction should be similar to 3D cases.

8.3.4 From hairpin vortex to one-leg vortex

Consider the interaction of shear and Liutex in 3D flows. The tensor formula could be:

$$R = \begin{bmatrix} 0 & -\frac{1}{2}R & 0 \\ \frac{1}{2}R & 0 & 0 \\ 0 & 0 & \lambda_r \end{bmatrix}, \; S = \begin{bmatrix} 0 & 0 & 0 \\ \varepsilon & 0 & 0 \\ 0 & 0 & 0 \end{bmatrix}$$

and

$$R + S = \begin{bmatrix} 0 & -\frac{1}{2}R & 0 \\ \frac{1}{2}R + \varepsilon & 0 & 0 \\ 0 & 0 & \lambda_r \end{bmatrix}. \tag{8.6}$$

The conclusion should be the same as in 2D case. If ε is positive (the shear and rotation have the same directions), the magnitude of Liutex R will not change at all, which means the shear cannot change the strength of rotation. On the other hand, if ε is negative (the shear and rotation have the opposite directions), the magnitude of Liutex or the strength of rotation will be weakened or even disappeared.

Let us consider the interaction of a shear and a hairpin vortex. Both are symmetric. Assume the direction of shear is clockwise, which we think ε is positive, and the hairpin has two counterrotating legs with the right leg clockwise and the left leg counterclockwise. Interacted with a clockwise shear, the right leg will keep the same rotation strength, namely R, but the left leg may be weakened (becomes thinner) or even disappear. This process will make the original symmetric hairpin become an asymmetric hairpin or even a one-leg vortex. One example with the symmetric Liutex is shown in Fig. 8.12. Moreover, Fig. 8.13 shows the one-leg vortex ring appearing in the upper level of the boundary layer and Fig. 8.14 depicts a secondary vortex ring with one strong leg and one weak leg in the lower boundary layer.

Confirmed by DNS and experiments, there are forests of hairpin vortices in the flow transition and early turbulence, but the hairpin vortex could be deformed or degenerated in the lower boundary layer where the viscosity is large or in fully developed turbulence zones due to the shear interaction with legs (Fig. 8.13). Note that the condition for the deformation or degeneration of symmetric hairpin is the existence of symmetric shears. That is the reason why in the inviscid flow region, the hairpin vortex keeps symmetric for a long time but the hairpin vortex in the lower boundary layer could quickly lose one leg. The only condition is the existence of fluctuated shear. If the shear moves in a clockwise motion, the clockwise vortex leg will not be affected. However, the counterclockwise vortex leg will be weakened or even disappeared. The asymmetric shear—Liutex interaction will cause the asymmetries of the hairpin vortices and further generate more to two asymmetric legs with one strong and one weak or even a one-leg vortex. These could happen on the top of hairpin vortices (see Fig. 8.13) or secondary vortices located in the lower boundary layers (see Fig. 8.14).

8.3.5 Asymmetric Liutex and chaos

As confirmed by both DNS and experiment, there are many one-leg vortices inside the lower boundary layer and fully developed turbulent flows. The one-leg vortex cannot

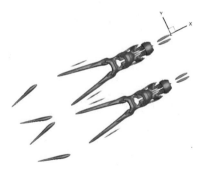

FIGURE 8.12

Symmetric Liutex.

Reproduced from Liu, J., et al., Mathematical foundation of turbulence generation-symmetric to asymmetric Liutex/Rortex, Journal of Hydrodynamics 31 (3), 2019, with the permission of Journal of Hydrodynamics.

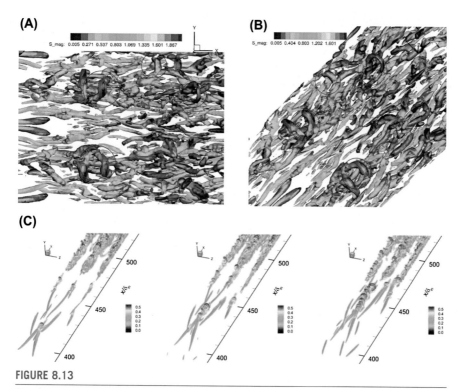

FIGURE 8.13

One-leg vortex or asymmetric vortices in the upper boundary layer (Liutex iso-surfaces colored by shear magnitude).

(A) and (B) reproduced from Liu, J., et al., Mathematical foundation of turbulence generation-symmetric to asymmetric Liutex/Rortex, Journal of Hydrodynamics 31 (3), 2019, with the permission of Journal of Hydrodynamics; (C) reproduced from Matsuura (2018).

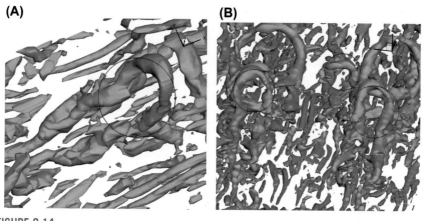

FIGURE 8.14

One-leg vortex in the second-level vortex rings near the wall (Liutex iso-surfaces).

Reproduced from Liu, J., et al., Mathematical foundation of turbulence generation-symmetric to asymmetric Liutex/Rortex, Journal of Hydrodynamics 31 (3), 2019, with the permission of Journal of Hydrodynamics.

keep static as the nature of Liutex, which keeps rigid rotation. The asymmetric Liutex will keep unbalanced and swinging, which will produce asymmetry with fluctuating, swinging, shaking, and chaos. As addressed early, there is no turbulence if there is no asymmetric Liutex. However, shear is always in the boundary layer, especially in the lower boundary layer and hairpin vortices always appear in the flow transition and early turbulence (see Fig. 8.15). Unfortunately, the interaction of hairpin vortex and shear will cause nonsymmetry according to the bias Liutex definition (smaller part of off-diagonal element) due to the nature of shear and vortex interaction. Therefore, asymmetry, the one-leg vortex, shaking of asymmetric vortices, and chaos are doomed. In other words, turbulence is doomed and that is the nature.

8.3.6 Section summary

According to the description earlier, the following conclusion can be achieved:

(1) Shear will not change the rotation strength R if shear and $\vec{R}$ have the same directions, but shear may reduce R if they have opposite directions and the shear magnitude is larger than the original shear contained in the original velocity gradient tensor.

(2) The symmetric hairpin vortex may lose its symmetry when it interacts with symmetric shear.

(3) One leg in symmetric hairpin vortex can be weakened or can disappear due to the shear-vortex interaction. Therefore, a hairpin vortex is unstable in boundary layer.

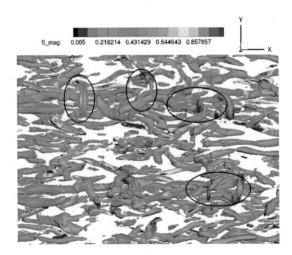

FIGURE 8.15

Liutex iso-surfaces colored by shear magnitude.

Reproduced from Liu, J., et al., Mathematical foundation of turbulence generation-symmetric to asymmetric Liutex/Rortex, Journal of Hydrodynamics 31 (3), 2019, with the permission of Journal of Hydrodynamics.

(4) One-leg or asymmetric vortices are shaking, swinging, and then cause turbulence.

(5) The nature of Liutex magnitude definition (smaller element of a pair) and interaction of shear and vortex are the mathematical foundation of turbulence generation; therefore, the symmetry loss and chaos are doomed. The fact that the hairpin vortex is unstable and one-leg vortex is doomed to be generated is due to the asymmetry of the Liutex strength definition. Therefore, the symmetry of vortex structure is finally lost and flow will become turbulent.

8.4 Summary

In this chapter, the Liutex similarity in a fully developed turbulent boundary layer, the turbulence structure with multiple level co-rotating vortices and mathematical foundation of turbulence asymmetry caused by the hairpin vortex/Liutex and shear interaction are introduced to reveal the secret of turbulence generation and sustenance and the turbulence structure.

Liutex and proper orthogonal decomposition for hairpin vortex generation

9

9.1 Proper orthogonal decomposition for vortex ring formation (Charkrit and Liu, 2019)

9.1.1 Short review of the hairpin vortex formation in early flow transition

Vortices are considered as the building blocks of turbulent flows. It is well known that spanwise vortex, Λ-shaped vortex, and hairpin vortex commonly appear in natural flow transition in both experiment and DNS. The hairpin vortex typically consists of three parts: (1) two counterrotating legs; (2) a ring-like vortex known as the vortex head; (3) necks that connect the head and two legs. The formation of hairpin vortex is an important topic to study the physics of vortex and turbulence generation. To study how one type of vortex becomes another type especially in flow transition is one way to get better understanding about turbulence. In a natural flow transition, a spanwise vortex is formed at the very beginning, then the Λ-vortex appears. It was found in the studies (Yan et al., 2014a; Liu et al., 2014) that the mechanism of a spanwise vortex becomes a Λ-vortex due to the 2-D T-S wave growth and interaction with the 3-D T-S waves. The two roots of a Λ-vortex known as two legs and two counterrotating vortices are generated by 2-D and 3-D T-S wave interaction. After the Λ-shaped vortex is formed, the hairpin vortex appears. Later, multiple vortex rings are formed one by one. The evolution of vortex structures in flow transition is displayed in Fig. 9.1.

For the mechanism of the Λ-vortex development to the hairpin vortex, many researchers believe that the formation of hairpin vortex is caused by vorticity self-deformation. Hama (1960) and Hama and Nutant (1963) studied the mechanism of the formation and development of Λ-shaped vortices in water. Knapp and Roache (1968) also concluded that the Λ-vortex to hairpin vortex is caused by self-deformation observed by means of smoke visualizations in air. Moin et al. (1986) gave a similar self-deformation mechanism by using Biot–Savart law for 2-D and DNS for 3-D. The results demonstrated that "perturbation of layer of vorticity leads to its roll-up into filaments of concentrated vorticity" or, equivalently, the Λ-vortex

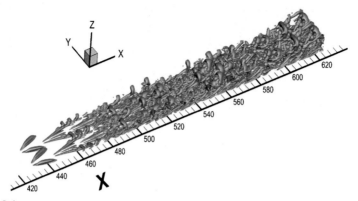

FIGURE 9.1

Evolution of vortex structures in flow transition.

becomes the hairpin vortex by self-deformation as shown in Fig. 9.2. Although the calculation of DNS study could be correct, the definition of vortex is incorrect as it is emphasized again and again in this book that vortex is not vorticity tube, but is Liutex lines and Liutex tubes. This would lead to the misunderstanding on the mechanism of the hairpin vortex formation through the self-deformation of the Λ-vortex.

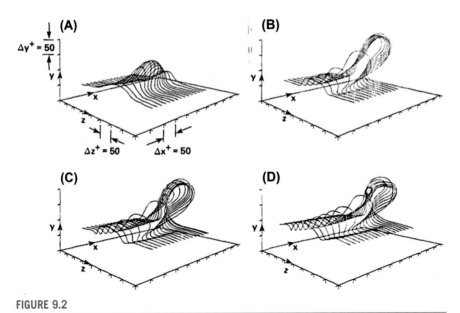

FIGURE 9.2

Sketch of vortex layer at different time from (A) to (D).

Reproduced from P. Moin, A. Leonard, J. Kim, Evolution of curved vortex filament into a vortex ring, Phys. Fluids 29 (4), 1986, with the permission of AIP Publishing.

Since Helmholtz introduced the concept of vortex filament and tube in 1858, which is actually vorticity tube, many textbooks adopt the Helmholtz vortex definition and his three theorems. The vorticity is mathematically defined by velocity curl, $\vec{\omega} = \nabla \times \vec{v}$, and the vorticity line is defined as the line everywhere tangent to the local vorticity vector $\vec{\omega}$. However, vortex is a rotational group of fluids and, as addressed earlier, has no correlation with vorticity. This misunderstanding becomes a major obstacle to turbulence research, and it is sure that vorticity tubes cannot represent vortex in general, especially in boundary layers where viscosity is important.

The combination of λ_2 and vortex filament was applied by Yan et al. (2014a) and Liu et al. (2014) to study the mechanism of the hairpin vortex formation in flow transition in a boundary layer at a freestream Mach number of 0.5 using DNS simulation. They gave a contradiction to the previous conclusions of self-deformation of the Λ-vortex to hairpin vortex. They conclude that there is no such a process that a Λ-vortex is self-deformed to a hairpin vortex as many literatures suggested. Actually, the vortex ring is not a part of the original Λ-vortex but is formed separately, and the formation of a Λ-vortex to a hairpin vortex is caused by the shear layer instability (K-H instability). As a new vortex definition or Liutex is given, this chapter further confirms the aforementioned new mechanism of hairpin vortex formation by using Liutex and proper orthogonal decomposition (POD). In addition, the modified Liutex Omega method is applied to capture the vortices of POD modes in this chapter in order to effectively visualize vortices and obtain more structures in both weak and strong vortices simultaneously.

9.1.2 **Proper orthogonal decomposition method**

(1) Proper orthogonal decomposition

In order to understand the mechanism of hairpin vortex formation and other complex turbulence structures, the POD is applied here. The POD has been widely applied to explore the complex flow fields since they can examine the dominant and coherent structures in fluid flow.

The POD is one of the most widely applied methods to analyze fluid flow. There are two versions of POD technique. The original POD was proposed by Lumley (1967). The other version is called snapshot POD, which was introduced by Sirovich in 1987. POD was first utilized in turbulent flows by Lumley (1967) to decompose the multiscale coherent structures into spatial modes with various energy levels, and its successful application was later confirmed in various areas of fluid mechanics, such as coherent structures in a turbulent boundary layer (Sen et al., 2007), Couette flow (Moehlis et al., 2002), circular cylinder wake (Bergmann and Cordier, 2008; Wang H., et al., 2014), jet flow (Meyer et al., 2007; Schlatter et al., 2011; Cavar and Meyer, 2012), laminar separation boundary layer (Jin and Ma, 2018), and micro-vortex generator (MVG)-controlled compression ramp flow (Chen et al., 2019), and so on. Meyer et al. (2007) analyzed the instantaneous velocity field of a turbulent jet in cross-flow from stereoscopic PIV snapshots by using POD and indicated that the wake vortex structures, rather than the jet shear layer vortices, are the dominant dynamic structures responsible for the strong interaction with the jet core. Cavar and Meyer (2012) identified the coherent structures of a turbulent jet in cross-flow through POD algorithm and revealed that the relationship between the counterrotating vortex pair, the hanging vortex, and the wake vortices could be illustrated by the first two POD modes. They also indicated that the shedding process involving oscillation of the jet core is responsible for the creation of wake vortices. There are also some POD studies focusing on exploring the essence of the complicated spatiotemporal structures in the supersonic or hypersonic flows. Yang and Fu (2008) extracted the flow structures in supersonic plane mixing layers from DNS data with the aid of the POD method and pointed out the main difference between the incompressible and compressible mixing layers in terms of frequency spectrum of the POD modes. The leading POD modes represent most of the turbulence energy for incompressible flow, while much more high-ordered POD modes are needed accounting for the total energy for compressible flow. For a NACA0015 foil wake flow obtained by time-resolved particle image velocimetry (TR-PIV), Prothin et al. (2014) applied POD analysis to highlight the unsteady nature of the wake flow using phase averaging based on the leading ordered POD coefficients to characterize the coherent process in the near wake of the rudder. Berry et al. (2017) applied POD on time-resolved schlieren to extract the canonical flow structures in supersonic multistream rectangular jets and indicated that it is an effective approach for the identification of the flow physics that dominates those modern military nozzles. The buzz phenomenon of a typical supersonic inlet was numerically analyzed by Luo et al. (2020) using POD, and their study suggested that the dominant flow patterns and characteristics of the buzzed flow are obtained by decoupling the computed pressure field into spatial and temporal subparts. Moreover, the first mode represents the mean features and dominates the global flow field, while the second mode reflects the dominant frequency characteristics of buzz.

In POD, the flow structure is decomposed into orthogonal mode ranking by their kinetic energy content. In recent years, there have been many applications by using POD in flow analyzation. For examples, POD was used to turbulent pipe flow (Duggleby et al., 2007a,b). In papers by Freno and Cizmas (2014) and Li and Zhang (2016), POD was applied to identify turbulent discontinuous and nonlinear flows. A mixing layer downstream on a thick splitter plate obtained by DNS was analyzed by POD in 2010 (Laizet et al.). Hellström et al. (2016) and Hellström and Smits (2017) used POD to analyze coherent structures in pipe flow and Gunes (2004) applied POD to a transitional boundary layer with and without control. POD was also used to investigate asymmetric structures of flow on the flat plate (Charkrit, 2019). The vortex structure in MVG wake was examined by Dong et al. (2019a). Jin et al. (2018) used POD to analyze entropy generation in a laminar separation boundary layer.

(2) Steps of POD

The POD is applied to extract the coherent structure of complex flows. POD modes are ranked by energy content.

The POD algorithm consists of the following steps.

Step 1: Collect the data

The n snapshots $x_j = p(\xi, t)$ are assembled columnwise in a matrix $P \in \mathbb{R}^{m \times n}$, which is set from the DNS data by

$$P = \begin{bmatrix} | & | & & | \\ x_1 & x_2 & \cdots & x_n \\ | & | & & | \end{bmatrix}_{m \times n} \tag{9.1}$$

over three-dimensional discrete spatial points ξ at times t. Here, m is the number of spatial points and n is the number of snapshots (time steps). In this case, $n \ll m$. Each column in the matrix P is a snapshot in time and it can be the components such as velocity, vorticity, pressure, or Liutex vector.

In POD method, the flow field is decomposed into a set of base functions and mode coefficients as

$$p(\xi, t) = \sum_{i=1}^{n} \phi_i(\xi) a_i(t) \tag{9.2}$$

or it can be written in the matrix form as

$$P = \Phi Q \tag{9.3}$$

where Φ contains the spatial modes $\phi_i(\xi)$ and Q contains the temporal amplitudes $a_i(t)$.

Then, the covariance matrix C is constructed by $C = P^T P$, and the eigenvectors ψ_i and eigenvalues λ_i of C can be obtained by

$$C \psi_i = \lambda_i \psi_i, \quad i = 1, 2, \ldots, n. \tag{9.4}$$

Since C is symmetric and positive-semidefinite, λ_i are real and nonnegative with $\lambda_1 \geq \lambda_2 \geq \cdots \geq \lambda_n \geq 0$. Also, the eigenvectors ψ_i are orthogonal. Then, the POD spatial mode ϕ_i can be recovered by

$$\phi_i = P \psi_i \frac{1}{\sqrt{\lambda_i}}, \quad i = 1, 2, \ldots, n \tag{9.5}$$

or it can be written in the matrix form as

$$\Phi = P \Psi W \tag{9.6}$$

where W is an $m \times n$ matrix with all elements zero except for those along the diagonal. The diagonal elements of W consist of $\frac{1}{\sqrt{\lambda_i}}$.

As the DNS data is a discrete set, the POD is applied as a matrix decomposition called the singular value decomposition (SVD). The POD algorithm proposed by Schmid (2013) and Mendez et al. (2017) is processed in the same way as SVD.

Step 2: Compute the SVD of matrix P

By Eq. (9.6), the matrix P can be represented by

$$P = \Phi S \Psi^T \tag{9.7}$$

where

(a) T stands for the transpose of the matrix.
(b) Φ is an $m \times m$ orthogonal matrix

$\boldsymbol{\Psi}$ is an $n \times n$ orthogonal matrix.

The matrix $\boldsymbol{\Phi}$ is called the left matrix. It is the matrix of spatial structure.

The matrix $\boldsymbol{\Psi}$ is called the right matrix. It is the matrix of temporal structure.

(c) S is an $m \times n$ matrix with all elements zero except for those along the diagonal.

The diagonal elements of S consist of $S_{ii} = \sigma_i \geq 0$. σ_i is called a singular value of P with $\sigma_1 \geq \sigma_2 \geq \cdots \geq \sigma_n$. In CFD, σ_i stands for the kinetic energy of fluid flow. The matrices $\boldsymbol{\Phi}, \boldsymbol{\Psi}$, and S are demonstrated as follows.

$$\boldsymbol{\Phi} = \begin{bmatrix} | & | & & | \\ \phi_1 & \phi_2 & \cdots & \phi_m \\ | & | & & | \end{bmatrix}, S = \begin{bmatrix} \sigma_1 & 0 & \cdots & 0 \\ 0 & \sigma_2 & \cdots & 0 \\ \vdots & \vdots & \ddots & \vdots \\ 0 & 0 & 0 & \sigma_n \\ 0 & 0 & 0 & 0 \\ \vdots & \vdots & \vdots & \vdots \\ 0 & 0 & 0 & 0 \end{bmatrix},$$

$$\boldsymbol{\Psi} = \begin{bmatrix} | & | & & | \\ \psi_1 & \psi_2 & \cdots & \psi_n \\ | & | & & | \end{bmatrix} \tag{9.8}$$

Thus, Eq. (9.7) can be written as

$$P = \begin{bmatrix} | & | & & | \\ \phi_1 & \phi_2 & \cdots & \phi_m \\ | & | & & | \end{bmatrix}_{m \times m} \begin{bmatrix} \sigma_1 & 0 & \cdots & 0 \\ 0 & \sigma_2 & \cdots & 0 \\ \vdots & \vdots & \ddots & \vdots \\ 0 & 0 & 0 & \sigma_n \\ 0 & 0 & 0 & 0 \\ \vdots & \vdots & \vdots & \vdots \\ 0 & 0 & 0 & 0 \end{bmatrix}_{m \times n} \begin{bmatrix} - & \psi_1 & - \\ - & \psi_2 & - \\ & \vdots & \\ - & \psi_n & - \end{bmatrix}_{n \times n}.$$

$$\tag{9.9}$$

Next, the matrices $\boldsymbol{\Phi}, S$ and $\boldsymbol{\Psi}$ can be evaluated from Eq. (9.7). The procedure to obtain $\boldsymbol{\Phi}, S$ and $\boldsymbol{\Psi}$ is as follows.

(1) Solve for the right matrix $\boldsymbol{\Psi}$ and the singular matrix $\boldsymbol{S}$ by computing $\boldsymbol{P}^T\boldsymbol{P}$.

Since $\boldsymbol{P}^T\boldsymbol{P} = \boldsymbol{\Psi}\boldsymbol{S}^T\boldsymbol{S}\boldsymbol{\Psi}^T = \boldsymbol{\Psi}\boldsymbol{S}^2\boldsymbol{\Psi}^T$, then $\boldsymbol{P}^T\boldsymbol{P}\boldsymbol{\Psi} = \boldsymbol{\Psi}\boldsymbol{S}^2$, which is the eigen decomposition of matrix $\boldsymbol{P}^T\boldsymbol{P}$. Then we obtain

$$\boldsymbol{\Psi} = \begin{bmatrix} | & | & & | \\ \psi_1 & \psi_2 & \cdots & \psi_n \\ | & | & & | \end{bmatrix} \tag{9.10}$$

where ψ_i are eigenvectors of $\boldsymbol{P}^T\boldsymbol{P}$ corresponding to eigenvalues, and

$$\boldsymbol{S}^2 = \begin{bmatrix} \lambda_1 & 0 & \cdots & 0 \\ 0 & \lambda_2 & \cdots & 0 \\ \vdots & \vdots & \ddots & \vdots \\ 0 & 0 & 0 & \lambda_n \\ 0 & 0 & 0 & 0 \\ \vdots & \vdots & \vdots & \vdots \\ 0 & 0 & 0 & 0 \end{bmatrix} \tag{9.11}$$

where λ_i are eigenvalues of $\boldsymbol{P}^T\boldsymbol{P}$.

Then, the singular matrix $\boldsymbol{S}$ is obtained by

$$\boldsymbol{S} = \begin{bmatrix} \sigma_1 & 0 & \cdots & 0 \\ 0 & \sigma_2 & \cdots & 0 \\ \vdots & \vdots & \ddots & \vdots \\ 0 & 0 & 0 & \sigma_n \\ 0 & 0 & 0 & 0 \\ \vdots & \vdots & \vdots & \vdots \\ 0 & 0 & 0 & 0 \end{bmatrix} \tag{9.12}$$

where $\sigma_i = \sqrt{\lambda_i}$.

(2) Solve for the left matrix $\boldsymbol{\Phi}$ by plugging S and $\boldsymbol{\Psi}$ into Eq. (4.8). Thus, the matrix $\boldsymbol{\Phi}$ can be obtained by

$$\boldsymbol{\Phi} = P\boldsymbol{\Psi}S^{-1}. \tag{9.13}$$

(3) Dimension reduction and the matrix reconstruction

If we use S and $\boldsymbol{\Psi}$ in (9.8) to solve for $\boldsymbol{\Phi}$, the matrix $\boldsymbol{\Phi}$ will be a huge $m \times m$ matrix. This is not a good idea to obtain $\boldsymbol{\Phi}$ with the high dimension matrix. To avoid this, we can obtain $\boldsymbol{\Phi}$ with the lower dimension. Since the rank of S is n, we can write S in terms of $n \times n$ matrix,

$$S = \begin{bmatrix} \sigma_1 & 0 & \cdots & 0 \\ 0 & \sigma_2 & \cdots & 0 \\ \vdots & \vdots & \ddots & \vdots \\ 0 & 0 & 0 & \sigma_n \end{bmatrix} \tag{9.14}$$

Then, Eq. (9.9) becomes

$$P = \begin{bmatrix} | & | & & | \\ \phi_1 & \phi_2 & \cdots & \phi_n \\ | & | & & | \end{bmatrix}_{m \times n} \begin{bmatrix} \sigma_1 & 0 & \cdots & 0 \\ 0 & \sigma_2 & \cdots & 0 \\ \vdots & \vdots & \ddots & \vdots \\ 0 & 0 & 0 & \sigma_n \end{bmatrix}_{n \times n} \begin{bmatrix} - & \psi_1 & - \\ - & \psi_2 & - \\ & \vdots & \\ - & \psi_n & - \end{bmatrix}_{n \times n} \tag{9.15}$$

Similarly, we can solve for $\boldsymbol{\Phi}, S$, and $\boldsymbol{\Psi}$. The matrix $\boldsymbol{\Phi}$ is reduced to a lower dimension $m \times n$, which is much more convenient than a higher dimension $m \times m$. Moreover, the matrix $\boldsymbol{\Phi}$ can be reduced to a lower dimension $m \times r$ with $r \leq n$ by truncating the last $m - r$ columns. Then by truncating the last $n - r$ rows and columns of S, the dimension of S becomes $r \times r$. For the matrix $\boldsymbol{\Psi}$, to truncate the last $n - r$ rows of $\boldsymbol{\Psi}$, the dimension of matrix $\boldsymbol{\Psi}$ becomes $r \times n$. Thus, Eq. (9.16) will be an approximated form of Eq. (9.15) as the following expression.

$$P \approx \begin{bmatrix} | & | & & | \\ \phi_1 & \phi_2 & \cdots & \phi_r \\ | & | & & | \end{bmatrix}_{m \times r} \begin{bmatrix} \sigma_1 & 0 & \cdots & 0 \\ 0 & \sigma_2 & \cdots & 0 \\ \vdots & \vdots & \ddots & \vdots \\ 0 & 0 & 0 & \sigma_r \end{bmatrix}_{r \times r} \begin{bmatrix} - & \psi_1 & - \\ - & \psi_2 & - \\ & \vdots & \\ - & \psi_r & - \end{bmatrix}_{r \times n} \tag{9.16}$$

Note that the approximated matrix P in Eq. (9.16) is called the matrix reconstruction.

(4) Reduction criterion:

To choose the size r of the reduced dimension of matrix, it can be computed from the relative energy of the snapshots by the first r POD basis vectors as the following formula

$$\varepsilon(r) = \frac{\sum_{i=1}^{r} \sigma_i^2}{\sum_{i=1}^{n} \sigma_i^2} \tag{9.17}$$

r is usually chosen as the minimum integer such that

$$1 - \varepsilon(r) \leq tol \tag{9.18}$$

where tol means a given tolerance with $0 < tol < 1$. For examples, $tol = 0.0011$.

(5) Linear combination of POD mode

By Eqs. (9.7) and (9.9), the matrix P can be written in the form of the superposition of basis as follows.

$$P = \sum_{k=1}^{n} \sigma_k \phi_k \psi_k^T, \tag{9.19}$$

or

$$P = \sigma_1 \phi_1 \psi_1^T + \sigma_2 \phi_2 \psi_2^T + \cdots + \sigma_n \phi_n \psi_n^T \tag{9.20}$$

The vector u_i is called a POD mode. The modes are the bases. The whole structure can be extracted into n coherent structures. Dimensional reduction can keep the most important mode as a basis. The first mode will be the most dominant energy structure and the last mode will be the least dominant energy structure. If P is reduced to the rank $r < n$ as in Eq. (9.16), the approximated P is expressed by

$$P \approx \sum_{k=1}^{r} \sigma_k \phi_k \psi_k^T \tag{9.21}$$

or

$$P \approx \sigma_1 \phi_1 \psi_1^T + \sigma_2 \phi_2 \psi_2^T + \ldots + \sigma_r \phi_r \psi_r^T, \tag{9.22}$$

which represent mode1, mode 2, ..., mode r.

9.1.3 Vortex structures in early transition by vortex core line identification

There have been many vortex identification methods used as convenient tools to capture vortices. However, they are all strongly dependent on the selection of threshold. However, the Liutex core line method will give a unique vortex structure, which is threshold-free.

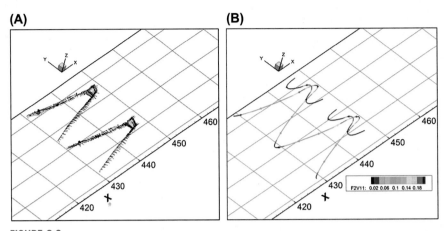

FIGURE 9.3

The Λ-vortex at $t = 6.0T$, where T is the period of T-S wave. (A) Liutex magnitude gradient lines without iso-surface. (B) Vortex core lines colored (red is strong and blue is weak) by Liutex magnitude.

Fig. 9.3A demonstrates Liutex magnitude gradient lines and the vortex core lines, are illustrated in Fig. 9.3B. Moreover, the visualization of vortex with different thresholds can be represented by the unique vortex core lines as shown in Fig. 9.4A–C. Note that Fig. 9.4A–C give totally different vortex structures, but they have the same vortex core lines, which are unique.

9.1.4 Ring-like vortex formation

The evolution of vortex formation in flow transition is displayed in Fig. 9.5 by the modified Liutex Omega method. It can be observed that the vortex formation starts with the spanwise vortex, and then the Λ-vortex is generated. After the Λ-vortex is formed, the ring-like vortex will be generated. After that, the multiple hairpin

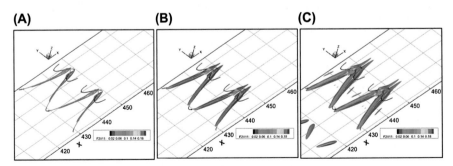

FIGURE 9.4

The Λ-vortex at $t = 6.0T$, where T is the period of T-S wave, visualized by Liutex magnitude and vortex core line. (A) Iso-surface of $R = 0.1$ (B) Iso-surface of $R = 0.05$. (C) Iso-surface of $R = 0.02$.

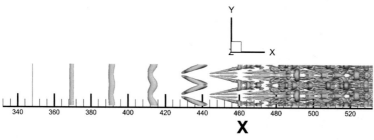

FIGURE 9.5

The evolution of vortex formation on flow transition by the modified Omega Liutex method with iso-surface of $\widetilde{\Omega}_L = 0.51$.

vortices appear. In the process, the Λ-vortex becomes the hairpin vortex, which can be found in the area between $x = 450$ and $x = 470$ in the streamwise direction as shown in Fig. 9.5. As addressed earlier, it was widely believed that the Λ-vortex self-deforms to the hairpin vortex (Hama, 1960, Hama et al., 1963, Knapp et al., 1968, Moin et al., 1986).

However, Liu et al. (2014) and Yan et al. (2014a) have found the mechanism of self-deformation does not exist, which is a misunderstanding caused by mistreating vortex as vorticity tube. By the DNS observation, they concluded:

(1) The Λ-vortex and the multiple ring-like vortices are formed separately.
(2) The multiple ring-like vortices are not part of Λ-vortex.
(3) There is no process that the Λ-vortex self-deforms to the hairpin vortex.

Since the Λ-vortex is not self-deformed to the hairpin vortex, there must be another mechanism for the hairpin vortex formation, which can be descried as follows.

(1) A momentum deficit zone (low-speed zone) is formed above the Λ-vortex and further generates a Λ-shaped high shear due to the Λ-vortex root ejection.
(2) The first vortex ring is strengthened by the high shear layer through the Kelvin−Helmholtz (K-H) type instability near the tip of the Λ-structure.
(3) Multiple vortex rings are all formed by shear layer instability, which is generated by momentum deficit.
(4) If there is a shear layer inside the flow field, the shear must transfer to rotation and further to turbulence when the Reynolds number is large enough.
(5) All small vortices are generated by shear layers.
(6) The multiple-level shear layers are generated by vortex sweeps and ejections. The sweep brings high-speed flow down (positive spike) to the lower boundary layer and the ejection brings the low-speed flow up (negative spike) to the upper boundary layer. They form the multiple-level shear layers.
(7) The shear layer instability (K-H type instability) plays an important role to transferor shear to Liutex.

From the earlier analysis, they conclude that the self-deformation theory of hairpin vortex formation is produced by misunderstanding that considers "a vortex" as a "vorticity tube."

After Liutex and Liutex core line methods are discovered, Charkrit et al. (2019) restudied the mechanism of hairpin vortex formation by using POD, Liutex, and Liutex core line methods. Therefore, no specified threshold is required. As shown in Fig. 9.6 (left), the evolution of the Λ-vortex to the hairpin vortex is captured by Liutex magnitude of $R = 0.05$ with the convenient tool. To uniquely represent the vortex, the vortex core lines are visualized and colored by Liutex magnitudes as shown in Fig. 9.6 (right). The Λ-vortex is shown with iso-surface at $t = 6.0T$ and starts to form the first ring in Fig. 9.6 (Left). It is clearly shown by the vortex core lines that the rings, which are on top and in the upstream of the tips, are separated from the legs of the Λ-vortex as shown in Fig. 9.6 (right). Similarly, at $t = 6.3T$ and $6.5T$, the second and third rings are generated separately above the leg of Λ-vortex by the visualization of vortex core lines shown in Fig. 9.6 (Right). The Λ-vortex is likely connected with ring-like vortex at the upstream of the tip part when it is observed with iso-surface. However, the unique vortex core demonstrates that the Λ-vortex is apparently separated from the ring-like vortex.

9.1.5 Kelvin—Helmholtz instability

The Kelvin—Helmholtz instability occurs when there is a velocity shear at the interface between two fluid layers. A 2-D vortex ring formation, which is caused by the K-H type instability, can be explained as shown in Fig. 9.7. A low-speed (on the bottom) and a high-speed region (on the top) generate a shear layer as shown in Fig. 9.7A—B. Then, the shear layer instability (K-H instability) forms the ring-like vortex by transferring the shear to a pair of rotations as in Fig. 9.7C—D. In other words, it is a process to transfer nonrotational vorticity or shear to rotational vorticity or transfer shear to Liutex. After that the pair of rotations becomes one but still have a pair of cores inside the rotation as shown in Fig. 9.7E—H. Therefore, pairing of vortices is an outstanding symbol of K-H instability. Note that although K-H instability is for the inviscid flow, it still could be used to describe about "the shear layer instability" for the viscous flow as an approximation.

In the case of pairing in K-H instability, Charkrit et al. (2019) propose a hypothesis that the vortical structures and some features in the same pairs should be similar. Hence, the POD is applied to show results that support the hypothesis and analyze the ring-like vortex formation.

9.1.6 POD analysis on K-H instability

By the DNS data, the vortex structure of flow on a flat plate is investigated. Fig. 9.8 shows one example of vortex in flow transition at $t = 13.00T$, where T is a period of

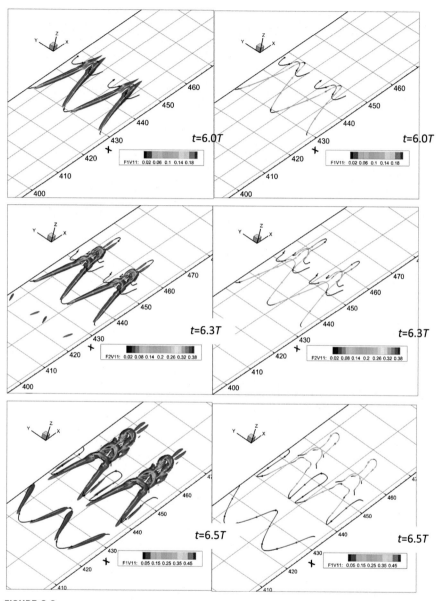

FIGURE 9.6

The evolution of Λ-vortex and ring-like vortex at $t = 6.0T$, $t = 6.3T$, and $t = 6.5T$. (Left) Iso-surface of Liutex magnitude $R = 0.05$ with the vortex core lines. (Right) The vortex core lines without iso-surface colored by Liutex magnitudes.

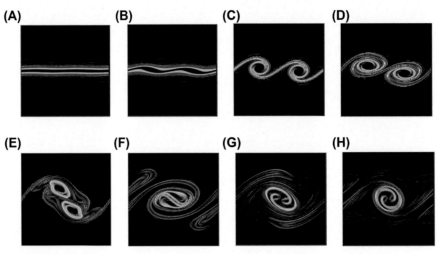

FIGURE 9.7

Numerical simulation in 2-D on K-H instability (Start from (A) to (H), respectively).

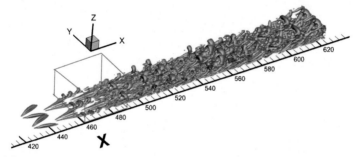

FIGURE 9.8

Vortex structures by modified Omega Liutex method with $\widetilde{\Omega}_L = 0.52$ at $t = 13.00T$.

T-S waves along the streamwise direction between $x = 430$ and $x = 630$, visualized by using the modified Liutex Omega method with iso-surface of $\widetilde{\Omega}_L = 0.52$.

In order to apply POD in the analysis of K-H instability, the domain of study is the area where a Λ-vortex becomes a hairpin vortex. Therefore, the specific area of POD analysis is from $x = 460$ (index = 316) to $x = 490$ (index = 376) as shown in Fig. 9.8. The POD is applied over 100 snapshots (i.e., 100 time steps) at time between $t = 12.51T$ and $t = 13.50T$ to investigate the principal components of the coherent structures. A subzone is extracted to reduce the computation cost. The parameters of the subzone are given in Table 9.1.

Table 9.1 Parameters of subzone.

	Start index	End index
I (in x direction)	316	376
J (in y direction)	1	128
K (in z direction)	1	200

The snapshot x_i is defined to install into the matrix $\boldsymbol{P}$ as expressed in Eq. (9.1) by

$$
x_i =
\begin{pmatrix}
u^{(i)}_{316,1,1} \\
\vdots \\
u^{(i)}_{376,1,1} \\
u^{(i)}_{316,2,1} \\
\vdots \\
u^{(i)}_{376,2,1} \\
\vdots \\
u^{(i)}_{I,J,K} \\
\vdots \\
u^{(i)}_{316,128,200} \\
\vdots \\
v^{(i)}_{I,J,K} \\
\vdots \\
w^{(i)}_{I,J,K} \\
\vdots \\
w^{(i)}_{376,128,200}
\end{pmatrix}
\quad \text{for } i = 1, ..., 100,
\tag{9.23}
$$

where $u^{(i)}$, $v^{(i)}$, and $w^{(i)}$ are velocity fields at $t = (12.50 + 0.01i)T$, for $i = 1, ..., 100$. Here, i is the time index.

The POD is applied over 100 snapshots in this study. Therefore, 100 POD modes are obtained. The eigenvalues in POD method are evaluated and ordered. The POD modes are ranked by the energy content represented by the singular values or the eigenvalues. The results of eigenvalues of all modes are investigated. It is found that two eigenvalues are similar to each other in pairs. The eigenvalues of mode 1 and the pairs of some other modes are presented as shown in Table 9.2.

Table 9.2 Eigenvalues of some modes.

Pair no.	Eigenvalues of modes	
	Mode 1: 87949612.7013267	
Pair 1	Mode 2: 200288.009080135	Mode 3: 194889.125828145
Pair 2	Mode 4: 39234.2518602893	Mode 5: 38676.5370753112
Pair 3	Mode 6: 34451.0788998119	Mode 7: 34089.7551347020
Pair 4	Mode 8: 18673.8222690667	Mode 9: 18322.4804874360
Pair 5	Mode 10: 17169.6248912740	Mode 11: 17018.6436759498
Pair 6	Mode 12: 11631.8322897197	Mode 13: 11494.4484482407
Pair 7	Mode 14: 8146.26652550679	Mode 15: 8090.41774555481
Pair 8	Mode 16: 2844.71672727352	Mode 17: 2795.57150355193
Pair 9	Mode 18: 1452.34341004029	Mode 19: 1397.08863389295
Pair 10	Mode 20: 1201.55853843886	Mode 21: 1196.85608390518
Pair 11	Mode 34: 6.10591736207254	Mode 35: 5.27626505327566

In Fig. 9.9, mode 1, which represents the mean velocity (streamwise velocity), is taken out and only the fluctuation modes, which are mode 2 to mode 100, are considered. The eigenvalue of each mode except for mode 1 and the pairings of eigenvalues can be seen in Fig. 9.9.

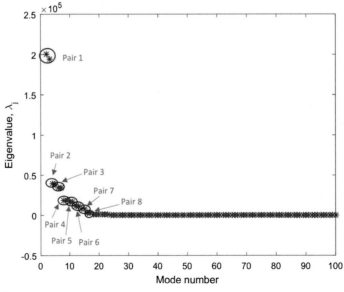

FIGURE 9.9

Eigenvalues of mode 2 to mode 100 indicated the pairs of modes.

By dimensional reduction, the number of modes, r, is chosen by Eq. (9.18) and shown in Fig. 9.10. The proper number of modes to reconstruct the vortex structure is 20 modes since it can keep the most cumulative energy at almost 100%.

The reconstruction performs very well with the first 20 modes comparing to the original flow data. Fig. 9.11 demonstrates that the vortex structures of the original flow and the reconstructed flow of first 20 modes in three different time steps.

The vortex structures in both original DNS results and reconstructed ones are some kind different in shape at different time steps as shown in Fig. 9.11. However, the same shape of vortex structures of each POD mode in different time step from $t = 12.51T$ to $13.50T$ is obtained, which is really the eigenvector, since the POD mode represents a spatial structure of the flow with the time average.

Mode 1 represents the most dominant mode in terms of kinetic energy content as it has the most principal component with $E(r) = 99.25\%$. In Fig. 9.12, mode 1 is shaped to streamwise vortex structure with the iso-surface of $\widetilde{\Omega}_L = 0.52$. The spatial shapes of the other modes are also illustrated in pairs, which can be seen in Fig. 9.12. The shapes of modes in the same pair are similar as they have similar eigenvalues and similar eigenvectors.

Modes in pair 1, which is a pair of modes 2 and 3, are likely in streamwise characteristic visualized by modified Liutex Omega method with iso-surface of $\widetilde{\Omega}_L = 0.52$. For the other modes, the vortex shapes are dominated by the spanwise structures. As seen in Fig. 9.12, the higher-order modes have smaller spanwise vortex structures and more numbers of rings visualized by the modified Liutex-Omega method.

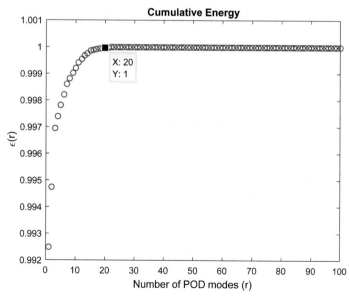

FIGURE 9.10

Cumulative energy of all modes.

(A)

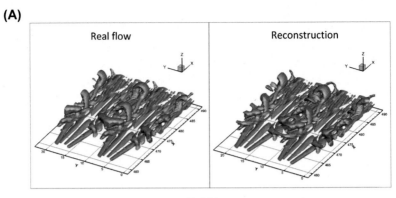

The vortex structure at $t = 12.51T$

(B)

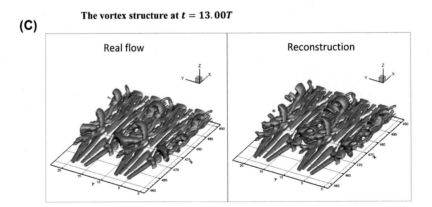

The vortex structure at $t = 13.00T$

(C)

The vortex structure at $t = 13.50T$

FIGURE 9.11

The vortex structures of real flows (left) and reconstructions of the first 20 modes (right) with iso-surface of $\widetilde{\Omega}_L = 0.52$ with $\varepsilon = 0.001\left(\beta^2 - \alpha^2\right)_{max}$.

POD Mode Shape

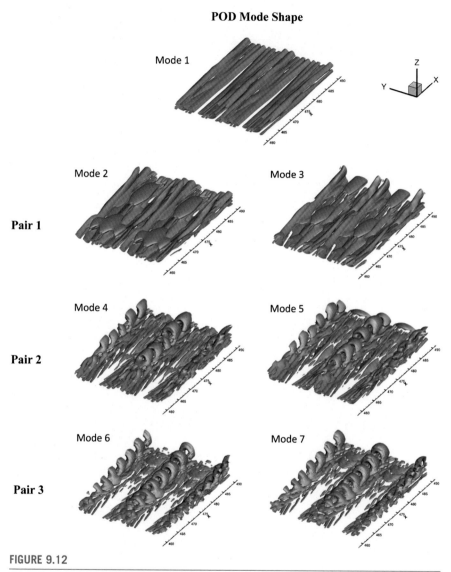

FIGURE 9.12

Vortex structures of some modes with iso-surface of $\widetilde{\varOmega}_L = 0.52$ with $\varepsilon = 0.001\left(\beta^2 - \alpha^2\right)_{max}$.

To investigate the fluctuation of flows, the POD time coefficients are applied to demonstrate periodicities of POD modes. The POD time coefficients are obtained as described earlier. As seen in Fig. 9.13, there is a little fluctuation in mode 1 since the graph of POD time coefficient is likely constant. Thus, mode 1 can represent the mean flow. For the other modes, time coefficients are illustrated in pairs. The similar

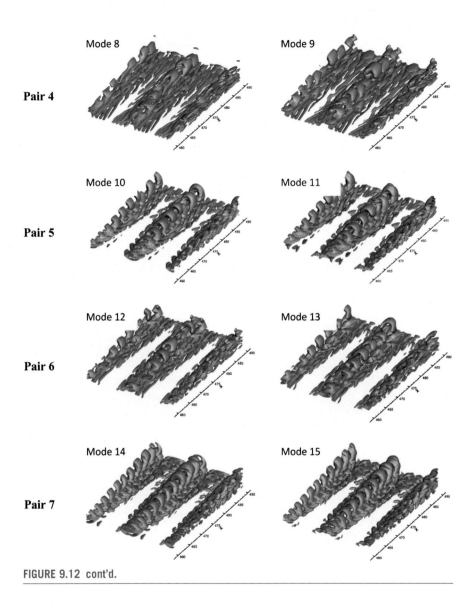

FIGURE 9.12 cont'd.

fluctuation features of modes as periodicities and amplitudes are demonstrated in the same pairs. It can be seen that the higher-order modes have higher fluctuations as shown in Fig. 9.13.

9.1.6.1 Summary

In Charkrit's work, the Liutex core line method, which represents the rotation axis, is applied to show the formation of hairpin vortex. The results show that the ring-like vortex is not part of the Λ-vortex and they are formed separately. Second, the POD is

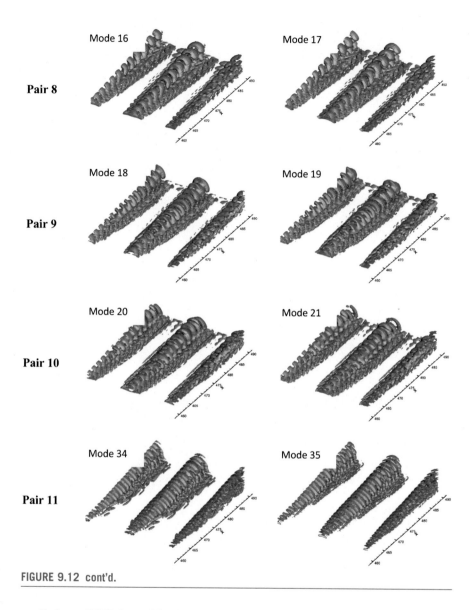

Mode 16   Mode 17

Pair 8

Mode 18   Mode 19

Pair 9

Mode 20   Mode 21

Pair 10

Mode 34   Mode 35

Pair 11

FIGURE 9.12 cont'd.

applied to a DNS data with 100 snapshots to analyze the vortex structure and K-H instability characteristics in the early boundary layer flow transition where the area of a Λ-vortex becomes a hairpin vortex. Mode 1 is the most energetic mode with leading streamwise structure and a little fluctuation. The pair 1 is dominated by the stremiwise vortices and other modes are in spanwise shapes with fluctuations. The higher-order modes have smaller spanwise vortex structures as they are ranked

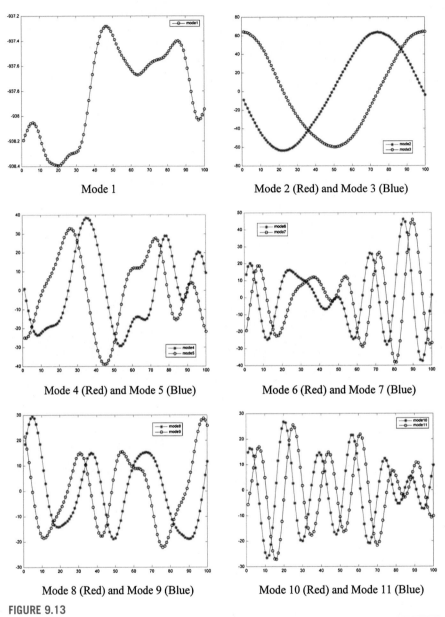

FIGURE 9.13

POD time coefficients (y-axis) with mode number (x-axis).

by the eigenvalues. The smaller eigenvalues in higher-order modes get the smaller-scale structures. The higher-order modes, which are dominated by the spanwise structures, show more fluctuations. Moreover, the important result shows that the

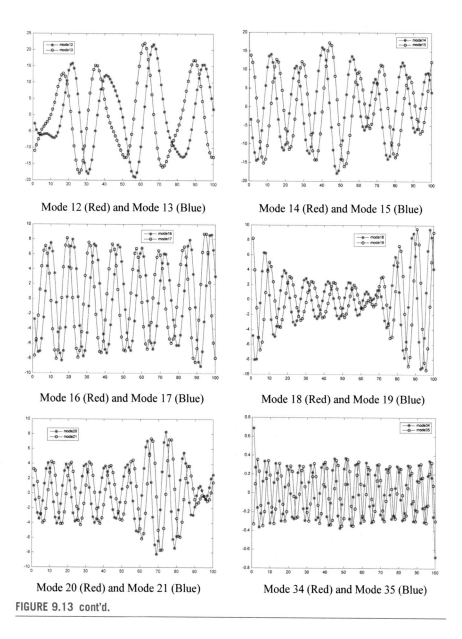

Mode 12 (Red) and Mode 13 (Blue)

Mode 14 (Red) and Mode 15 (Blue)

Mode 16 (Red) and Mode 17 (Blue)

Mode 18 (Red) and Mode 19 (Blue)

Mode 20 (Red) and Mode 21 (Blue)

Mode 34 (Red) and Mode 35 (Blue)

FIGURE 9.13 cont'd.

eigenvalues and eigenvectors of POD modes are shown in pairs, which is typical sign of the K-H instability. The same characteristics such as mode shapes and fluctuations are also established in pairs. This evidence is a strong proof that the K-H instability plays a key role in hairpin vortex formation. Therefore, it can be strongly confirmed that the K-H instability is the main factor of the formation of the hairpin vortex from the Λ-vortex, which transforms the nonrotational vorticity or shear to the rotational vorticity or Liutex. The theory of self-deformation of the Λ-vortex to hairpin vortex has not been observed and maybe has no scientific foundation. Note that the current input for POD is velocity but Liutex can be input as well as Liutex is a vector same as velocity. However, Q cannot be used as input since it is a scalar. Vorticity is a vector, but cannot be used for the vortex structure analysis.

9.2 Liutex and POD analysis on vortical structures in MVG wake (Dong et al., 2020)

9.2.1 Brief introduction

For supersonic flows, the shock waves and turbulence interactions may lead to the flow distortion, flow separation, or aerodynamic force fluctuations and then deteriorate the performance of flight vehicles. As a convenient and effective passive device for control of shock-induced flow separation, micro vortex generator (MVG) is particularly interested by many engineering applications. Due to its importance in both scientific research and engineering applications, many efforts on study of the control mechanism of the MVG wake and the distinct coherent structures (vortices) have been studied over the past decades. Coherent structures are organized with large-scale structures that persistently appear, disappear, and reappear with a characteristic temporal lifespan, although they are not explicitly steady (Berkooz et al., 1993). They also play a significant role in chaotic fluctuations with small-scale structures, due to the redistribution and partitioning of the turbulent kinetic energy. Therefore, a complete and precise description of the characteristics of canonical coherent structures is obviously vital to a deep understanding of predicting or even controlling the evolution in a supersonic flow. Babinsky et al. (2009) proposed a schematic illustration of the various vortices in MVG wake through experiments that a small horseshoe vortex, a pair of primary streamwise vortices, a pair of secondary vortices generating from the side wall, and another pair of secondary vortices originating from the top edges of MVG are typical coherent structures in MVG wake. On the other hand, a new model of the vortex organization behind MVG was given by Li and Liu (2010), which is different from the conventional expressions. They suggested that the streamwise vortices could be generated initially at the trailing edge of MVG, but rapidly decay in a short time and become rather weak when developing further downstream. However, the arc-shaped shear layer around the wake quickly becomes unsteady and then induces the K-H vortices, and eventually develops into a train of vortex rings when moving further downstream.

Recently, high-order DNS and LES together with TR-PIV have been widely applied to capture the unsteady flows with the reasonable spatial and temporal resolutions (Zhang et al., 2014). However, the accurate vortex identification methods are still the essential issue for identifying these vortex structures in shock and boundary layer interaction (SBLI). Liutex method is applied in this chapter (Liu et al.,

2018a; Gao and Liu, 2018; Wang Y., et al., 2019a). In order to overcome the sensitivity of the vortex identification methods, a normalized Liutex method or Liutex-Omega method (Dong et al., 2019b; Liu and Liu, 2019d), which combines the ideas of the Liutex and the Ω method (Liu et al., 2016), was proposed as a relative strength of the fluid rotation to overcome the sensitivity of the threshold selection. Furthermore, the Liutex core line method is defined as a line where the Liutex magnitude gradient vector is aligned with the Liutex vector, being able to extract core features manually (Gao et al., 2019b) and automatically (Xu et al., 2019). In order to get deep understanding of the vortex evolution in the MVG wake, the Liutex system is applied in a supersonic MVG wake flow with multiscale vortical structures for the vortex identification as well as the quantitative analysis. Furthermore, a systematic POD analysis on the MVG wake together with Liutex core line method is carried out to give some revelations.

Low-frequency oscillations of shock and turbulent boundary layer interaction (SBLI) are very harmful and major obstacle in design and flight of supersonic commercial aircrafts due to the unacceptable low-frequency noises. To find the source of the low-frequency noise and SBLI control, researchers have spent almost one century to find the mechanism but with fruitless outcome. It is still a mystery if the low-frequency noise of SBLI is caused by inflow turbulence, shock-induced boundary layer separation, or others? After several years of research, it has been found that the SBLI is really shock–vortex interaction or shock–Liutex interaction and the noise is overwhelmingly caused by shock oscillation, and the low-frequency noise is determined by the vortex size and moving speed or by the Liutex spectrum. And using MVG can reduce the shock oscillation and then remove the low-frequency noise but keep the high-frequency oscillations, which are less harmful. These important discoveries have paved ways for people to understand the source of SBLI that caused low-frequency noise and pave way for commercial supersonic airplane to become reality.

9.2.2 Case description

Fig. 9.14 gives the dimensions of the compression ramp as well as the MVG, The axes x, y, and z respectively represent the spanwise, normal, and streamwise directions, which is unusual. The corner of MVG is located at $x = 0$ along the streamwise direction, and the compression ramp angle α is 24 degrees. The configuration of MVG follows the experimental study performed by Babinsky et al. (2009) with $\alpha = 24$ degrees, $c = 7.2h$, $s = 7.5h$, where h is the height of MVG and s denotes the distance between MVGs. In order to alleviate the difficulty of grid generation, the trailing edge declining angle behind MVG is set as 70 degrees. The grid of the computational domain is $n_{streamwise} \times n_{normal} \times n_{spanwise} = 1600 \times 192 \times 137$.

For the inflow boundary condition, 20,000 instantaneous turbulent profiles of the DNS simulation results (Liu et al., 2014) are used for obtaining a turbulent inflow boundary. The far-field boundary condition is utilized on the top boundary and the outlet boundary is set as an outgoing flow without the reflection. The adiabatic,

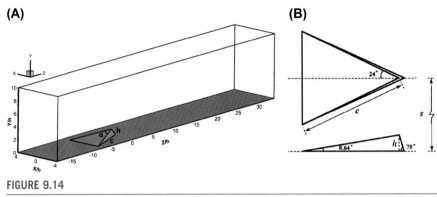

FIGURE 9.14

Configurations of (A) the computational domain and (B) MVG geometry.

Reproduced from Dong, X., et al., POD analysis on vortical structures in MVG wake by Liutex core line identification, Journal of Hydrodynamics 32, 2020, with the permission of Journal of Hydrodynamics.

zero-pressure gradient and nonslipping conditions are adopted for the wall boundary conditions. The periodic boundary conditions are applied in the front and rear boundaries in spanwise direction. Moreover, the initial and reference parameters in the turbulent flow are given in Table 9.3, where Re_θ is the Reynolds number, which is based on the momentum thickness θ, and δ is the undisturbed turbulent boundary layer thickness.

9.2.3 Liutex vector and Liutex core line method

As Liutex is a vector, all the iso-surface, the Liutex vector field, and the Liutex lines have been successfully applied to illustrate the hairpin vortex. The Liutex vector field and the Liutex lines distributions on $R = 0.05$ iso-surface for a hairpin vortex of the turbulent flow simulated by DNS are shown in Fig. 9.15A and B. In Fig. 9.15B, R concentrates at the center of the vortex head and legs; it however, becomes zero in the nonrotational region.

   According to Gao et al. (2019b), the vortex rotation axis is the concentration of Liutex magnitude gradient lines as the local Liutex maxima on the plane perpendicular to this vortex rotation axis line, and the concentration line is a special Liutex line where the Liutex vector $\boldsymbol{R}$ is aligned with the Liutex magnitude gradient ∇R, where $\nabla R = \left[\partial R/\partial x \quad \partial R/\partial y \quad \partial R/\partial z\right]^{\mathrm{T}}$. The distribution of Liutex magnitude gradient lines concentrated to a single line in a hairpin vortex is shown in Fig. 9.15C. Thus, the vortex core center or the vortex rotation axis line is defined as a line consisting of the points in the flow field satisfying $\nabla R \times \boldsymbol{r} = 0$ & $R > 0$. Consequently, the Liutex core line is displayed in Fig. 9.15D.

Table 9.3 Initial and reference parameters of the turbulent flow.

Ma_∞	Re_θ	T_∞	T_W	δ	U_∞
2.5	5760	288.15 K	300 K	9.44 mm	850 m/s

Reproduced from Dong, X., et al., POD analysis on vortical structures in MVG wake by Liutex core line identification, Journal of Hydrodynamics 32, 2020, with the permission of Journal of Hydrodynamics.

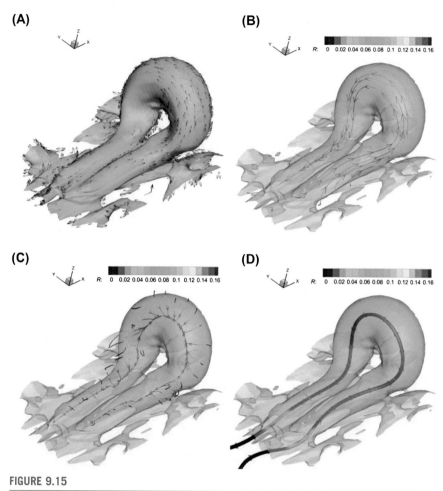

(A)

(B)

(C)

(D)

FIGURE 9.15

$R = 0.05$ iso-surface of hairpin vortex in a turbulent boundary layer shown with (A) Liutex vector field, (B) Liutex lines, (C) Liutex magnitude gradient lines, (D) Liutex core line.

Reproduced from Dong, X., et al., POD analysis on vortical structures in MVG wake by Liutex core line identification, Journal of Hydrodynamics 32, 2020, with the permission of Journal of Hydrodynamics.

9.2.4 Vortex structures in MVG wake

(1) Liutex identification

Fig. 9.16 shows the vortex structures in MVG wake identified by Liutex iso-surface. In the figure, the vortices A, B, and C are obtained from the wake downstream, which are spanwise-, wall-normal-, and streamwise-dominant vortex structures respectively with relatively stronger rotation strength than those in upstream. It shows that the visualization effect is altered by the selection of threshold. To verify the accuracy of the Liutex core line method, the vortex core line and the 2D surface streamlines of the vortices A, B, and C are shown in Fig. 9.17. In Fig. 9.17A,

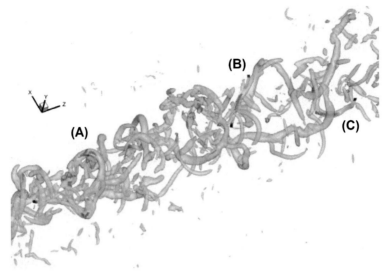

FIGURE 9.16

Vortex structures in MVG wake identified by Liutex magnitude iso-surface.

Reproduced from Dong, X., et al., POD analysis on vortical structures in MVG wake by Liutex core line identification, Journal of Hydrodynamics 32, 2020, with the permission of Journal of Hydrodynamics.

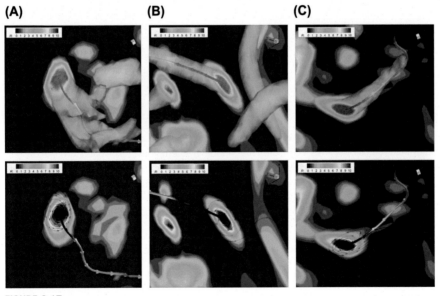

FIGURE 9.17

Three vortices (A), (B), and (C) identified by surface streamlines and the vortex core line.

Reproduced from Dong, X., et al., POD analysis on vortical structures in MVG wake by Liutex core line identification, Journal of Hydrodynamics 32, 2020, with the permission of Journal of Hydrodynamics.

the streamlines distribute a spiral pattern around the Liutex core line on the cross section of the vortex A, and similar phenomenon can be found from Fig. 9.17B and C. As is found, the Liutex core line is a robust and efficient tool to extract the rotation axis of a vortex structure without user-specified threshold.

(2) POD for multiscale coherent structures

For further investigating the principal components of multiple vortical structures in the MVG wake, POD analysis is applied over 200 snapshots of the flow field around MVG in this chapter, and the velocity is chosen to be the input data and also the Liutex core line method is utilized to illustrate the characteristics for each POD mode.

The idea of snapshot POD (Lumley et al., 1967) is to obtain a proper orthonormal basis $\{\psi_i\}_{i=1}^r$ from a real-valued $M \times N$ matrix $\mathbf{U} = (\boldsymbol{u}_1, \boldsymbol{u}_2, ..., \boldsymbol{u}_N)$ of rank $r \le \min\{M, N\}$ with columns $\boldsymbol{u}_j \in \mathbb{R}^M$, $1 \le j \le N$ and N denotes the number of the snapshots. As a conventional input, the matrix $\mathbf{U}$ consists of the velocity components $\boldsymbol{u} = (u, v, w)^{\mathrm{T}}$, where u, v, and w are velocity components in x, y, and z directions. The covariance matrix of velocity $\mathbf{R} = \mathbf{U}^{\mathrm{T}}\mathbf{U}$ has the following expression:

$$\mathbf{R}\mathbf{A}_i = \lambda_i \mathbf{A}_i, \ i = 1, \ 2, ..., N \tag{9.25}$$

where λ_i and $\mathbf{A}_i$ are the eigenvalues and corresponding eigenvectors of auto-covariance matrix $\mathbf{R}$, and the rank of the eigenvalues in a descending order is $\lambda_1 \ge \lambda_2 \ge \cdots \ge \lambda_N$. Then the characteristic function ϕ_i can be obtained by projecting the matrix $\mathbf{U}$ onto each eigenvector $\mathbf{A}_i$, and the normalized form, which is called POD mode, can be written as

$$\phi_i = \sum_{i=1}^r \frac{1}{\sqrt{\lambda_i}} \langle \mathbf{A}_i^j, \ \boldsymbol{u}_j \rangle_{\mathbb{R}^M}, \ j = 1, 2, ..., N \tag{9.26}$$

where $\langle ., . \rangle_{\mathbb{R}^M}$ denotes the canonical inner product in $\mathbb{R}^M$. After the POD modes are obtained, the reconstruction of each snapshot of the original flow field can be performed,

$$\boldsymbol{u}_j = \bar{\boldsymbol{u}} + \sum_{i=1}^N a_i^j \phi_i, \ j = 1, \ 2, ..., N \tag{9.27}$$

$$a_j = \boldsymbol{\Phi}^{\mathrm{T}} \boldsymbol{u}_j, \ \boldsymbol{\Phi} = [\boldsymbol{\phi}_1, \boldsymbol{\phi}_2, ..., \boldsymbol{\phi}_N] \tag{9.28}$$

where a_i represents the time coefficient of each POD mode ϕ_i. The total energy of the flow can be obtained from the sum of all eigenvalues $\sum_{i=1}^N \lambda_i$. The size r of the reduced dimension of matrix can be determined by the relative energy in terms of the first r POD basis vectors according to the following formula,

$$\varepsilon(r) = \frac{\sum_{i=1}^r \lambda_i}{\sum_{i=1}^N \lambda_i} \tag{9.29}$$

The MVG wake flow field is identified by iso-surface of $R = 0.6$ (Fig. 9.18). A subzone is selected for the following POD analysis to reduce the computation complexity. The index information of the subzone is listed in Table 9.4.

The snapshot u_j is defined as follows:

$$u_j = \begin{bmatrix} u^{(j)}_{34,15,621} \\ \vdots \\ u^{(j)}_{104,15,621} \\ u^{(j)}_{34,16,621} \\ \vdots \\ u^{(j)}_{104,16,621} \\ \vdots \\ u^{(j)}_{I,J,K} \\ \vdots \\ u^{(j)}_{104,115,1020} \\ \vdots \\ v^{(j)}_{I,J,K} \\ \vdots \\ w^{(j)}_{I,J,K} \\ \vdots \\ w^{(j)}_{104,115,1020} \end{bmatrix} \quad \text{for } j = 1, ..., 200 \qquad (9.30)$$

where $u^{(j)}$, $v^{(j)}$, and $w^{(j)}$ are velocity components at $t = (188 + 0.5j)T$. Thus, the dimension of the original flow field matrix $U = (u_1, u_2, ..., u_{200})$ should be 8605200×200. To determine a proper number of first r modes to reconstruct the snapshots, the accumulation energy $\varepsilon(i)$ for $i = 1, ..., 200$ and the eigenvalues λ_i

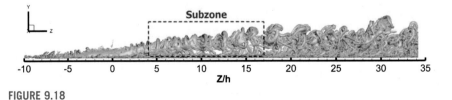

FIGURE 9.18

MVG wake flow field identified by $R = 0.6$ iso-surface.

Reproduced from Dong, X., et al., POD analysis on vortical structures in MVG wake by Liutex core line identification, Journal of Hydrodynamics 32, 2020, with the permission of Journal of Hydrodynamics.

Table 9.4 Parameters of subzone.

Start/End index	Subzone
I (In *x* direction)	34 ~ 104
J (In *y* direction)	15 ~ 115
K (In *z* direction)	621 ~ 1020

Reproduced from Dong, X., et al., POD analysis on vortical structures in MVG wake by Liutex core line identification, Journal of Hydrodynamics 32, 2020, with the permission of Journal of Hydrodynamics.

of the covariance matrix R in descending order except for the mean mode, which is named as the mode 0, are shown in Fig. 9.19. $\epsilon(i)$ represents the energy proportion of the first i modes, and thus the first 45 modes contain more than 80% of the total energy, which hints that there is a wide range of characteristic energy containing multiscales contributing to the velocity fluctuations in the MVG wake flow.

The spatial organization of the first few characteristic modes is presented based on Liutex identification in Fig. 9.20, which gives four characteristic POD modes with descend order identified by Liutex magnitude iso-surface (left) and the Liutex core line (right). In Fig. 9.20A, a pair of primary streamwise vortices with the largest scales is identified by $R = 4.0$ and it indicates that mode 0 is contributed by the time-averaged velocity flow field. In addition, comparing with the primary one, another pair of counterrotating streamwise vortices with the rather weak rotating strength and an opposite rotation axis direction is found below according to the Liutex

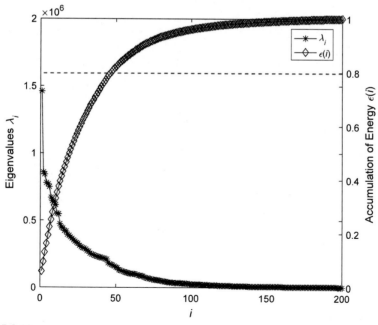

FIGURE 9.19

Eigenvalues of the covariance matrix and the accumulation energy for 200 POD modes.

Reproduced from Dong, X., et al., POD analysis on vortical structures in MVG wake by Liutex core line identification, Journal of Hydrodynamics 32, 2020, with the permission of Journal of Hydrodynamics.

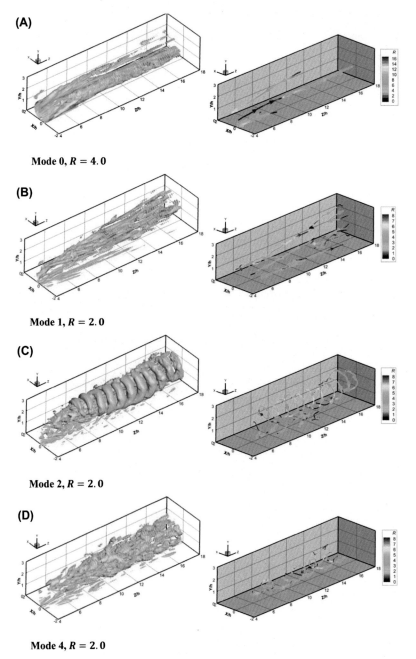

(A)

Mode 0, $R = 4.0$

(B)

Mode 1, $R = 2.0$

(C)

Mode 2, $R = 2.0$

(D)

Mode 4, $R = 2.0$

FIGURE 9.20

Vortical structures of characteristic POD modes in MVG wake identified by Liutex magnitude iso-surfaces (left) and by Liutex core lines (right). (A) Mode 0, $R = 4.0$; (B) Mode 1, $R = 2.0$; (C) Mode 2, $R = 2.0$; (D) Mode 4, $R = 2.0$.

Reproduced from Dong, X., et al., POD analysis on vortical structures in MVG wake by Liutex core line identification, Journal of Hydrodynamics 32, 2020, with the permission of Journal of Hydrodynamics.

core line distribution. It has been discovered by many experimental researches in the mean flow and confirmed to be the secondary vortex.

Mode 1 in Fig. 9.20B, which is a lower-order POD mode, accounts for 6.0% of the total kinetic energy. It can be observed from the $R = 2.0$ iso-surface in Fig. 9.20B that mode 1 is similar to mode 0 since it has the streamwise-oriented vortical structure distribution. It indicates that the streamwise component of the MVG wake is dominant in terms of kinetic energy contribution. According to the Liutex core line as well as the distribution of the Liutex magnitude, the vertical structures of mode 1 lift or descend when developing downstream and therefore becomes less organized. This suggests that the mode 1 is featured by the fluctuated roll-up motion of streamwise vortex structures.

Being different from the mode 1, the mode 2 in Fig. 9.20C has most unsteady feature of the wake structures and accounts for 3.6% of the total kinetic energy. Mode 2 and mode 3 are pairing and thus display the same vortex characteristics. According to the distribution of the Liutex core line, mode 2 presents the organized unclosed vortex rings, which shed with a relatively high frequency. Furthermore, another unsteady feature is shown by the upright vortices, which connect the primary wake and the wall vortices.

Fig. 9.20D gives the vortex structures for mode 4. It is not easy to get new findings from the iso-surface of $R = 2.0$ (left figure), while an amount of streamwise–spanwise-oriented flow patterns are discovered to be contained in the small-scale structures by Liutex core line of the right figure. These streamwise–spanwise-oriented vortices actually contribute to the low-frequency oscillations of the wall vortices.

The POD coefficients and the corresponding power spectra are shown in Figs. 9.21 and Fig. 9.22, respectively. In Fig. 9.21A, the time coefficients of the mode 0 show extreme small amplitude without any periodic characteristics and thus demonstrate their contributions to the steady part of the instantaneous flow field. Although $a(t)$ of mode1 is nonperiodic, it shows unsteady temporal characteristics. The coefficients $a(t)$ of mode 2 and mode 3 in Fig. 9.21B display extreme fluctuated and periodic curves along with the number of snapshots of the flow field N and match the behavior of the vortex ring shedding. Meanwhile, the peak Strouhal

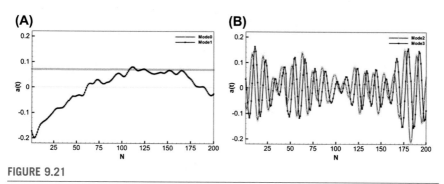

FIGURE 9.21

Time coefficients of the characteristic POD modes.

Reproduced from Dong, X., et al., POD analysis on vortical structures in MVG wake by Liutex core line identification, Journal of Hydrodynamics 32, 2020, with the permission of Journal of Hydrodynamics.

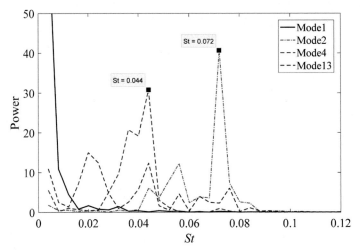

FIGURE 9.22

Spectra of the time coefficients of POD modes.

Reproduced from Dong, X., et al., POD analysis on vortical structures in MVG wake by Liutex core line identification, Journal of Hydrodynamics 32, 2020, with the permission of Journal of Hydrodynamics.

number $St = 0.072$ (Fig. 9.22), is the dominant high frequency of K-H vortices in MVG wake, where $St = f \cdot h / U_\infty$. Mode 4 takes a relative lower frequency ($St = 0.044$) due to low-frequency oscillations of the wall vortices.

9.2.5 Summary

The Liutex core line method and the snapshot POD are utilized in a supersonic wake flow of MVG at $Ma = 2.5$ and $Re_\theta = 5760$ to reveal the physical meaning for each POD eigenmode of the flow field. Compared with other vortex identification methods, the identification of the Liutex core line is verified to be the most appropriate method, which provides full information of a fluid rotation motion. The physical mechanism of each POD mode for multiple vortical structures is investigated by Liutex core line identification to give some revelations. Some conclusions are obtained and expressed as follows:

(1) Compared with other scalar-based vortex identification methods, the Liutex core line identification is applied in MVG wake flow and verified to be an appropriate method to provide various information of the motion of the fluid rotation.

(2) The mechanism of each POD mode for multiple vortical structures in MVG wake is investigated by the Liutex core line method. Mode 0 is contributed by the time-averaged velocity flow field. The mode 1 is featured by a fluctuated roll-up motion of streamwise vortex with large-scale structures, and the streamwise component of the MVG wake is demonstrated to be dominant in terms of the total kinetic energy contribution. Mode 2 reveals the most unsteady feature of the MVG wake and a dominant higher frequency of $St = 0.072$. In addition, another unsteady feature is obtained from the upright vortices, which connect the wake flow and the wall vortices. Mode 4 with the distribution of streamwise–spanwise-oriented vortices subjects to the low-frequency oscillations ($St = 0.044$) of the wall vortices.

9.3 **POD analysis by using Liutex vector input**

The input in the last section is velocity such as u, v, w in a xyz system. However, vortex can be exactly represented by Liutex and the input of Liutex, such as L_x, L_y, L_z, is natural to use. Fig. 9.23 and Table 9.5 provide a subdomain for the POD study on the vortex structure in the MVG wake.

The snapshot x_j is defined as

$$
x_j = \begin{pmatrix}
L^{(j)}_{x571,1,35} \\
\vdots \\
L^{(j)}_{x770,1,35} \\
L^{(j)}_{x571,2,35} \\
\vdots \\
L^{(j)}_{x770,2,35} \\
\vdots \\
L^{(j)}_{xI,J,K} \\
\vdots \\
L^{(j)}_{x771,1,35} \\
\vdots \\
L^{(j)}_{yI,J,K} \\
\vdots \\
L^{(j)}_{zI,J,K} \\
\vdots \\
L^{(j)}_{z770,110,105}
\end{pmatrix} \quad \text{for } j = 1, \dots, 120 \tag{9.31}
$$

where $L^{(j)}_x$, $L^{(j)}_y$, and $L^{(j)}_z$ are fluctuation Liutex vector fields at $t = (1512 + 2j)T$ and $X = (x_1, x_2, x_3, \dots, x_{120}) \in R^{m \times n}$ ($m = 4686000$, $n = 120$).

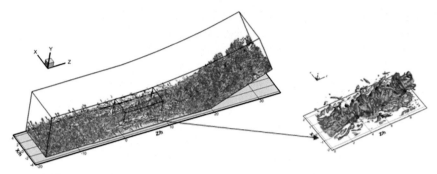

FIGURE 9.23

Vortex structure of Liutex in MVG wake.

Table 9.5 Parameters of subzone.

	Start index	End index
I (in *x* direction)	35	110
J (in *y* direction)	1	110
K (in *z* direction)	571	770

From the singular value of matrix X as shown in Fig. 9.24, there are two pairs of POD modes among the first sixth POD modes. Because of the significant difference between singular value of mode 1 and mode 2 (as shown in Table 9.6), those two modes are not considered as a pair. Modes 3−4 and modes 5−6 have instead similar singular values, which are then considered as paired. Modes 3−4 have strong

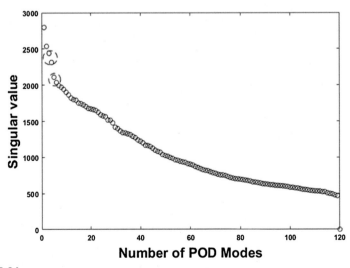

FIGURE 9.24

Singular value of matrix X.

Table 9.6 Singular value of the first six POD modes using Liutex vector directly (L_x, L_y, L_z) as an input.

	Mode 1	Mode 2	Mode 3	Mode 4	Mode 5	Mode 6
Singular value	2795	2535	2439	2314	2110	2037

primary vortices and modes 5—6 are secondary vortices. Both pairs are caused by shear layer, which generate spanwise vortex rings through K-H type instability or K-H modes. The pairing of POD modes is a common sign of the K-H instability.

Mode 1 has the highest energy compared to other POD modes (see Fig. 9.25). The low-order POD modes have higher energy than high-order ones, indicating that the difference of cumulative energy between two consecutive modes is decreasing.

It can be clearly observed that modes 1—2 are streamwise vortices while modes 3—4 and modes 5—6 have characteristics of spanwise vortex structure. Modes 3—4 and modes 5—6 display staggered array structure like vortex street rolled from K-H instability, which is caused by the fluctuation motion induced by vortex rings. These findings are similar to the ones shown in Section 9.2. There are six time coefficients of POD modes and two time coefficients are demonstrated in pairs like modes 3—4 and modes 5—6. All the three pairs have different fluctuation as shown in Fig. 9.26.

The vortex structure in Fig. 9.27 in the MVG wake subzone shows evidently that the head shape of K-H vortices is arc-shaped. It also illustrates that there is a vortex ring generated by MVG attributable to K-H instability. Kelvin—Helmholtz vortices lie on the top and circle around streamwise vortices along the Z-direction (Z-direction is the streamwise direction). K-H vortices are getting larger downstream

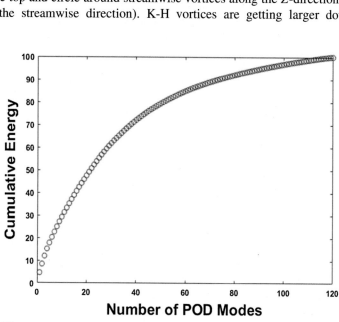

FIGURE 9.25

Cumulative energy.

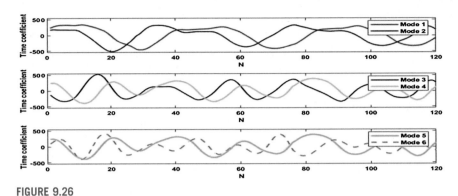

FIGURE 9.26

Time coefficient of the first six POD modes.

because of the growth of vortex pair spacing, which agrees with the result of Sun et al. (2012) and Sun (2015).

The results of loading data directly from velocity (u, v, w) by using $\widetilde{Q}_R$ and Liutex methods and loading data directly from Liutex vector (L_x, L_y, L_z) both show that there are pairs of the first four POD modes (3−6) over 120 snapshots, which is a strong sign of K-H instability. This supports Li and Liu's theory (2010) discovered through large eddy simulations that the mechanism of vortex ring generation should be K-H instability and MVG generates vortex rings to destroy the shock. Therefore, although Babinsky's theory addressed that the velocity profile getting fuller after applying MVG through experimental methods, which is well-known, the theory may not have scientific foundation and cannot be confirmed by either LES results or PIV observation. The POD analysis using Liutex input has got same conclusions as the velocity input has. It shows that mode 1 has the highest energy among all POD modes. There are two pairs of POD modes: modes 3−4 and modes 5−6. Modes 1−2 are dominated by streamwise vortices. However, mode pairs of 3−4 and 5−6 are strongly dominated by spanwise vortices. Actually, the pairing of POD modes is a common case, which is a sign of K-H instability. The absolute strength of rotation, or Liutex, is correlated with cumulative energy of rotation.

9.4 Summary

Mechanism of hairpin vortex formation is important in turbulence research. In this chapter, the new Liutex core line method and the proper orthogonal decomposition (POD) are applied to support the hypothesis that the ring-like vortex is formed by the Kelvin−Helmholtz (K-H) instability. The new modified Liutex-Omega method is efficiently used to visualize the shapes of vortex structures. The POD analysis shows that the eigenvalues and eigenvectors of POD modes are presented in pairs, which is typical in the K-H instability. This evidence is a strong proof that K-H instability plays a key role in the hairpin vortex formation from the Λ-vortex, which transforms

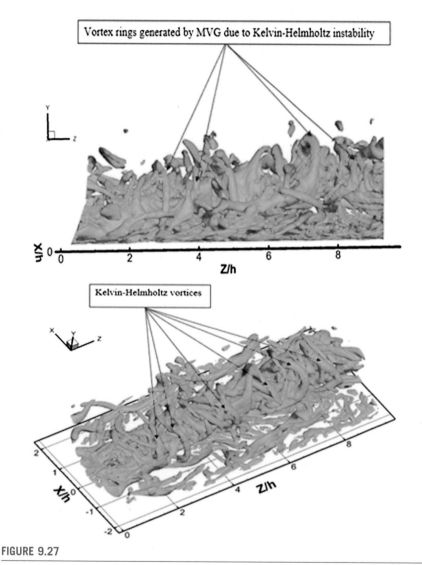

Vortex rings generated by MVG due to Kelvin-Helmholtz instability

Kelvin-Helmholtz vortices

FIGURE 9.27

Subzone of vortex structures in MVG wake.

the non-rotational vorticity or shear to the rotational vorticity or Liutex. This further confirms that the theory of Lambda vortex self-deformation to hairpin vortex has no solid scientific foundation. And then, the Liutex core line method and the snapshot POD are utilized in a supersonic wake flow of MVG. The K-H instability is confirmed again as the driven force of vortex generation. This study clearly shows that shock wave and boundary layer interaction (SWBLI) is really shock-Liutex interaction and the frequency of shock oscillations is determined by the Liutex

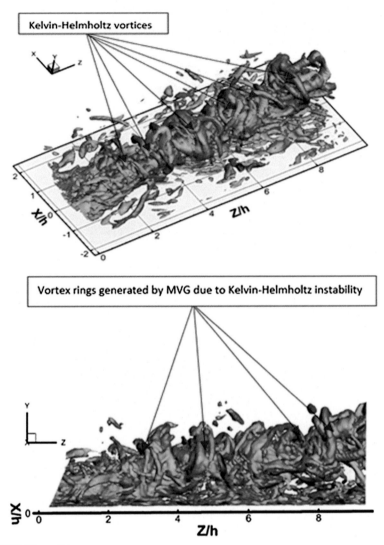

FIGURE 9.27 cont'd.

spectrum, which may pave the way to reduce the low frequency noises generated by the supersonic flight vehicles. Also, the Liutex vectors are used as the direct input for POD analysis. The result shows the same conclusions as the velocity input.

Liutex iso-surface and LXC-core line methods in DNS analyses

10

10.1 Visualizations of representative data and graphical highlights of the 3rd-G VI applications and findings

The introduction of Liutex system and the associated 3rd-G VI techniques provide an opportunity to visualize and to analyze the vortex structures from a brand-new perspective for fluid mechanics research. Toward this end, the current chapter aims at applying the Liutex system to revisit a number of popularly studied fundamental flow configurations within the fluid mechanics community. The studies are conducted by making use of the direct numerical simulation (DNS) and experimental database that is currently built through the joint collaborations among the co-author's research groups.

So far, the DNS database includes a number of representative configurations with simple geometries, such as the turbulent boundary layer (TBL) flows in a zero-pressure-gradient flat plate (ZPGFP), pressure-gradient-driven channel, square duct and square annular duct (PGDCHN, PGDSQD, and PGDSAD), and flows around single- or multicylinders, rib tabulators, airfoils, and wings, and so on. Albeit simple, these DNS data are fraught with very rich digital information of fluid physics and an abundance of vortex-related phenomena, such as the instability, transition, fully-developed turbulence, Karman vortex street, flow separations, and TBL, and so on. Therefore, revisiting these data based on the brand-new 3rd-G VI of Liutex theory is expected to discover and resolve a variety of unknown or unexplained fluid mechanics challenges. Within the context, the current section briefs a number of configurations currently available at http://simplefluids.fudan.edu.cn/publications/ book-elsevier with the representative ones being selected to be discussed in detail in the book, which include (1) flow around single and double cylinders confined in channel; (2) flow around rib tabulators confined in channel, (3) turbulence in Type-A (ZPGFP) and Type-B (PGDSAD) TBLs.

10.1.1 Flow around single and double cylinders

The flows over circular cylinders are one of classical research topics in fluid mechanics. Starting from the single-cylinder wake, many related configurations and

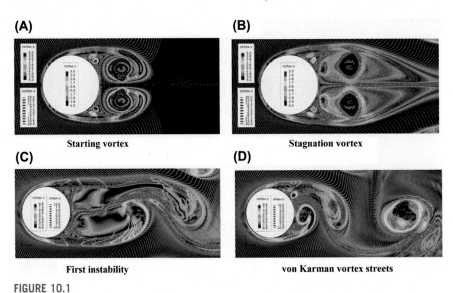

FIGURE 10.1

Two-dimensional flow around single cylinder confined in a channel.

Animations available at http://simplefluids.fudan.edu.cn/publications/book-elsevier/.

phenomena were intensively investigated, such as single and double cylinders with flow separation, vortex shedding, instability and Karman vortex streets, and so on (Robinson, 1991; Leweke and Williamson, 1998). The recent interests and attentions have been shifted to and focused on the more complicated secondary instability in the wake region because of the rapid progress of research methods based on both experiments and computations, which permits to detect the complicated dynamics in the three-dimensional (3D) vortical structures fundamental for understanding turbulences (Wu et al., 2019). The physical domain of such studies commonly involved a single cylinder (Fig. 10.1) or double cylinders (Fig. 10.2) in tandem placed in a confined or unconfined space with a uniform incoming flow. These configurations are important in understanding the fluid dynamics and revisiting these flows based on the Liutex system can not only reveal much richer vortex structures, but also lead to more discoveries of a variety of instability mechanisms. The current databases contain both the 2D and 3D visualizations of all the detailed transient vortex phenomena mentioned above.

(1) Visualization of vortex evolutions in the flow around 2D single cylinder confined in a channel.

(2) Visualization of vortex evolutions in the transition of 3D double cylinders in tandem confined in a channel

(A)

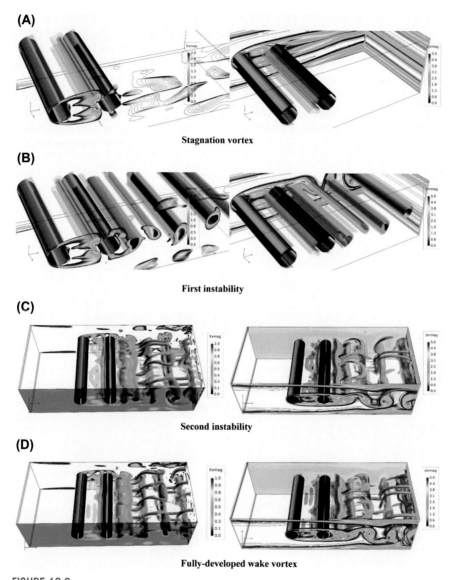

Stagnation vortex

(B)

First instability

(C)

Second instability

(D)

Fully-developed wake vortex

FIGURE 10.2

Three-dimensional transition of double cylinders in tandem confined in a channel.

Animations available at http://simplefluids.fudan.edu.cn/publications/book-elsevier/.

10.1.2 Flow around rib tabulators confined in channel

Regarding the flows around rib confined in a channel, the configurations are commonly applied in the cooling structures in a high-temperature turbine blade of aero-engine. Therefore, a variety of research attentions are recently paid to the configurations and the major interests are particularly concentrated on the thermal turbulence inside the channel with a strongly coupled flow and heat interactions. On the other hand, the surface roughness becomes an important issue from both manufacturing and fluid-flow heat-transfer perspectives. Since the channel and the roughness are at the scales of milli- and micron-meters, respectively, the fluid flow and heat transfer are significantly affected by the roughened surface in the channel. In addition, the roughened structures are located in the inner surface of turbine blade, which is difficult to be altered after the blade is casted and even more difficult to directly measure its heat-transfer effect by experiment. Therefore, the DNSs of thermal turbulence, as seen in Fig. 10.3, are in extremely high demands for these roughened channel with rib-tabulator structures by engine industry and our DNS database can provide the strongly-coupled thermal turbulence data for these configurations,

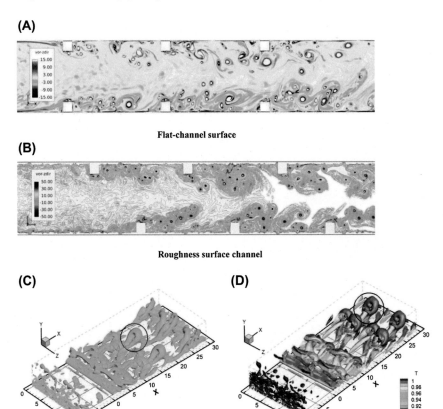

(A)

Flat-channel surface

(B)

Roughness surface channel

(C)   **(D)**

(c) 3-D bypass transition induced by rib with hairpin vortex structures presented by Liutex magnitude iso-surfaces

(d) Hairpin Liutex-iso-surface coloured by temperature to demonstrate convective heat-transfer enhancement mechanisms

FIGURE 10.3

Two- and three-dimensional flows around multiple rib tabulators: vortical structures in fully developed turbulence.

Animations available at http://simplefluids.fudan.edu.cn/publications/book-elsevier/.

which enables the hairpin flow structure formation study and the associated heat-transfer enhancement investigations mentioned in the following analysis.

10.1.3 Turbulence in Type-A (ZPGFP) and Type-B (PGDSAD) TBLs

As mentioned earlier in Chapter 7, the LXC-core line capability developed based on the Liutex theory opens the door to study the vortex kinematics and dynamics, as demonstrated by the current Chapter and Chapter 7, respectively. Moreover, the transient behaviors of these LXC-core lines illustrated by Fig. 10.4 at an arbitrary time instant

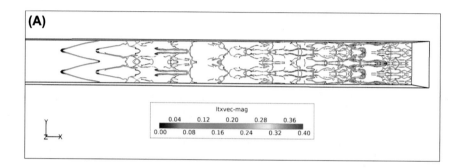

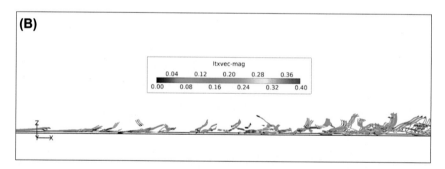

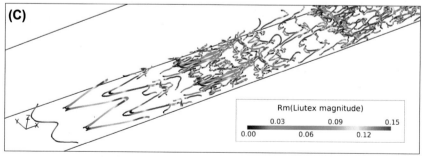

FIGURE 10.4

Type-A (ZPGFP) TBL LXC-core lines in transition stage from (A) top, (B) side, (C) 3D views; Type-B (PGDSAD) TBL LXC-core lines in fully-developed turbulence from (D) front, (E) side, (F) 3D view (colored by Liutex magnitudes).

Animations available at http://simplefluids.fudan.edu.cn/publications/book-elsevier/.

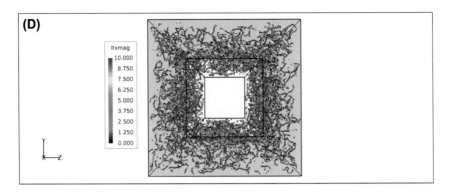

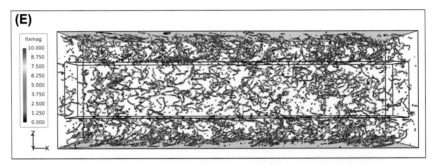

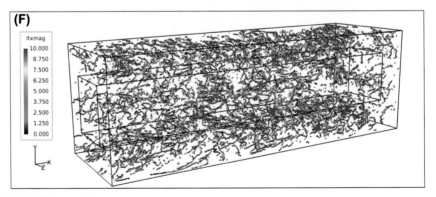

FIGURE 10.4 cont'd.

provide the fluid mechanics community and other vortex science—related communities with a very useful and reliable tool to study the vortex patterns and their evolutions since the structures defined by Liutex-core line are mathematically unique, which guarantees that the represented vortical structures are not contaminated and distorted by any threshold issue. Within the context, the time-sequencing LXC-core line evolutions for the two typical TBLs of ZPGFP(Type-A) and PGDSAD(Type-B) are provided at the website of http://simplefluids.fudan.edu.cn/publications/book-elsevier/, as a demo to promote and share the research in this aspect.

10.2 Hairpin-vortex formation based on 3rd-G VI of Liutex theory

10.2.1 Short review

Transition of boundary layer is a classical and an important phenomenon originally attracting popular attentions from the fluid mechanics community. From the physical perspective, as seen in Robinson (1991) and Liu et al. (2014), researchers were interested in investigating and clarifying the mechanisms giving rise to a transition. Regarding the engineering application, as demonstrated in Liu et al. (2018) and Wu et al. (2019), many fluid engineers were keen to fundamentally understand and actively control the transition process. Historically, the boundary-layer transitions can be classified into two distinctive types, namely the natural and the bypass transitions. Three typical stages were found in a natural transition, which included the stages of perceptibility, linear and nonlinear growths, leading to the phenomena of boundary-layer instability and breakdown. In a bypass transition reported in Buzica et al. (2019), the perceptibility and linear growth can only be maintained in a very short period of time and consequently are often overlooked in comparison with a natural transition.

A bypass transition is usually induced by the disturbances imposed on the inlet boundary. Lu et al. (2012a) applied the TS waves to generate an accelerated natural transition and further, the hairpin vortices in a forest pattern were obtained by Liu et al. (1995). On the other hand, Wu et al. (2009) introduced the periodical intermittent localized disturbances in the freestream to trigger the transition of a perturbed boundary layer. The hairpin packets were found, which played an important and active role in the breakdown of boundary-layer bypass transition. Moreover, the inception of transitional turbulent spot was found by Wu et al. (2015b), which was considered similar to the secondary instability of natural transition in a boundary layer, or more specifically a process from the spanwise vortex filament changing into a Λ-vortex and thereafter into a hairpin packet. In addition, the surface roughness effects were widely introduced to induce and study a bypass transition, including a step in the two-dimensional form by Le et al. (1997) or a rib structure in Miura et al. (2010), a three-dimensional microramp in Lu and Liu (2012) and Yan et al. (2012). These microvortex generators can give rise to a series of vortex pairs with the general pattern characterized by a variety of hairpin vortices being imposed along the interface between freestream and boundary layer (see Ye et al. 2016, 2018).

The discovery of coherent structure is a milestone work in turbulence research, which was traditionally explained as a cascade of vortices in the streak patterns. Within this regard, the hairpin-shaped vortices at one scale level were, for the first time, postulated and introduced by Theodorsen (1952) with a presence of forward-leaning pattern, which, later on, turned out to be a valuable finding that provided the explanation for the turbulence generation and sustainment in the boundary layers. Subsequently, Head and Bandyopadhyay (1981) conducted an important research, which obtained the vortical structures at a certain distance away from the wall and visualized these structures by finding out that these structures tended to generally incline at an approximate angle of 45° over a wider range of Reynolds numbers. Smith et al. (1991) carefully reviewed these early studies on the detailed dynamics of near-wall turbulence and concluded that the gradient of shear forces played an important role inside the boundary layer, which dominated the generation, growth, and deformation of the hairpin vortical structures.

So far, the hairpin structures are much more clarified in terms of their formation processes. Based on the observations of the characteristics of vorticity iso-surfaces, firstly, the stretch of a spanwise vortex, or filament, transforms the structures into a Λ-shaped structure in the streamwise direction. Secondly, the Λ-shaped structure is further elongated and lifted, giving rise to an eventual hairpin pattern. However, the paths leading to the hairpin pattern can be in multiple ways, as exhibited by Liu et al. (1995, 2014) and Zhou et al. (1999) in terms of the generation process leading to the hairpin structures or the hairpin packets. Based on their findings, the starting vortical structures, with their threshold strength exceeding a certain value relative to the mean stream, can lead to the formations of new hairpin vortices upstream of the primary vortical structures. Moreover, the younger vortical structures were also induced downstream by the primary structures. Consequently, the new-born hairpins can be found both upstream and downstream of the primary one. Although a hairpin-vortex study often deals with single structure, the hairpin vortices are usually found in packets. It is confirmed by Eitel-Amor et al. (2015) that the laminar-turbulent transition is fraught with these packets. Meanwhile, the hairpin structures were evidently found decaying with the temporal evolution, which were more often observed in the low Reynolds number or transition flow. However, the quasi-streamwise streak-shaped vortices, instead of hairpin vortical structures, were found dominating in a fully-developed turbulence. Moreover, the lifespan of these large-scale streaks were claimed by Jiménez (2015) and Lozano-Durán and Jiménez (2014) proportional to their sizes and the structures tended to be geometrically and temporally self-similar.

The hairpin-vortical structures were recently generated by a new way, as reported in Wang et al. (2015), in which the hairpin structures were found originating from a near-wall streamwise vortex pair. The hairpin head with the arch-shaped structures first grew from the downstream end of a structure strongly spiraling in streamwise and then attached to the weaker streamwise rotating structure with an opposite sign in their overlaping zone, which thereby formed a complete structure of hairpin shape.

Liu et al. (2014) found that the Λ-shape and ring-like vortical structures tended to take their formations separately and independently. When the Λ-shaped structure becomes stronger, a shear layer with higher strength always occurs above the roots of Λ-structure, which is induced by an ejection of the Λ-shaped rotation that brings the low momentum fluid from the bottom of boundary layer. Consequently, the instability due to the shear layer gives rise to the formation of a new-born vortex with the ring-shaped structure. The ring-shaped vortex explained by Liu et al. (2014) is a similar hairpin-shaped structure first introduced by Robinson (1991).

As a matter of fact, many possibilities exist in the formations of various types of vortical-structure patterns when vortices merging with each other in different amalgamation scenarios. However, as demonstrated by Tomkins and Adrian (2003), three idealized schemes in the vortex merging processes were schematically depicted and proved by either experiments or DNS data. The hairpin merging process is one of the idealized models, within which the inner legs of vortices were typically annihilated. These idealized possible schemes were proved by experiments conducted at the higher Reynolds numbers of $Re_\theta = 1015$ and 7705 by Tomkins and Adrian (2003). Based on the DNS data from Li et al. (2020), it was found that the hairpin vortex was possible to evolve from a pair of arch-shaped vortices at the low Reynolds numbers of $288 < Re_\theta < 425$. This merging process was observed by the data visualization using the 3rd-G VI approach, specifically the LXC-core lines, with the detailed processes being presented and analyzed in this chapter. The inherent mechanisms are subsequently studied and in-depth clarified, which promote the current understandings of boundary-layer transition and the vortical structure evolution in a wall-bounded turbulence. In the following analyses, the nomenclatures conform with the presentations set by Head et al. (1981) and Robinson (1991) in the descriptions of arch- and hairpin-shaped vortical structures.

Several existing researches conducted the coherent structure coupled heat-transfer simulations. Zhao et al. (2016), for instance, studied the thermal fields affected by the coherent structures, such as the vortices and velocity streaks. However, the details were not provided in terms of the streak characteristics for the temperature field. Although Zhao et al. (2016) mentioned the temperature streak structures, the detailed distribution patterns and their connections to the vortex streaks were yet to be analyzed and reported until Li et al. (2020). The chapter provides the inherent correlations between the temperature and velocity streaks and the quantified and much more clarified mechanisms behind these vortical phenomena in the wall-bounded turbulence enabled by the current 3rd-G VI techniques. These 3rd-G VI approaches applied to extract the vortex structures include the iso-surfaces (Liutex magnitude scalar), rotational vectors (Liutex vectors), and LXC-core lines poineered by Xu H. et al. (2019) and Gao et al. (2018).

10.2.2 Physical model, boundary conditions, and DNS data validation

The computational domain is schematically plotted in Fig. 10.5 under the Cartesian coordinate system. The domain dimensions are set at $L = 132e$, $H = 24e$, and $S = 16e$, with e being the square rib height. The rib is located at a distance of $L_l = 11e$ from the inlet and is place on the surface of lower wall.

The no-slip velocity condition was set at all the wall surfaces, including the upper and lower walls and the rib surface. The nondimensional temperature was given at $T_w = 1.1$. The pressure gradients normal to the walls were set to zero and the density

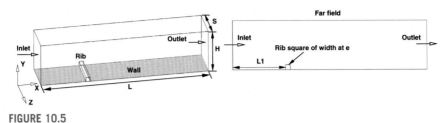

FIGURE 10.5

Schematic drawing of the physical model and computational domain.

was calculated by the state equation of the ideal-gas model. The back pressure was prescribed as the environmental pressure on the outlet where all the other variables were determined by the zero-gradient condition in the streamwise direction or, equivalently, the Neumann boundary condition. On the inlet, the nondimensional temperature was set at $T_i = 1.0$ and the reference pressure P_{ref} was determined based on the ideal-gas state equation with a Mach number at $Ma = 0.2$. The pressure on inlet was given by $P_i = 1.002\, P_{ref}$, which took into the account the pressure losses induced by the rib.

To mimic the flow in reality and to quickly bypass the transition, the disturbances were imposed on the inlet. The inlet conditions were generated by a separate DNS data of the fully-developed turbulence in channel. The instantaneous turbulent velocities on a crosssection in between the channel plates were extracted out and the data between $0 < Y < 1$ were then normalized and imposed onto the inlet of the computational domain. Fig. 10.6A provides the instantaneous vortices represented by the Liutex scalar magnitudes, which stands for the rigid rotational strength on the inlet. The corresponding velocities in the cross-streamwise directions are given in Fig. 10.6B and C.

Turbulence intensity defined by $I = \left[\left(\overline{u'u'} + \overline{v'v'} + \overline{w'w'} \right) \middle/ 3 \right]^{0.5}$ is an important and characteristic quantity with the overbar denoting the time-averaging operation for the relevant velocity components given in Hosseinverdi et al. (2019). The time-averaged distribution of turbulence intensity is plotted in Fig. 10.7 along the direction (y) normal to the wall, which indicates that the turbulence intensity, if averaged in y direction, is roughly at 5.28% relative to the incoming velocity. Generally speaking, the turbulence intensity is considered in a lower range if less than 1% and higher if more than 10%. Therefore, the current turbulence inlet condition is in the range of a medium level and the highest turbulence intensity occurs at a distance of $y = 0.07$ away from the wall. The time-averaged velocity is symmetrical with respect to the plane of $z = 8.0$ in the spanwise direction.

The DNS simulation was performed with the criterion numbers at $Re = 1200$ and $Pr = 0.71$. The time advancement was conducted at the time step of $\Delta t = 4 \times 10^{-4}$, which guaranteed the stable computation. After the turbulence being fully-developed, the statistics, including the time-averaged velocity and Reynolds stresses, are then taken over another 300,000 time steps and are analyzed in the following sections.

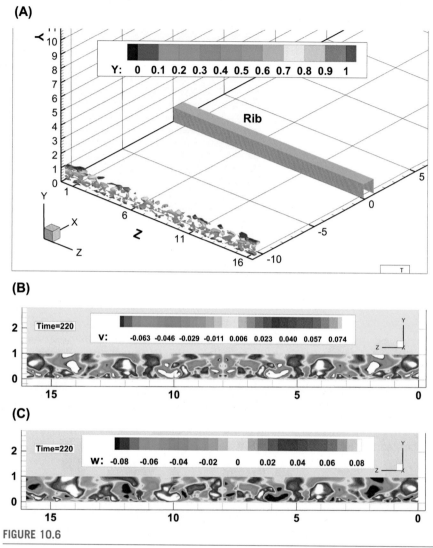

FIGURE 10.6

Inlet conditions of (A) the three-dimensional view of Liutex iso-surfaces near the inlet; contour plot of velocities of (B) v-component and (C) w-component on the inlet.

The code validation and grid independence must be conducted, which is important for the numerical simulation being claimed as DNS. The paper from Li et al. (2019) carefully examined the convergence of grids in a channel (CHN) with rib at $Re = 1200$. For the channel, it was confirmed that the grids of $428 \times 135 \times 64$ were sufficient to resolve the flow at $Re = 1200$, with the grid resolution in each direction at $\Delta x_{min}^+ = 0.31$, $\Delta y_{min}^+ = 0.302$, and $\Delta z^+ = 7.29$. For the current investigation, the grids of $1420 \times 110 \times 128$ were selected to provide that $\Delta x_{min}^+ = 0.31$,

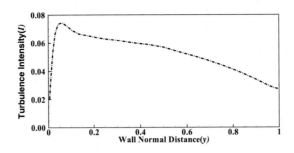

FIGURE 10.7

Time-averaged turbulence intensity distribution on the inlet boundary.

$\Delta y_{min}^+ = 0.302$, and $\Delta z^+ = 7.29$. Tables 10.1 and 10.2 list the detailed parameters and the grid resolution in y direction is presented in Fig. 10.8 with the minimum wall distance of $\Delta y_{min}^+ = 0.302$.

10.2.3 Analyses of the DNS data based on 3rd-G VI approach

(1) Data presentations and visualizations

The vortical structures with two different thresholds of $|R| = 0.05$ and 0.2 are plotted in Fig. 10.9A and B, which clearly indicate that the traditionally so-called vortex breakdown phenomena are, in fact, due to the structure's strong dependency on the selected thresholds. The same drawbacks exist for all the other traditional VI methods based on iso-surfaces, such as Q, λ_2, or λ_{ci} criteria. On the contrary, the LXC-core lines, as colored by the Liutex scalar magnitude of $|R|$, present the

Table 10.1 Parameters of domain ($L \times H \times S$), mesh ($Nx \times Ny \times Nz$), Re number, and Pr number in the DNS simulations.

	$L \times H \times S$	$Nx \times Ny \times Nz$	Re	Pr
TBL	132 × 24 × 16	1420 × 110 × 128	1200	0.71
CHN	42 × 10 × 8	428 × 135 × 64	1200	0.71

Table 10.2 Grid resolutions in the streamwise (x), spanwise (z), and wall-normal (y) directions with $^+$ denoting the quantities measured by the TBL or frictional velocity scale.

	$\Delta x_{min}^+ (\Delta x_{min})$	$\Delta x_{max}^+ (\Delta x_{max})$	$\Delta y_{min}^+ (\Delta y_{min})$	$\Delta z^+ (\Delta z)$
TBL	0.31(0.0053)	5.83(0.10)	0.30(0.0052)	7.29(0.13)
CHN	0.31(0.0053)	8.45(0.15)	0.30(0.0052)	7.29(0.13)

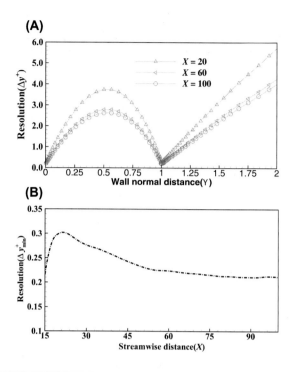

FIGURE 10.8

Grid resolution Δy^+ distributions (A) in the wall-normal direction at three representative streamwise locations of $x = 20$, $x = 60$, $x = 100$, and (B) in the streamwise direction of x.

strongly nonuniform distributions of $|R|$ along the strings, which are therefore definitively independent on any thresholds of $|R|$ as seen in Fig. 10.9C and D. The comparisons between the Liutex iso-surface and LXC-core lines clearly demonstrate that a unique tool is enabled by the LXC-core lines to investigate the vortex structures and their kinematic and dynamic evolutions.

Fig. 10.9C and D produce the typical LXC-core lines structures evolving downstream at a specific time instant. The spanwise LXC-core line can be pretty clearly seen right behind the rib. When traveling downstream, these spanwise-dominated LXC-core lines start to oscillate in the spanwise and therefore, the lines are twisted, which first start to take the arch shape and eventually evolve into the typical Λ- and hairpin-shaped structures downstream. An important observation, as seen in Fig. 10.9D, can be made that the rotating strengths of LXC-core lines, as represented by the colors along the lines, are increased significantly in the bypass transition regime, which roughly occurs in $5 < X < 20$ and then the strengths start to decay in the regions downstream. After traveling a long distance downstream away from the rib, the colors of LXC-core lines tend to approach to lower rotating strength regime, suggesting that the vortical strengths become weaker surrounding the core lines.

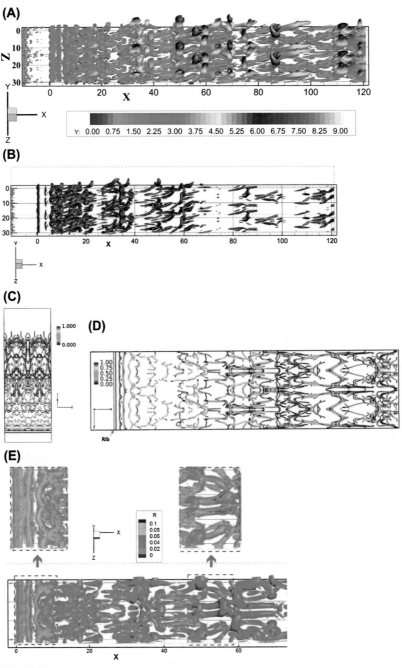

FIGURE 10.9

Three-dimensional (3D) views of Liutex magnitude iso-surfaces with thresholds at (A) |RI = 0.05 and (B) IRI = 0.2; LXC-core lines (C) 3D view and (D) Top view; (E) Local LXC-core lines overlapped by the vortical structures represented by the Liutex iso-surfaces visualized with a threshold at IRI = 0.05.

The global view of vortices in the TBL, exhibited by the Liutex magnitude iso-surfaces, is presented in Fig. 10.9A, where the Liutex iso-surfaces are plotted at |R| = 0.05 by Li et al. (2020). The rib block is located in between $0.0 \leq X \leq 1.0$. The vortices in a fully-developed turbulence were introduced into the flow field through imposing a separate time-sequential DNS solution in a channel onto the inlet, which provided the vital inflow disturbances. Before reaching the rib, these vortices were dissipated to some degrees, which might be attributed to the incompatibility between the inflow disturbances and the mainstreams. However, when the stream flows past over the rib as seen in Fig. 10.9D, the large-scale vortical structures in spanwise were generated and stimulated immediately beyond the rib. Comparing to the natural transition as given by Liu et al. (1995, 2014), the current bypass transition took place within a fairly shorter evolution distance. The transition time and distance were extremely shortened due to the existence of rib structure. However, various vortices, including both small- and large-scale vortical structures, were found and visualized downstream as seen in Fig. 10.9. The distinctive phenomenon is the vortex lift-up in the process of migrating downstream, which is clearly presented by the red-colored hairpin vortex heads depicted by the Liutex iso-surfaces. It is worth to mention that the vortex scales in the bypass-transition regime, i.e., approximately $5 < x < 20$, tend to be smaller and denser comparing to those in the fully-developed region downstream, as demonstrated in Fig. 10.9. The instantaneous vortical structures were examined carefully at a variety of time instants and the phenomena were observed repeating consistently.

The DNS data presented in Fig. 10.9 can be visualized to contain the vortices at large scale with only a single layer occupying the fully-developed TBL. However, the results given by Liu et al. (1995, 2014) demonstrated the multilayer vortices residing in the entire domain of fully-developed TBL. Liu et al. (2014) systematically analyzed the physical mechanisms of the turbulence generation and maintenance based on the multilayer vortical structures. Wu et al. (2009) also suggested the existence of multilayer vortices in the fully-developed TBL (2009). The sizes of vortices in the current study tend to be large in terms of the TBL thickness and clearly these vortices are sparser in the fully-developed region comparing to those given by Wu et al. (2009). The major reason can be attributed to the effects of momentum Reynolds number, which is in the range of $250 < Re_\theta < 438$ for the current study. However, the remarkably higher momentum Reynolds numbers of $750 < Re_\theta < 940$ were reported by Wu et al. (2009). Within the context, the present DNS data suggest that the turbulence with a lower Re tends to nurture the larger-sized and sparser-spaced vortices in the TBL.

The instantaneous streamwise velocity contours are presented in Fig. 10.10A and B on the spanwise slices of $z = 1$ and $z = 8$, respectively. The streamwise velocity is found increasing in these figures when the flow passes over the rib. In addition, the recirculation zone with pretty large area exists at the rear of rib, which qualitatively conforms with the experimental observation in Fouladi et al. (2017). The instantaneous temperature contours are given on the slices of $z = 1$ and $z = 8$ in Fig. 10.10C and D, respectively. The upward lifting motions of the vortices induce

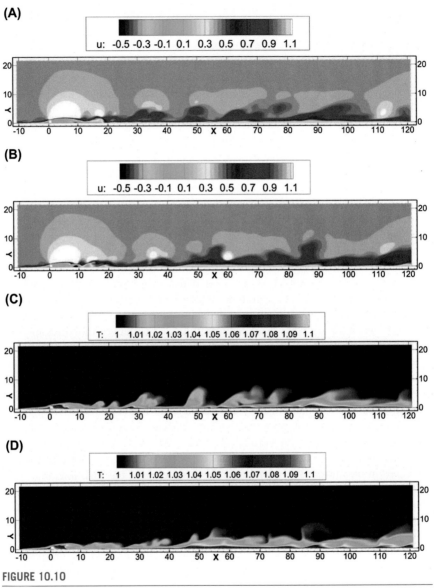

FIGURE 10.10

Contours distributions of: (A) the streamwise velocity component at spanwise slice of $z = 1$ and (B) $z = 8$; the temperature at spanwise slice of (C) $z = 1$ and (D) $z = 8$.

the local high-temperature zone being lifted up and inclined. The vortical flow structures pump the near-wall high-temperature fluids into the low-temperature stream located in the upper layer away from the wall. Natrajan and Christensen (2006) pointed out that these up-inclined hairpin vortices essentially dominate the energy transfer process in the semi-log layer of TBL near wall, as explained in detail by Wang D. et al. (2019).

(2) Validations based on the statistics of flat-wall-bound thermal turbulence

The variation of boundary-layer thickness (δ) with the rib height (e) is plotted Fig. 10.11A with the horizontal axis (x) being the streamwise direction. The corresponding variations in Fig. 10.11B are presented for the momentum thickness Reynolds number (Re_θ) and the frictional Reynolds number (Re_τ) in a region of $30 < x < 120$. Re_θ based on momentum thickness is defined as $Re_\theta = \theta U_\infty / v$ with θ being the momentum thickness, U_∞ being the free-stream velocity, and v being the kinetic viscosity. Similarly, Re_τ based on the boundary-layer thickness δ, the frictional velocity u_τ, and the kinetic viscosity v is given by $Re_\tau = \delta u_\tau / v$, as explained by Schlatter and Örlü (2010) and Pirozzoli and Bernardini (2013). Both Re_θ and Re_τ were found monotonically increasing with the streamwise distance of x by Schlatter

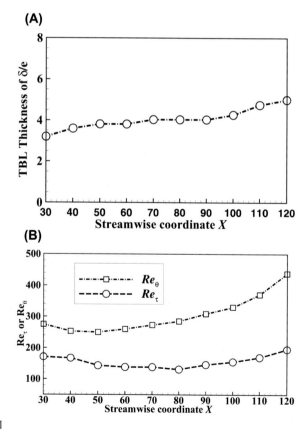

FIGURE 10.11

(A) Variation of boundary-layer thickness δ with the rib height e and (B) Re_θ and Re_τ variations in the streamwise direction.

and Örlü (2010) in a ZPGFP TBL. However, for the present study, Re_θ and Re_τ provide a performance of firstly decreasing and thereafter increasing, which can be attributed to the effect of rib. The rib height was arranged at the same order of magnitude with the TBL thickness. Consequently, the recirculation zone behind the rib presents a remarkable influence on the TBL downstream, which causes the behaviors of Re_θ and Re_τ in Fig. 10.11B.

To study the flows in the region downstream the rib, five representative locations are selected, where the data are extracted out to examine and validate the relation between Re_θ and Re_τ. The current results, as demonstrated in Fig. 10.12, are in a good consistence with the power-law relation of $Re_\tau = 1.13 \times Re_\theta^{0.843}$ given by Schlatter and Örlü (2010), which confirms the validity of power-law relation for lower Reynolds number since the Re_θ and Re_τ in the current study, as seen in Fig. 10.12, are much lower than the existing cases from Wu et al. (2009), Schlatter and Örlü (2010), and Liu et al. (2014).

The streamwise velocities at three representative locations within the recirculation zone are given in Fig. 10.13A−C. At the streamwise location of $x = 3$, a negative velocity zone is found, which is roughly at one rib height in the wall-normal direction of y. However, the back-flow velocity is small and not quite evident compared to the reverse velocity at the location of $x = 6$. For the flow at $x = 10$, the recirculation zone is found existing, but with a height being less than that at the location of $x = 6$. The recirculation zone vanishes when increasing x beyond the location of $x = 10$. Thereafter, the velocity profiles are gradually recovered as presented in Fig. 10.13D−F. The tangible accelerations are found near $y = 2$ on the top of the rib. Due to the conservation of mass and momentum in the streamwise direction, the acceleration effects persist beyond the rib, which causes an evidently elongated recirculation zone behind the rib.

As well known, the skin-friction coefficient is a key flow parameter defined as the nondimensional wall shear stress τ_w and calculated by

$$C_f = \frac{\tau_w}{(\rho_\infty U_\infty^2)/2} = \frac{\mu(\partial u/\partial y)_w}{(\rho_\infty U_\infty^2)/2}, \quad (10.1)$$

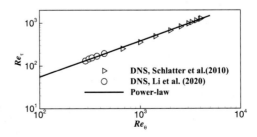

FIGURE 10.12

Variations of Re_τ with Re_θ for the five representative locations equally distanced from $x = 80$ ($Re_\theta = 285$) to $x = 120$ ($Re_\theta = 438$).

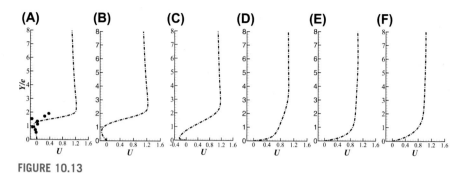

FIGURE 10.13

Streamwise velocity profiles at the selected six locations of (A) $x = 3$; (B) $x = 6$; (C) $x = 10$; (D) $x = 20$; (E) $x = 40$; (F) $x = 80$ with the solid dots representing the experiments by Fouladi et al. (2017) for a ribbed plate.

In Fig. 10.14A, C_f is found negative in the zone of recirculation behind the rib. When $x > 20$, C_f gradually descends downstream along the streamwise direction of x. The skin-friction coefficient in Fig. 10.14B suggests that the fully-developed turbulent zone tends to have a larger C_f than the one in laminar region based on the Blasius theory elaborated in the books from Chen (2002) and White (2012). However, the current data underpredict the solution given by the power-law theory in the books, which may be attributed to the lower Reynolds number in the current study since the power-law theory was introduced for the fully-developed TBL on the semi-infinite plate with a higher Reynolds number. On the other hand, the rib can exert a non-negligible effect on the skin-friction coefficient. The effects of rib impose an acceleration in the transition of boundary layer, which is essentially similar to the natural transition on a flat plate with sufficient length. However, the rib effect is actually equivalent to imposing a bypass by adding a certain length of flat plate. Within the context, the Re_x axis is equivalently shortened in Fig. 10.14B in terms of the friction coefficient variations. Or equivalently, the C_f variation curve in Fig. 10.14B is elongated to the right to some extent, which yields a C_f curve well in conformity with the power-law predictions.

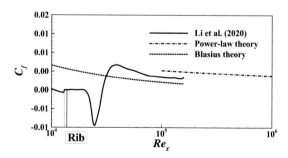

FIGURE 10.14

Time-averaged skin-friction coefficient variations with (A) the streamwise coordination and (B) the Re_x defined as $Re_x = x_d U_\infty / v$ with x_d being the distance measured from the inlet (Chen, 2002).

Similar to C_f, another key flow parameter of pressure coefficient C_p is commonly defined as

$$C_p = \frac{P - P_0}{\left(\rho_\infty U_\infty^2\right)/2} \qquad (10.2)$$

with the reference pressure of P_0 being chosen at $x = -4.5$ and the inlet density and velocity being set at $\rho_\infty = 1$ and $U_\infty = 1$ as suggested by Miura et al. (2012) and Matsubara et al. (2015). Fig. 10.21 presented a significant pressure drop right behind the rib and the current results present a quite similar trend in line with the data reported by Miura et al. (2012).

On the thermal aspect, the well-known Nusselt number is the key parameter to measure the heat-transfer intensity on the wall surface, which is defined as

$$Nu = -\frac{L}{T_w - T_\infty}\left(\frac{\partial T}{\partial y}\right)_w \qquad (10.3)$$

with $T_\infty = 1.0$ being the inlet temperature and $T_w = 1.1$ being the temperature on the wall. $L = 10e$ is chosen as the reference length in Miura et al. (2012). Fig. 10.22 further verifies that the current results coincide with the data reported by Miura et al. (2012) in terms of the nondimensional wall-heat flux.

The Reynolds normal stresses, or turbulence intensities, are well known given by $\overline{u\prime u\prime} = \overline{uu} - \overline{u}\cdot\overline{u}$, $\overline{v\prime v\prime} = \overline{vv} - \overline{v}\cdot\overline{v}$ and $\overline{w\prime w\prime} = \overline{ww} - \overline{w}\cdot\overline{w}$ in x, y and z directions, respectively, which yields the turbulent kinetic energy being defined as $k = (\overline{u\prime u\prime} + \overline{v\prime v\prime} + \overline{w\prime w\prime})/2$. The Reynolds shear stress between mainstream and wall-normal directions is calculated as $-\overline{u\prime v\prime} = -(\overline{uv} - \overline{u}\cdot\overline{v})$.

The Reynolds stresses and the turbulent kinetic energy, given in Fig. 10.17, are quite intensively distributed inside the TBL beyond $x = 5$ and in particular, the zone in $5 < x < 20$ presents the significant turbulence intensities and Reynolds shear stress, which indicates that the turbulence is quite active in the region. On the other hand, Fig. 10.17 suggests that the vortices be pretty younger in the region of $5 < X < 20$ with the sizes of vortices being smaller and the number of vortices being denser comparing to the region downstream. Therefore, it is concluded that the bypass transition takes place approximately in the region of $5 < X < 20$. By checking the C_f and C_p in Figs. 10.14 and 10.15, respectively, it is obvious that these

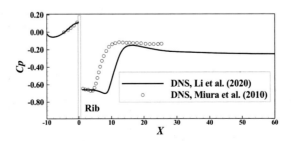

FIGURE 10.15

Mean pressure coefficient distributions.

coefficients are peaked with their maxima in the region and correspondingly, the Nuseelt number also is highlighted by its maximum, which clearly indicates that the bypass transition process is quite energy concentrated and involves a remarkable turbulence energy convertion and transfer and thereby, the friction drag is enhanced and the heat transfer is promoted, as demonstrated in Figs. 10.14 and 10.16.

10.2.4 Hairpin vortex generation mechanisms based on 3rd-G VI

Hairpin-shaped vortex is probably the most distinctive and featured structure being depicted in many experimental observations and was visualized by a variety of simulation data of wall-bounded turbulence as comprehensively reviewed by Adrian (2007). The closely-related phenomena also include the ejection, sweep, streak, and coherent structures in a wall-bounded turbulence as commented by Liu et al. (2014). Within the context, it is necessary and valuable to thoroughly understand the hairpin kinematic formation mechanisms and thereafter to further control the dynamics of these vortical motions. In principle, the hairpin formation can be induced by many possibilities due to the complexity of the wall-bounded turbulence. However, in general, the existing researches on the hairpin vortices can be conceptually classified into the three essential paths that can lead to the formation of a hairpin. Based on the current DNS data, the existing three ways leading to a hairpin are comprehensively reviewed. Moreover, to demonstrate the importance of 3rd-G VI method, a careful vortex identification practice enabled by the LXC-core lines was conducted in Li et al. (2020) and successfully revealed a fourth path giving rise to a hairpin-shaped vortex, which is here considered an appropriate topic to be introduced and elucidated in more details by the current chapter as following.

(1) Hairpin vortex originated from spanwise vortex

The first and classical evolution path leading to a hairpin vortex was reported by Liu et al. (2014) where a spanwise vortex was locally stretched in the mainstream direction by an existing disturbance. Usually, the disturbance was caused by a large velocity gradient near the wall, which imposed a strong shear effect to stretch the vortex in the streamwise direction. Consequently, a Λ-pattern vortex took place in

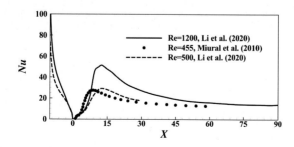

FIGURE 10.16

Mean Nusselt number distributions.

the evolving process, as seen in Fig. 10.18. Meanwhile, an upward-induced velocity became dominant, which gave rise to a lift-up for the head of Λ-shaped vortex explained by Zhou et al. (1999). Eventually, a hairpin vortex took the formation as illustrated in the schematic drawing presented in Fig. 10.18.

Fig. 10.19A provides the spanwise vortex in large scale behind the rib at the time instant of (A) $t = 10$. From then, the spanwise vortex is locally stretched in the streamwise direction and the three-dimensional structures are emerging in Fig. 10.19B at the time instant of (B) $t = 20$. Thereafter, the Λ-shaped vortex takes place at the time instant of (C) $t = 30$ in Fig. 10.19 with the nonlinear instability being enhanced. Meanwhile, the head of Λ-shaped vortex is gradually lifted up in the process of moving downstream. With the nonlinear instability being further promoted by the strong effects of shearing and stretching in the streamwise direction, the hairpin vortices are grown up and come into formation exhibited in

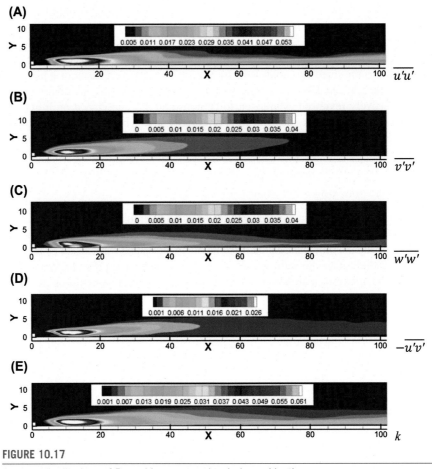

FIGURE 10.17

Contour distributions of Reynolds stress and turbulence kinetic energy.

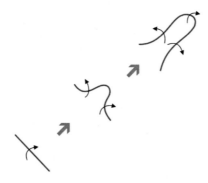

FIGURE 10.18

Schematic evolution path from a spanwise vortex to a hairpin vortex.

Fig. 10.19D at the time instant of (D) $t = 40$. The hairpin head is further lifted due to the upwardly induced velocity. In the current study, the existence of rib is equivalent to an effect of imposing strong disturbances that accelerate the transition process and thereafter, the classical evolution path is commonly taking place from a spanwise to a hairpin-shaped vortex as aforementioned.

The evolution process of a hairpin vortex downstream the rib is further demonstrated in Fig. 10.20 by plotting and visualizing the LXC-core lines, which apparently gets rid of the multi-structure issue due to the threshold selection in Fig. 10.19 and limpidly exhibits the vortex evolution history discussed earlier. A spanwise vortex, as represented by a straight spanwise LXC-core line in Fig. 10.20A, comes into existence behind the rib due to the rib's wall and blockage effects. The spanwise LXC-core line thereafter starts being stretched into a Λ-shaped vortex exhibited in Fig. 10.20B and C and then further develops into a hairpin pattern presented in Fig. 10.20D. The vortex evolutions are more cleanly and uniquely depicted here in Fig. 10.20, which is obviously attributed to the reason that the magnitudes of Liutex vector are the local rigid spinning speeds of fluid, the tangential direction of the core line is the local spinning axis, as explained by Xu H. et al. (2019). Consequently, the LXC-core line structures, strictly speaking, are more appropriate to represent the vortical structures than either the Liutex magnitude iso-surfaces or the vorticity magnitude iso-surfaces in the traditional VI.

(2) Hairpin vortex originated from streamwise vortex pair

As well known, streak structures and streamwise vortices are frequent words and common phenomena in wall-bounded turbulence and the interactions among these coherent structures can be quite complex. It was found, until recently, that a younger hairpin vortex was originated from a pair of streamwise vortices as reported by Lu et al. (2012a) and Yan et al. (2014a) and came into formation through a path schematically illustrated in Fig. 10.21. The streamwise vortices near wall can be generated through the transient growth of the streaks initially subjected to the sinuous disturbances in the spanwise velocity components. In the evolution process, the

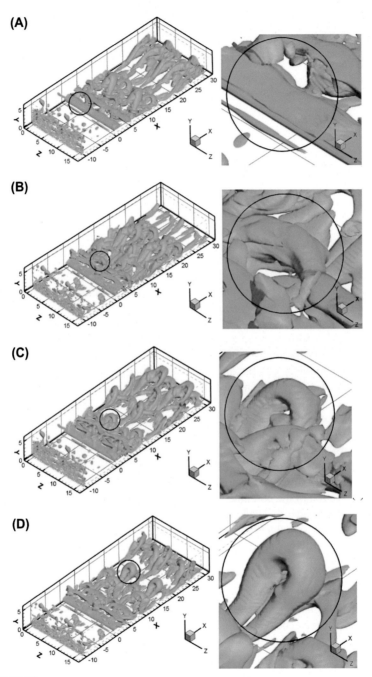

FIGURE 10.19

Evolution of Liutex magnitude (IRI) iso-surfaces with IRI = 0.1 at the time instants of (A) $t = 10$; (B) $t = 20$; (C) $t = 30$; (D) $t = 40$.

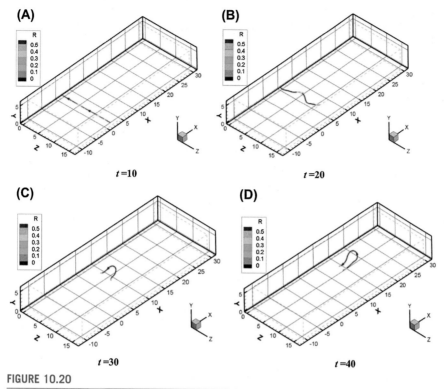

FIGURE 10.20

Evolution process of the LXC-core lines colored by the magnitudes of Liutex vector, with the framed regions being enlarged and highlighted that are corresponding to the marked zones in Fig. 10.19.

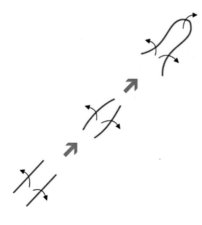

FIGURE 10.21

Schematic evolution path from a counterrotating streamwise-vortex pair to a hairpin vortex.

downstream end of a stronger streamwise vortex is first growing into the arch-shaped head through an ejection and then connects to its weaker counterpart in their overlapping region. The detailed mechanisms can be found in the articles from Lu et al. (2012a) and Yan et al. (2014a), which eventually lead to a complete hairpin structure as given in Fig. 10.21.

(3) Hairpin vortex originated from a primary hairpin vortex

As demonstrated in Fig. 10.22, a parent hairpin vortex traveling downstream can also give birth to a younger hairpin upstream itself. The phenomenon was reported in the existing researches from Zhou et al. (1996, 1999) in which the hairpin vortices, when exceeding a certain threshold strength relative to the mainstream, gave rise to a new-born hairpin vortex upstream the original one. The younger hairpin was also named as the second hairpin, which was possible to produce another lifting-up motion and thereby led to the formation of a tertiary hairpin vortex, so on and so forth. As explained by Liu et al. (2014), the strong shearing effects were playing an extremely important role in generating the new-born vortex and thereafter, sequentially evolving into the hairpin packets.

The phenomenon and process are well captured by the current DNS data. As exhibited in Fig. 10.23, a younger hairpin vortex is generated upstream and then is shed off from the primary vortex. The data visualization based on the current DNS data is well in consistence with the observations provided by Smith et al. (1991) and Zhou et al. (1999). Meanwhile, the current data further confirm that the wall-bounded turbulence is abundant with a varieties of hairpin breeding and reproducing processes, which makes the hairpin-patterned vortices be almost omnipresent and ubiquitous. However, to precisely and quantitatively identify these hairpin evolutions, a more mathematically rigorous quantity and physically limpid

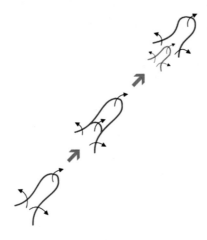

FIGURE 10.22

Schematic evolution path from a primary hairpin vortex to child hairpin vortices in tandem.

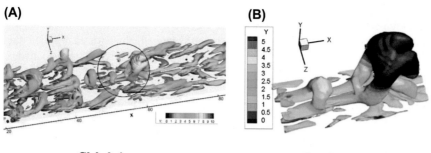

(A)

(B)

Global views   Local enlarged views

FIGURE 10.23

Secondary hairpin vortex generating from a primary hairpin at the time instant of $t = 120$, visualized by the Liutex magnitude iso-surface with the threshold at $|R| = 0.05$.

presentation is in high demand, which calls for the new generation of VI, i.e., the 3rd-G VI. Within the context, the current book demonstrates the advantage of the 3rd-G VI by making use of the findings in Li et al. (2020), which successfully utilized the Liutex quantity and the associated vortex presentation of Liutex-core line to successfully discover a new hairpin generation path.

(4) Hairpin vortex originated from arch-shaped vortex pair

In addition to the aforementioned three distinctive paths leading to the hairpin vortices, a brand-new formation process of a hairpin vortex was found and first detected in Li et al. (2020) based on the LXC-core lines and the turbulence big data established at Department of Aeronautics and Astronautics, Fudan University at Shanghai, PR China, with the website at http://simplefluids.fudan.edu.cn/publications/book-elsevier/, which are considered an appropriate topic to be elaborated in detail by the current chapter.

As schematically plotted in Fig. 10.24, the hairpin vortex can be conceptually developed and generated from a pair of arch-shaped vortices in a fourth path in

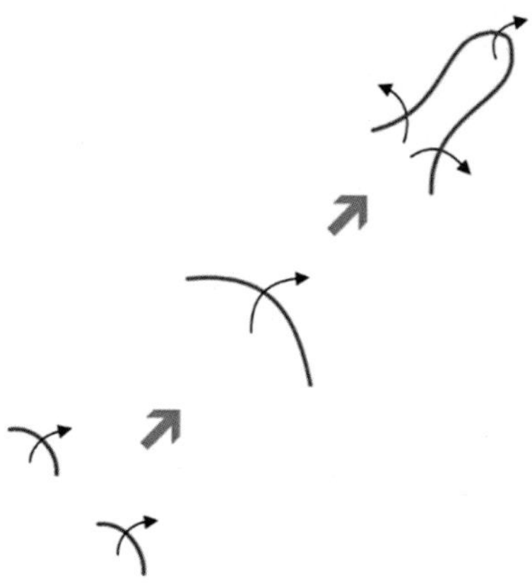

FIGURE 10.24

Schematic evolution path from a pair of arch-shaped vortices to hairpin vortex.

addition to the three ones aforementioned. From the beginning, the two counterrotating arch-shaped vortices travel downstream side by side. Near the inner ends of the arch-shaped vortices, the rotational strengths cancel out with each other due to their opposite rotational directions. Consequently, the two arch-shaped spanwise vortices tend to approach to and eventually connect with each other as the self-induced and spontaneous motions, which gives rise to a larger arch-shaped vortex. When further migrating downstream with the shearing and rotating effects, the enlarged arch-shaped vortex is stretched gradually and is eventually evolved into a hairpin-patterned vortex, which can be seen step by step in Fig. 10.25 by capturing the formation event from the time instant of (A) $t = 150$ to (F) $t = 200$.

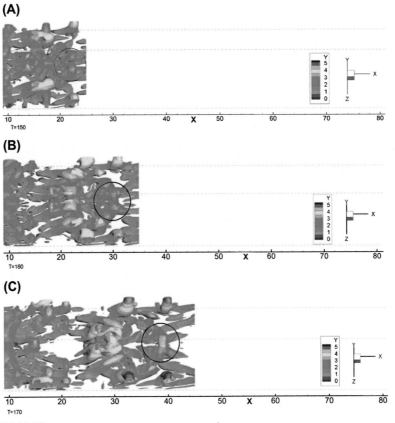

FIGURE 10.25

Formation process of the hairpin vortex from a pair of arch-shaped vortices side by side as highlighted by the dashed box and visualized by the Liutex magnitude iso-surface with a threshold at $|R| = 0.05$ at the time instant of (A) $t = 150$; (B) $t = 160$; (C) $t = 170$; (D) $t = 180$; (E) $t = 190$ and (F) $t = 200$.

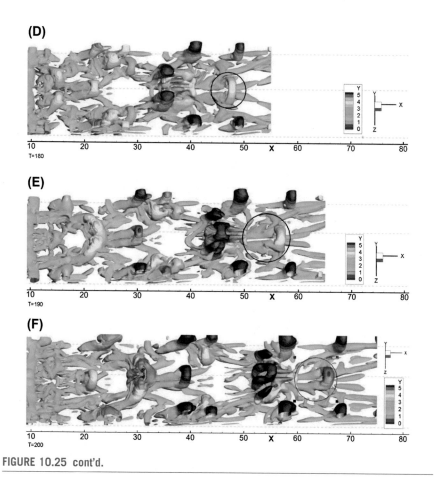

FIGURE 10.25 cont'd.

From the beginning, a pair of arch-shaped vortices in Fig. 10.25A at the time instant of $t = 150$ is captured migrating downstream side by side at the streamwise location of $x = 22$. At the next time instant of $t = 170$, Fig. 10.25C exhibits that an enlarged arch-shaped vortex takes place near the location of $x = 39$, indicating that the pair of arch-shaped vortices are connected with each other and therefore are merged into an arch-patterned vortex with an enlarged size comparing to the original side-by-side vortex pair. When the time advancing to the instant of $t = 200$, Fig. 10.25F presents a hairpin-patterned vortex arises, taking place near the streamwise location of $x = 65$, which is originated from the pair of arch-shaped side-by-side vortices with a counterrotation.

Fig. 10.26 further provides the enlarged local views for the three important steps in the evolution. Evidently, a pair of separated arch-shaped vortices first migrates side by side with the mainstream. Subsequently, these arch-shaped vortices are

(A)   **(B)**   **(C)**

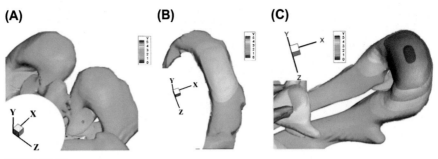

FIGURE 10.26

Enlarged local views to highlight the arch-shaped vortex pair merging event in Fig. 10.31 from (A) $t = 150$; to (B) $t = 175$; and up to (C) $t = 200$ with the Liutex magnitude iso-surfaces being extracted out at a threshold of IRI $= 0.05$ and with the same color legend as in Fig. 10.31.

merged into an enlarged single arch-shaped vortex. Thereafter, a hairpin-patterned vortex comes into formation. The evolution process is clearly exhibited in Fig. 10.27. Based on the colored Liutex magnitude iso-surfaces in Fig. 10.27, it can be judged that the vortices in Fig. 10.27A are originally located in a lower layer (closer to wall) in the TBL. With downstream migration, the vortices are gradually merged and meanwhile are lifted into an upper layer (far away from wall) in the TBL, nearly up to $y = 5$ in Fig. 10.27, which is obviously due to the TBL-induced uplifting velocity.

As elucidated in Chapter 7, the LXC-core lines given by Xu H. et al. (2019) make the 3rd-G VI equipped with a capability to more limpidly depict the vortex evolution and thereby, to more precisely probe and detect the kinematic mechanisms of the hairpin vortex formation. Toward this end, the LXC-core lines are carefully extracted out from these vortices and are plotted in Fig. 10.27, which provides a more accurate and uniquely defined evolution for these vortex systems. The side views of these LXC-core lines at the time instants of $t = 150, 170, 180$, and 200 are presented in Fig. 10.27 with the same reference length scale as in Fig. 10.27, and meanwhile, Fig. 10.28 plots the top views of these LXC-core lines at the

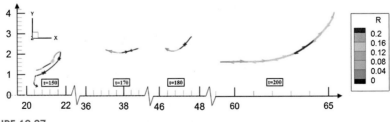

FIGURE 10.27

Side view of LXC-core lines evolution at the time instants of $t = 150, 170, 180$, and 200 for the same process in Fig. 10.31 with the color legend standing for the Liutex magnitude, i.e., the local rigid spinning speed around the core lines.

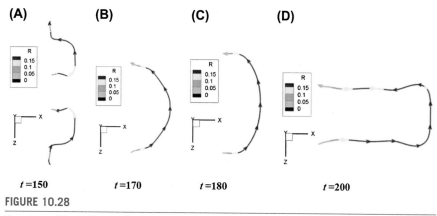

FIGURE 10.28

Top view of the LXC-core lines evolution for the same process in Fig. 10.25 with the color legend standing for the Liutex magnitude, i.e., the local rigid spinning speed around the core lines.

same time instants by the enlarged reference length scales. Therefore, the evolution process of vortices is more clearly and comprehensively presented, which provides the convincing evidence to validate and support the aforementioned statements. It is worthy to note that the rigid-rotation strengths, as represented by the colors on these LXC-core lines, remain almost the same in the evolution, but are redistributed significantly when developing into the hairpin-patterned vortex.

After presenting the detailed vortex evolution as the fourth path leading to a hairpin generation, the current study further conducts a quantified analysis on the velocity, the rotational strength, and the spacings of the arch-shaped vortex pair from the time instant of $t = 150$ when the vortex connection event occurs. Starting from the location of $x = 22$, the center locations of the arch-shaped vortices and the spanwise profiles are taken out from the DNS data at the three representative stations normal to the wall, i.e., $y = 2.0$, $y = 2.5$, and $y = 2.8$. The profiles of the streamwise velocities and the rotational strengths of Liutex magnitude $|R|$ are plotted in Fig. 10.29A where a pair of valleys are presented near the locations of $z = 6.75$ and $z = 9.25$ highlighted by the dash-dot lines in red color. These valleys represent the low-speed streaks induced by the arch-shaped vortices and the gap in between the two valleys provides a good measure of the spanwise spacing for the vortex pair. Based on these reasonings, the spacing at the starting location of $x = 22$ is measured roughly at about 2.5 rib heights in Fig. 10.29A. The height of the vortex pair is estimated comparable to the TBL thickness and the vortex pair is traveling downstream at approximately the mainstream speed. Fig. 10.29B presents the corresponding vortex strength distributions, i.e., the local rigidly-spinning speed as represented by the Liutex magnitude of $|R|$. The rotational strength drops to a near zero at $z = 8$ in between the arch-shaped vortex pair, which suggests that the pair be detached and disconnected. Within each the vortex, the maxima, standing for by the

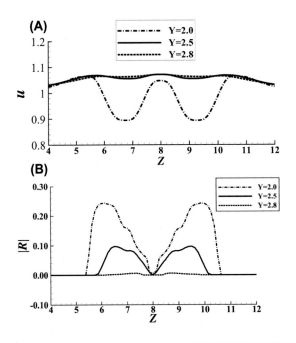

FIGURE 10.29

Profiles of streamwise velocity and Liutex magnitude IRI, with the data being extracted out at the streamwise location of $x = 22$ and the three representative wall-normal locations of $y = 2.0$ (dash-dot line), $y = 2.5$ (solid line), and $y = 2.8$ (dashed line) at the time instant of $t = 150$.

two peaks, exceeds 0.2, i.e., $|R| > 0.2$, at the wall-normal distance of $y = 2.0$ and descends quickly to zero with the increase of y, as presented in Fig. 10.29B.

Eitel-Amor et al. (2015) attempted to conduct a quantified analysis for the temporal evolution of vortex structures by using the spanwise vorticity, i.e., the 1st-G VI quantity, in which the volume-averaged rms (root-mean-square) was taken from the entire computation domain. Similar to the research by Eitel-Amor et al. (2015), a local framed domain with a certain size moving with the vortices is selected in the book as the zone to take the averaging information for the vortical structures. However, considering the complexity of the vortical structures, it is more appropriate that the moving domain is chosen locally for the integration to quantify the vortex evolution, instead of the entire computational domain being used for the integration by Eitel-Amor et al. (2015). Moreover, due to the strong shearing effect induced by the wall, it is, nowadays, well known that the vorticity is no longer proper to be chosen as the VI quantity to study the near-wall vortex structures. Instead, the domain-averaged spanwise component of $|R_z|$, i.e., the 3rd-G VI quantity, is more appropriately selected to quantify the temporal evolution of these vortices, which effectively gets rid of the contamination of shearing effects induced by the wall.

Therefore, the integral average of the spanwise rigid rotation strength $|R_z|$ is defined as the quantity to investigate the temporal evolution of vortex structures, which reads

$$R_{ave} = \frac{\int |R_z| ds}{S_t} \tag{10.4}$$

with S_t being the surface area surrounding the local moving frame domain for integration around the hairpin vortex in the plane of $z = 8$. Fig. 10.30 provides the time history for the R_{ave} evolution, which exhibits a gradual increase from $t = 140$, and a peak near $t = 190$ and then a descend afterward. These variations suggest that the strength of the hairpin head is developing from a zero at the beginning, and then the vortex being generated and the vortical strength being enhanced until maximized near $t = 190$. Thereafter, the hairpin head gradually loses its strength when the hairpin vortex is fully developed, which is qualitatively conforming with the observation reported by Eitel-Amor et al. (2015).

Due to the counterrotation effects around the inner ends of the Liutex vector lines in Fig. 10.31, the spinning strengths become weaker because of the vortex cancellation with each other. Consequently, the outer ends of the arch-shaped vortices present stronger rotation strengths than their inner counterparts. The two vortices tend to be attracted and gradually get closer to each other due to the inwardly induced velocity. Toward this end, a larger-sized arch-shaped vortex comes into formation

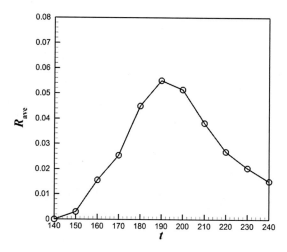

FIGURE 10.30

Temporal developing history of R_{ave} with the instantaneous spatial integration being taken over the domains of: (1) $17 \leq X < 19$ and $1 \leq X < 6$ at $t = 140$; (2) $22 \leq X < 24$ and $1 \leq Y < 6$ at $t = 150$; (3) $29 \leq X < 31$ and $1 \leq Y < 6$ at $t = 160$; (4) $38 \leq X < 40$ and $1 \leq Y < 6$ at $t = 170$; (5) $47 \leq X < 49$ and $1 \leq Y < 6$ at $t = 180$; (6) $55.7 \leq X < 57.7$ and $1 \leq Y < 6$ at $t = 190$; (7) $64.7 \leq X < 66.7$ and $1 \leq Y < 6$ at $t = 200$; (8) $74 \leq X < 76$ and $1 \leq Y < 6$ at $t = 210$; (9) $83 \leq X < 85$ and $1 \leq Y < 6$ at $t = 220$; (10) $92.3 \leq X < 94.3$ and $2 \leq Y < 7$ at $t = 230$; (11) $102 \leq X < 104$ and $2 \leq Y < 7$ for $t = 240$.

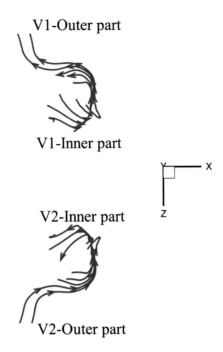

FIGURE 10.31

Inner and outer parts of the Liutex vector lines near the arch-shaped vortex pair at time instant of $t = 150$.

after the arch-vortex pair being merged with each other and then further evolves into a hairpin-shaped vortex after the amalgamation.

Historically, the Biot–Savart law was used to qualitatively interpret the evolution behaviors of vorticity in the 1st-G VI. Within the context, more quantitatively rigorous investigation of the vortex rollup and stretching, particularly in the TBL where both viscosity and turbulence fluctuations are non-negligible, has to be conducted based on the Liutex transport equation, instead of the traditional vorticity transport equation or Helmholtz equation. In this regard, the R-S decomposition in the Liutex theory of 3rd-G VI given by Liu et al. (2018a) provides a feasible way to systematically and quantitatively study the vortex dynamics as demonstrated in Chapter 7.

In the aforementioned four conceptually representative paths generating a hairpin vortex, the initial pattern and orientation of the perturbation play a critically important role in the subsequent vortex evolution. The vortex evolution takes the first path if the pattern of perturbation is dominant in streamwise. If the perturbations are prominent in spanwise, for example, the distinctive sinusoidal disturbance in the spanwise velocity, the hairpin vortex formation then takes the second path. In addition, the perturbation strength also needs to be considered as an important factor. The third-path formation can only take place when the strength exceeds a critical threshold as demonstrated by Wang et al. (2015). For the third path leading to a hairpin, the reproducing mechanism of parent-breeding mode with a positive feedback cycle takes place in the environment of sufficiently strong shear. The lifespan of the breed offspring is strongly dependent on the strength of the parent hairpin as illustrated by Eitel-Amor (2015). The second and tertiary hairpins can be reproduced by a parent hairpin under a strong local shearing condition, and the breeding process can even give rise to the hairpin packet as observed in Wu et al. (2009). However, as presented by Eitel-Amor et al. (2015), the new-born hairpins may disappear quickly if the local shear effect is too weak. With regard to the fourth path for hairpin generation, if the distinctive pair of disturbances are located within a sufficiently short spanwise distance, the arch-shaped vortices are likely to take place and thereafter merge into a single arch-shaped vortex, which eventually evolves into a hairpin-patterned vortex in Fig. 10.28. Needless to say, the initial intensity of disturbance is critical for the occurrence of a hairpin, although the precise threshold for the disturbance strength is still a subject for further investigation.

In summary, it is worthy to mention that a number of key parameters are involved in these four major evolutionary paths leading to a hairpin, which include (1) the strength threshold of initial disturbance, the location of disturbance normal to the wall, and the spanwise spacing between the pair of arch-shaped disturbances, and so on. The chapter demonstrates a systematic way to make use of the 3rd-G VI, particularly the Liutex iso-surface and LXC-core line, to more quantitatively and rigorously study the kinematic evolutions of vortex behaviors in a complex turbulence. Toward this end, the current research specifically proposes to clarify the roles of these key parameters played in the generating hairpin vortex based the 3rd-G VI quantity of Liutex and thereby more profoundly understanding the laminar-turbulent transition process. However, it has to be claimed that the transition is extremely complex due to the many factors of strong nonlinearity, which must not preclude any other possibilities of hairpin vortex generating paths and mechanisms. As commented by Adrian (2007), "With the growth of hairpins in the spanwise direction, various vortex reconnection events are possible, depending upon the shapes of the hairpins and their relative positions at the time of encounter." Therefore, an in-depth investigation is always a subject of study for the hairpin-vortex formation which turns out also be an important phenomenon and interesting topic even in the wake turbulence presented in Chapter 7.

Moreover, the streak structures are the closely-related topics with a hairpin generation. It is quite common that the streaks come into existence before the hairpin vortex is generated. For instance, the streaks are found existing on both sides of the streamwise vortex in the second path leading to a hairpin, which is depicted more in detail and in depth by Chapter 8. The instability due to the streak meandering can promote bridging the streamwise vortices and therefore, accelerate the hairpin formation. In the parent-breed model of the third path, the streak between vortex legs is pivotal for generating the child hairpin vortex, as seen in Adrian (2007) and Liu et al. (2014), and the streak upward movement can cause the hairpin being lifted away from the wall. Since the vortices are the elementary blocks in all these turbulence phenomena, the definition of Liutex with its scalar, vector, and even tensor forms provides a meaningful quantity and essential tool to quantitatively detect, interrogate, depict and eventually to understand the mechanisms behind these turbulence kinematic structures and their dynamic evolutions.

10.2.5 Summary

The section provides the analyses of the DNS data for the flat-plate TBL with a rib structure at a low Reynolds number ($Re = 1200$), which includes the detailed coherent and vortical structure visualizations, kinematic evolution process, and quantitative depictions based on the 3rd-G vortex identification method.

The bypass transition occurring in the region of $5 < X < 20$ is characterized by the small-sized and densely distributed vortices comparing to the fully-developed zone downstream. The Reynolds stresses and turbulence kinetic energy are quite intensified and promoted in the early transition, which suggests that the vortex generation be quite active and dynamic in the region, giving rise to a pretty high local friction coefficient, pressure coefficient and the closely-related heat-transfer Nusselt number. In the fully- developed regime downstream, i.e., $X > 80$, the vortical structures are comparatively larger in scale and sparser in number as demonstrated by the 3rd-G VI methods.

For the first time in history, the 3rd-G VI approaches, particularly the Liutex magnitude iso-surface and LXC-core lines, are applied to the DNS data to detect and to reveal the formation mechanisms of hairpin vortex. The first and the third paths leading to a hairpin are confirmed by the visualization of the current DNS data. Although the second path given by Wang et al. (2015) is not observed in the current study, a fourth path and mechanism are detected for generating a hairpin vortex from a pair of side-by-side arch-shaped vortices and the entire kinematic processes leading to a hairpin vortex are evidently presented for the first time. Moreover, the quantitative analyses, enabled and effected by the latest Liutex iso-surface and LXC-core lines, are successfully conducted, which not only limpidly exhibits the entire evolutionary path from the arch-shaped vortex pair to the hairpin-patterned vortex, but also quantitatively investigates the merging and hairpin-formation processes by tracking the time-history variations of the spatially averaged vortex strength in the mid-plane of the hairpin vortex. These VI practices prove that the 3rd-G VI methods truly open the door to reliably study the turbulence by quantitatively defining its elementary blocks of vortices.

CHAPTER

Comparison of Liutex and other vortex identification methods

11

As discussed in the previous chapters, vorticity is a vector, but is in general not correlated to the vortex and thus cannot be used to identify vortices. The second-generation methods such as Q, λ_{ci}, λ_2, and so on are scalar-based, strongly threshold-dependent and contaminated by shear and stretching (or compression). Therefore, the second generation is not appropriate to be applied to identify the vortex structure either. On the other hand, Liutex-based third generation of vortex identification methods, can represent six core elements of vortex and is a new and accurate tool to correctly identify the vortex structure. In particular, Liutex core line is unique and threshold-free, and the Liutex–Omega method is insensitive to threshold change. There have been many applications of Liutex and the third generation of vortex identification methods developed by the UTA team, which are reported by many literatures. In this chapter, a few computational examples are presented, which clearly show the superiority of Liutex over all other vortex identification methods.

11.1 Comparative assessment and analysis of Liutex in swirling jets (Gui et al., 2019a, b)

11.1.1 Short review

As well known, the identification of vortex plays an important role in studying the feature of the fluid flow and uncovering the mechanisms of turbulence. The most conventional definition of vortex is based on the vorticity $\overrightarrow{\varOmega_v}$ which is the curl of the fluid velocity $\overrightarrow{v}$, i.e.,

$$\overrightarrow{\varOmega_v} = \nabla \times \overrightarrow{v}, \ \varOmega_v = \left\|\overrightarrow{\varOmega_v}\right\|_2 \tag{11.1}$$

Hence, a connected fluid region with $\varOmega_v > 0$ is regarded as a vortex. More mathematically, based on the work of Chong et al. (1990), with $\nabla \overrightarrow{v}$ being the local velocity gradient tensor, its characteristic equation is

$$\lambda^3 + P\lambda^2 + Q\lambda + R = 0 \tag{11.2}$$

with three eigenvalues $\lambda_i (i = 1, 2, 3)$. Many vortex identification criteria were proposed based on Eq. (11.2), such as:

Liutex and Its Applications in Turbulence Research. https://doi.org/10.1016/B978-0-12-819023-4.00017-3
Copyright © 2021 Elsevier Inc. All rights reserved.

(1) Q criterion (Hunt et al., 1988): It is defined by the positive second invariant $Q = 0.5$ (b − a), where a and b are defined in Eqs. (1.8) and (1.9). Q represents the balance between the strain rate and the vorticity magnitude.

(2) λ_2 criterion (Jeong et al., 1995): The λ_2 vortex is defined by the second largest eigenvalue of $A^2 + B^2$. The connected region with the second largest eigenvalue λ_2 less than zero, i.e., $\lambda_2 < 0$, is regarded as a vortex.

(3) λ_{ci} criterion (Zhou et al., 1999): When the characteristic equation Eq. (11.2) has two complex eigenvalues, the imaginary part of the complex eigenvalue of the velocity gradient tensor is also used to quantify the local swirling strength of the vortex.

(4) Omega method (Liu et al., 2016)

$$\Omega_L = \frac{a}{a + b + \varepsilon} \tag{11.3}$$

where a and b are defined in Eqs. (1.8) and (1.9) and ε is a small positive number used to avoid division by zero.

(5) Liutex (Liu et al., 2018a, 2019; Gao et al., 2018) used a new decomposition $\nabla \vec{v} = A + R + S$, where R is regarded as a tensor representing rigid rotation with an angular speed of $R/2$.

$$\vec{R} = R\vec{r}$$

$$R = \left(\vec{\omega} \cdot \vec{r} \right) - \sqrt{\left(\vec{\omega} \cdot \vec{r} \right)^2 - 4\lambda_{ci}^2} \tag{11.4}$$

The vortex in the swirling flow is studied by different vortex identification methods. Note that the swirling flow is characterized by the motion of the fluid swirl imparted onto a directional jet flow or without the directional jet flow (Chen and Sun, 2010; Xing, 2014), and it might be one of the flow patterns mostly consistent with the definition of Liutex. Therefore, the direct numerical simulation data of the swirling jet flows in a rectangular container are utilized here, and the comparison of Liutex, Q, λ_2, Omega is carried out to assess the performance and the capacity of these vortex criteria.

11.1.2 Numerical simulation

11.1.2.1 Numerical setup and validation

In this work, a swirling air jet of diameter $d_i = 1$ cm is issued from the surface center of a rectangular box of dimension 0.10 m × 0.05 m × 0.05 m at the inlet velocity $U_0 = 1$ m/s. The flow domain is discretized by 512 × 256 × 256 Cartesian mesh grids. The Reynolds number is $Re = U_0 d_i/\nu = 3000$, where the kinematic viscosity $\nu = 1.43 \times 10^{-5}$ m²/s. The parameters used in the present simulation are listed in Table 11.1.

Table 11.1 Real parameters and corresponding dimensionless values used in simulation.

	Dimensionless value	Real value
Dimensions of bed, $L_x \times L_y \times L_z$	$10 \times 5 \times 5$	$0.10\,\text{m} \times 0.05\,\text{m} \times 0.05\,\text{m}$
Mesh numbers, $N_x \times N_y \times N_z$	$512 \times 256 \times 256$	$512 \times 256 \times 256$
Fluid density, ρ_f	1	$1.29\,\text{kg/m}^3$
Fluid density, μ	3.3×10^{-4}	$1.85 \times 10^{-5}\,\text{Pa s}$
Inlet diameter, d_i	1	$0.10\,\text{m}$
Inlet velocity, U_0	1	$4.3\,\text{m/s}$
Reynolds number, Re	3000	3000
Courant number, Co	0.0512	0.0512
Swirl number, S_n	0, 0.10, 0.20, 0.36	0, 0.10, 0.20, 0.36
Time step, Δt	0.001	$2.32 \times 10^{-6}\,\text{s}$
Total time, t	40	93 ms

Reproduced from Gui, N., et al., Comparative assessment and analysis of Rorticity by Rortex in swirling jets, Journal of Hydrodynamics 31 (3), 2019, with the permission of Journal of Hydrodynamics.

In the swirling flow, it is assumed that the rotational motion in the azimuthal direction is a translational streamwise motion. The inlet velocity profiles of the streamwise u and the azimuthal v velocities are shown in Fig. 11.1A and B, where the swirl number S_n is defined as the ratio of the flow rate of the rotational momentum to the translational momentum as

$$S_n = \frac{1}{r_i} \frac{\int_0^{r_i} \rho v u r^2 \, dr}{\int_0^{r_i} \rho u^2 r \, dr} \tag{11.5}$$

where $r_i = d_i/2$ is the radius of the jet inlet. As seen in Fig. 11.1, Table 11.1, three swirl numbers of $S_n = 0.10$, 0.20, and 0.36 are simulated in the present study by varying the azimuthal velocity. The inflow turbulence is not considered here. The nonreflecting boundary condition is used under the outlet condition (Frenander and Nordström, 2017), and the side walls are nonslipping wall boundaries. The configurations of the swirling flow are similar to earlier studies (Gui et al., 2010a, b, c), as simple configurations suitable for learning the swirl flows. With this spatial discretization ($\delta = 39\,\mu\text{m}$), the area of the jet inlet is discretized by about 50×50 mesh grids, which are both fine enough for the high resolutions of the direct simulation.

11.1.2.2 Numerical results
In this work, the features of the vorticity (Ω_V), Q, λ_2, Omega (Ω_L), and Liutex (R) are compared. The swirling flows of $S_n = 0, 0.10, 0.20, 0.36$ are compared and analyzed. The relations among the pseudo-rigid rotational motion (evaluated by Liutex R), the antisymmetrical shearing deformation (evaluated by the antisymmetric shear S), and the vorticity (evaluated by the vorticity Ω_V) are analyzed.

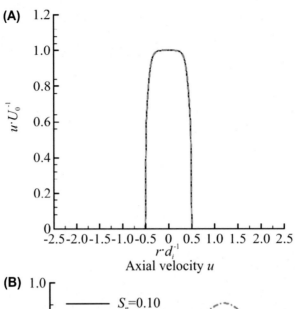

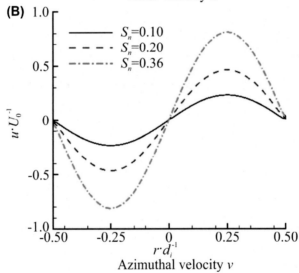

FIGURE 11.1

Velocity profiles at the inlet.

Reproduced from Gui, N., et al., Comparative assessment and analysis of Rortex vortex in swirling jets, Journal of Hydrodynamics 31 (3), 2019, with the permission of Journal of Hydrodynamics.

11.1.2.2.1 Vorticity identification

The vorticity identification in the case of $S_n = 0.36$ is shown in Fig. 11.2. The vortex is expanded in the lateral and spanwise directions after being initially issued into the flow domain at $t = 4$ (Fig. 11.2A) and expanded continuously to make a group of

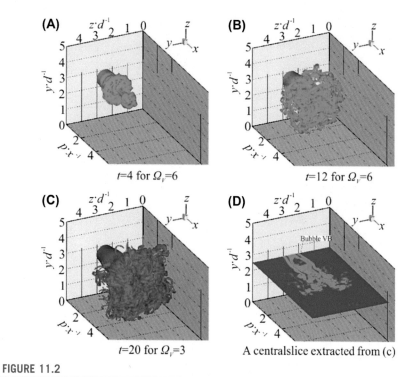

FIGURE 11.2

Snapshots of 3-D vorticity Ω_v for $S_n = 0.36$ and a central slice extracted from (c).

Reproduced from Gui, N., et al., Comparative assessment and analysis of Rortex vortex in swirling jets, Journal of Hydrodynamics 31 (3), 2019, with the permission of Journal of Hydrodynamics.

rather complicated twisted vortices ($t = 12$, Fig. 11.2B). The fully developed swirling vortices ($t = 20$) are shown in Fig. 11.2C with a clear bubble vortex breakdown (VB) (Li et al., 2008; Wang and Chen, 2009, Fig. 11.2D) and the turbulent motions downstream the VB region. These features of the VB and the complicated vortices are generally consistent with early results on the swirling flows (Gui et al., 2010a, b, c).

11.1.2.2.2 Comparison among Q, Ω_L, Ω_V, λ_2, and R

Let $t = 4$ (Fig. 11.3), $t = 20$ (Fig. 11.4) for the case study, to show the vortex presentations by Q (A), Ω_L (B), Ω_V (C), λ_2 (D), and R (E), respectively. It is clearly seen that all presentations have similar structures. They all indicate that the major ring-like vortex structure is formed perpendicular to the streamwise direction, where the interaction between such major streamwise vortex ring and the central axial direct vortex tube is through a circumferential array of secondary vortices. The secondary vortex rings follow a spiral distribution around the major vortex

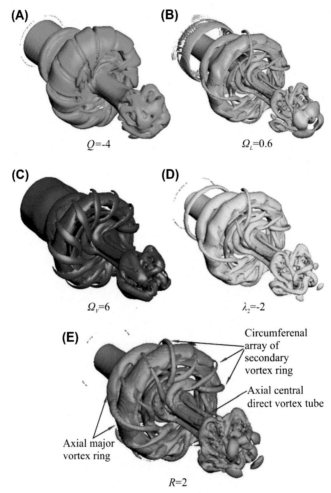

FIGURE 11.3

Vortex structures of $S_n = 0.36$ at $t = 20$.

Reproduced from Gui, N., et al., Comparative assessment and analysis of Rortex vortex in swirling jets, Journal of Hydrodynamics 31 (3), 2019, with the permission of Journal of Hydrodynamics.

ring. The structure can all be identified in all kinds of vortex identification. Thus, in such a free swirling flow, the Q, Ω_L, Ω_V, λ_2, and R criteria are all valid for the vortex study. It is also noticed that the diameter of the swirling jet visualized by Ω_V (vorticity) near the immediate inlet of the flow (Fig. 11.3C) is larger than those obtained by other kinds of vortex identifications, especially larger than that presented by Ω_L (Fig. 11.3B), λ_2 (Fig. 11.3D), and R (Fig. 11.3E). In fact, the diameter of the

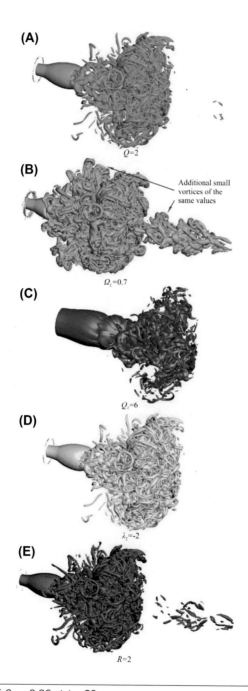

FIGURE 11.4

Vortex structures of $S_n = 0.36$ at $t = 20$.

Reproduced from Gui, N., et al., Comparative assessment and analysis of Rortex vortex in swirling jets, Journal of Hydrodynamics 31 (3), 2019, with the permission of Journal of Hydrodynamics.

jet presented by the latter is realistic. It means that the vortex presented by Ω_V has some shortcomings, for the possibility of making a fake recognition of the vortex cores.

Moreover, the strong vortex breakdown in the downstream of the bubble can be clearly seen from Fig. 11.4. The bubble shape can be more clearly recognized by Q (Fig. 11.4A), λ_2 (Fig. 11.4D), and R (Fig. 11.4E). The strongly kinked small-scale vortices in the downstream of the bubble are rather complex. Their structures are very similar, especially those presented by Q (Fig. 11.4A), λ_2 (Fig. 11.4D), and R (Fig. 11.4E). In Fig. 11.4B for Ω_L, the small-scale vortices are somewhat different from those presented by Q (Fig. 11.4A), λ_2 (Fig. 11.4D), and R (Fig. 11.4E), since more small-scale vortices can be seen immediately after the bubble VB, masked as the VB. Moreover, some additional small vortices can be observed in the central and further downstream regions (Fig. 11.4B). Notice that all the vortex criterion presentations are computed from the same data. They are at the same time point under the same flow condition, including the same level of the swirl number S_n. It means that the Ω_L definition can recognize the additional small-scale vortices, which other criteria cannot, except for Liutex.

In Fig. 11.4E, the additional small-scale vortices can also be recognized by Liutex very completely. In other words, Liutex can be used to clearly visualize the large-scale bubble VB and the strongly kinked small-scale vortices and the additional small-scale vortices in the further downstream region. It would be one of the best candidates for the vortex identification. Comparing Ω_L with Liutex, it is shown that the latter is somewhat better, since some additional small-scale vortices can still be observed in the immediate inlet of the flow domain, which are not recognized by other criteria except the Ω_V in Fig. 11.4C.

The Ω_V (Fig. 11.4C) may be not a good candidate for the vortex presentation although it has been used the most widely with a longest history. It is seen from Fig. 11.4C that the diameter of the jet near the inlet is larger than those obtained by other methods. It is in fact a fake since the jet inlet is not so large, and it is caused by the shearing (larger radial velocity gradient) around the periphery of the main jet. Moreover, the 3-D bubble VB is not so clear as presented in other cases.

11.1.2.2.3 Difference between Ω_V, R: the S vector

According to the former analysis, more attention is paid to the difference between the "worst" and "best" candidates of the vortex criterion, i.e., Ω_V, R, which are both vectors. At first, taking the case of $S_n = 0.36$ at $t = 16$ for example, the Ω_V, R vectors on the location of $R \neq 0$ are shown in Fig. 11.5. In Fig. 11.5A, the red and green vectors are the Ω_V, R vectors, respectively, at the same location at the same time. Particular attention is paid to the local region-A (Fig. 11.5B) and region-B (Fig. 11.5C) for a detailed inspection, and it is clear that Ω_V, R results are always different in both magnitudes and directions, although they sometimes look consistent with each other.

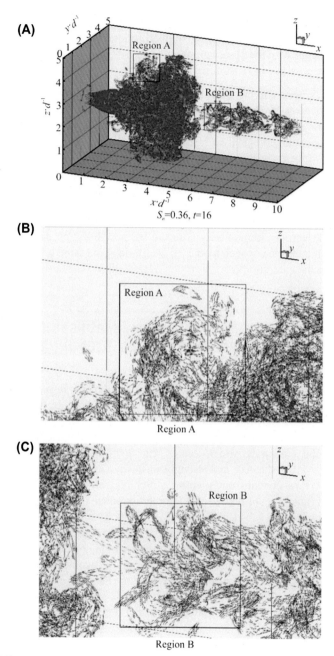

FIGURE 11.5

Visualization of Ω_V (red vectors), R (green vectors) at the locations of $R \neq 0$.

Reproduced from Gui, N., et al., Comparative assessment and analysis of Rortex vortex in swirling jets, Journal of Hydrodynamics 31 (3), 2019, with the permission of Journal of Hydrodynamics.

11.1.3 Summary

In this work, various vortex criteria are applied to identify the vortex structures in swirling flows. Based on the observations and analyses, the key features and conclusions of vortex criteria can be summarized as follows:

(1) All vortex criteria can be generally used to characterize the basic feature of the vortex structures, such as the bubble vortex breakdown, the axial major vortex ring, and the spiral distribution of the secondary vortex ring, in the swirling flows. A phenomenological comparison shows that the vorticity criterion Ω_V performs worse than others since it cannot correctly recognize the jet inlet diameter, and it cannot distinguish the shear from the pure rotational motion. Instead, the Ω_L, λ_2, Q, and Liutex criteria can identify the vortex core and the jet diameter correctly.
(2) The Ω_L, Liutex criteria have the ability to identify more additional secondary small-scale vortices immediately after the VB and in the far downstream, which cannot be clearly seen by the λ_2, Q criteria at the same levels of strongly kinked small-scale vortices.

11.2 Applications of different vortex identification methods in turbine rotor passages (Wang Y. et al., 2019)

11.2.1 Short review

The internal flow of the turbine passage is complex, and the flow losses are affected by many factors. Particularly, the flow losses near the end wall are obvious. Denton (1993) estimates that end-wall losses can account for 2/3 of total flow losses. Influenced by the end-wall boundary layer, a series of complex secondary flows are easily formed near the end wall, which causes the flow field to be strongly three-dimensional, strongly sheared, and strongly unsteady (Zou et al., 2014). The end-wall secondary flow contains many vortex structures, which interact strongly with each other and cause a lot of flow losses. In order to reduce the flow losses in turbine, researchers have done a substantial number of research on blade three-dimensional modeling (Filippov and Wang, 1963), end-wall modeling (Deich et al., 1960), leading edge modification (Zhang et al., 2012), and tip modeling (Zou et al., 2016, 2017). In the process of research, secondary flow structure and its evolution are an inevitable topic, and the accurate identification of vortex structure has been extremely important.

Since Helmholtz (1858) proposed "vortex motion" in his three vorticity theorems, more scholars began to pay attention to the definition and identification of vortices. Brachet et al. (1988) defined vortices as the region dominated by negative velocity gradient. Babiano et al. (1987) proposed that vortices are any region where vorticity is greater than a certain threshold. For a long time thereafter, vorticity has been regarded as a method that can reliably identify vortices. However, Robinson (1990a, b) pointed out that in the turbulent boundary layer, especially in the region near wall, the association between the large vorticity region and the actual vortices is rather weak. In recent studies, Dong et al. (2018a,b) also pointed out and verified that there is no direct relationship between vortices and vorticity through the study of boundary layer transition and that vorticity cannot be applied to identify vortices or rotation regions. With the recognition of this key issue, how to identify the vortices has become a controversial issue, which has also prompted the subsequent proposals of various vortex recognition rules. Since the eighties, researchers have proposed the Q criterion (Hunt et al., 1988), Δ criterion (Chong et al., 1990), λ_2 criterion (Jeong et al., 1995), and λ_{ci} criterion (Zhou et al., 1999). Up to now, a series of criteria based on the velocity gradient tensor have been formed for vortex identification, and various criteria for vortex identification are often used in the analysis of turbine cascade flow field. However, there are some common defects in the aforementioned vortices identification rules, such as the selection of threshold and the misidentification of strong shear zone, which are troublesome for the study of internal flow field in turbine cascade.

In 2016, a vortex identification method called Omega method, which has high robustness, was first proposed by Liu et al. (2016). Zhang et al. (2018a,b) applied this method to flow field analysis of reversible pump turbine and verified the robustness of Omega method and the simultaneous identification ability of strong and weak vortices. Recently, Liu and his UTA team (Liu et al., 2018a) further proposed the third-generation vortex identification method (Liu et al., 2019) − Liutex method whose identification parameter is a vector. Gao et al. (2018) compared Liutex with those vortex recognition methods based on eigenvalue of the velocity gradient tensor in mathematical and physical sense. The conclusion is that Liutex eliminates the contamination by shear compared with other methods and thus can accurately quantify the local rotational strength. Dong et al. (2019b) put forward a new normalized vortex identification method named Liutex−Omega method, which is based on the ideas of Liutex and the Omega method.

11.2.2 Numerical method

Steady-state RANS computations were performed in ANSYS CFX 12.0 employing the shear stress transport (SST) turbulence model. The N-S equation is discretized by the finite volume method and the "high precision" scheme in CFX is adopted. The wall boundary is set to be adiabatic, ignoring the effect of wall heat transfer.

In the numerical simulation, a single blade passage was modeled, including rotor and stator domains. Periodic conditions were assumed at the boundaries with the adjoining passages, while the mixing plane approach was employed at the stator rotor interface. The commercial software NUMECA Autogrid 5 is used to generate the static subdomain computational grid, while the ICEM is used to generate the rotor domain computational grid. The mesh elements in both domains are hexahedron meshes. The number of grids in rotor domain is about 1.92×10^6 and that in stator domain is about 6.4×10^5. The grids are shown in Fig. 11.6. In order to obtain the flow details of tip clearance region, the mesh of clearance region was specially refined in ICEM, and 32 cells were used in the radial direction of tip clearance.

Mass flow-averaged total pressure and circumferential average total temperature were imposed in the inlet plane, and the turbulence intensity was set to 5%. The exit static pressure was adjusted for the average total-to-total pressure ratio of 4.

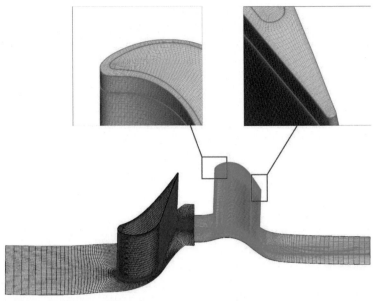

FIGURE 11.6

Computational mesh.

Reproduced from Wang, Y., et al., The applicability of vortex identification methods for complex vortex structures in axial turbine rotor passages, Journal of Hydrodynamics 31 (4), 2019, with the permission of Journal of Hydrodynamics.

11.2.3 Result and discussion

11.2.3.1 Identification of vortex structures in turbine rotor passages

In order to accurately identify the vortex structures in the turbine rotor passage, this section will adopt four methods including the Q criterion, Omega method, Liutex method, and Omega—Liutex method for vortex identification.

The Q criterion, proposed by Hunt et al. in 1988, is a commonly used method for vortex identification, which is based on the second invariant of the velocity gradient tensor. The Q criterion represents the extent to which the amplitude of local vorticity term exceeds that of local strain rate term, and can be expressed as:

$$Q = \frac{1}{2}\left(||\boldsymbol{B}||_F^2 - ||\boldsymbol{A}||_F^2\right) \tag{11.6}$$

where $||\cdot||_F$ represents the Frobenius norm and $\boldsymbol{A}$ is the symmetric part of the velocity gradient tensor and $\boldsymbol{B}$ is the antisymmetric part.

The Omega method was first proposed by Liu et al. in 2016 and was proved to be Galilean invariant (Wang Y. et al., 2018). This method has been applied in the boundary layer transition (Wang Y. et al., 2017), and the applicability of this method in the flow transition section has been verified. The Omega method defines the vortices as the region where the vorticity overtakes the deformation and can be calculated by the following formula:

$$\Omega = \frac{||B||_F^2}{||A||_F^2 + ||B||_F^2 + \varepsilon} \tag{11.7}$$

In practical calculation, ε is usually introduced into denominator to remove the noises caused by division by zero. In order to restrain the influence of ε on the calculation results, Dong et al. (2018c) have proved that $\varepsilon = 0.001\left(||B||_F^2 - ||A||_F^2\right)_{max}$ is better.

The main idea of the Liutex method is to define the vortex as a region where the vorticity overtakes the deformation in a plane perpendicular to the local rotation axis. Liutex is unique, Galileo invariant (Wang Y. et al. 2018), and systematic. The local rotation axis is determined as the real eigenvector of the velocity gradient tensor, and its strength is obtained by determining the size of the in-plane vorticity term and deformation term perpendicular to the rotation axis with two successive coordinate transformations. The explicit expression of Liutex strength (magnitude) R is:

$$\vec{R} = R\vec{r} = \left\{ \left(\vec{\omega}.\vec{r}\right) - \sqrt{\left(\vec{\omega}.\vec{r}\right)^2 - 4\lambda_{ci}^2} \right\}\vec{r} \tag{11.8}$$

where $\vec{\omega}$ is vorticity, $\vec{r}$ is the real eigenvector of the velocity gradient tensor, λ_{ci} is the imaginary part of the conjugate complex eigenvalue, and, represents the vector dot product, as shown in Eq. (11.8).

The Omega-Liutex method was proposed by Dong et al. (2019b), who applied the idea of the Omega method to Liutex. It is a normalized Liutex identification method with the threshold range of 0−1. According to the definition, the Omega-Liutex method can be obtained from the following formula:

$$\Omega_R = \frac{\beta^2}{\alpha^2 + \beta^2 + \varepsilon} \tag{11.9}$$

where α and β can be calculated by the following formulae:

$$\alpha = \frac{1}{2}\sqrt{\left(\frac{\partial V}{\partial Y} - \frac{\partial U}{\partial X}\right)^2 + \left(\frac{\partial V}{\partial X} + \frac{\partial U}{\partial Y}\right)^2} \tag{11.10}$$

$$\beta = \frac{1}{2}\left(\frac{\partial V}{\partial X} - \frac{\partial U}{\partial Y}\right) \tag{11.11}$$

In practical calculation, the Omega-Liutex method will also introduce $\varepsilon = b * (\beta^2 - \alpha^2)_{max}$ on denominator and b can be set to 0.001−0.002.

Fig. 11.7 shows the vortex structures of rotor passage identified by the aforementioned four vortex identification methods, and the iso-surfaces are colored with the

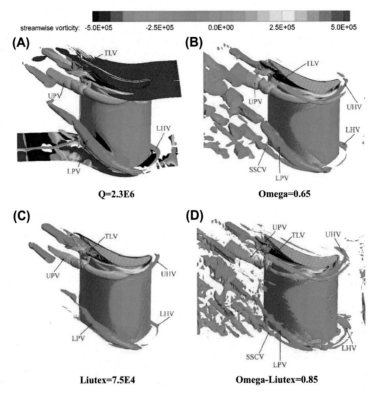

FIGURE 11.7

Vortex structure of rotor passage identified by different vortex identification methods and streamwise vorticity distribution (*LHV*, lower horseshoe vortex; *LPV*, lower passage vortex; *SSCV*, suction-side corner vortex; *TLV*, tip leakage vortex; *UHV*, upper horseshoe vortex; *UPV*, upper passage vortex).

Reproduced from Wang, Y., et al., The applicability of vortex identification methods for complex vortex structures in axial turbine rotor passages, Journal of Hydrodynamics 31 (4), 2019, with the permission of Journal of Hydrodynamics.

streamwise vorticity. The four vortex identification methods can identify the tip leakage vortex (TLV), the upper passage vortex (UPV), the lower passage vortex (LPV), and the horseshoe vortices (HV), but the relative sizes of the vortices identified by each method are different. The suction-side corner vortex (SSCV) is relatively weaker with respect to the TLV and the passage vortices. The identification results of corner vortex with the aforementioned four methods are quite different. There is no SSCV in the identification results of the Q criterion and the Liutex method, while the Omega method and the Omega-Liutex method can identify the suction-side corner vortex extremely clearly. This indicates that the Omega method and the Omega-Liutex method have better performance in identifying and visualizing weak vortices.

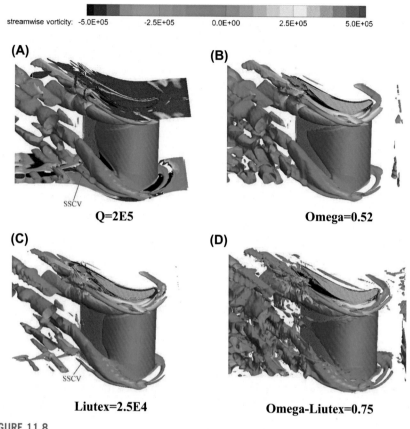

FIGURE 11.8

Vortex structure of rotor passage identified by different vortex identification methods with smaller threshold and streamwise vorticity distribution (*SSCV*, suction-side corner vortex).

Reproduced from Wang, Y., et al., The applicability of vortex identification methods for complex vortex structures in axial turbine rotor passages, Journal of Hydrodynamics 31 (4), 2019, with the permission of Journal of Hydrodynamics.

Fig. 11.8 shows the identification results of the four kinds of vortex identification methods with smaller thresholds. Compared with Fig. 11.7, it can be found that the identification results of the Q criterion and Liutex method are greatly affected by the threshold value. Increasing of the threshold leads to some vortices not being recognized. And the most obvious change of vortices in the passage is the SSCV, followed by the scraping vortices outside the gap. The HV structure near the casing identified by the Liutex method has also changed significantly. At a higher threshold, the pressure-side HV cannot be identified by the Liutex method since the Liutex magnitude is a scalar and thus strongly dependent on the selection of the threshold as well. Although the identification results of the Omega and Omega-Liutex method vary with the threshold, the key vortex structures can still be identified. Regardless of the threshold value, in Fig. 11.7B and D and Fig. 11.8B and D, the suction-side angle vortices can be clearly observed, and the evolution process of the UPV can still be shown. The sensitivity of these two methods to threshold is relatively low, and the difficulty of threshold selection will be reduced. Even if the most suitable threshold is not selected, some important vortex structures can still be identified. On the contrary, if the Q criterion or the Liutex method is applied without an appropriate threshold, it will easily lead to some vortices be omitted or misidentified. Therefore, the threshold selection of these two methods often requires users to have some professional experience.

A vortex is intuitively recognized as the rotational/swirling motion of the fluids. It is the intuitive and convenient method to judge the existence and location of vortices by streamlines. For the display consistency of vortex identification methods and streamlines, Fig. 11.9 focuses on the difference between vortex structures reflected intuitively by surface streamlines and vortex structures identified by the four vortex identification methods. In Fig. 11.9A, there are significant differences between vortices (e.g., Zone B & Zone C) identified by the Q criterion and surface streamlines. If the location of vortices identified by streamlines is taken as the standard, the Omega method is more accurate and the Liutex method is the most accurate one in identifying the upper passage vortex (in Zone B of Fig. 11.9B and C). However, the identification of strong vortices (e.g., Zone A & Zone B) by the Omega-Liutex method departure from the vortex structures visualized by streamlines. It is noteworthy that the Omega-Liutex method has high consistency with streamlines in the identification of weak vortices (e.g., Zone C of Fig. 11.9D). Accordingly, it is easy to find that the Liutex method is the most consistent with streamlines in identification of strong vortices, while the Omega-Liutex method coincides best with streamlines in identifying weak vortices.

Tip leakage loss accounts for a large proportion of flow losses in rotor passage. In order to further understand the relationship among losses, vortices, and shear, the local turbulent dissipation terms at 50% of chord and 90% of chord are shown in Fig. 11.10. And contour of streamwise vorticity is provided in Fig. 11.10. The turbulent dissipation term is defined as:

$$\Phi = \mu_{eff} \cdot \left\{ 2\left(\frac{\partial u_i}{\partial x_i}\right)^2 + \left(\frac{\partial u_j}{\partial x_i} + \frac{\partial u_i}{\partial x_j}\right)^2 - \frac{2}{3}\left(\sum \frac{\partial u_i}{\partial x_i}\right)^2 \right\} \tag{11.12}$$

where, $\mu_{eff} = \mu + \mu_\tau$, and μ_τ is eddy viscosity, μ is dynamic viscosity. Turbulent dissipation causes the loss of mechanical energy irreversibly converted to internal energy, which will inevitably lead to entropy production and can be used to measure the local loss. It is found that the high loss distributions at 50% of rotor chord and 90% of rotor chord coincide well with the high vorticity regions (Zone B in 11.10). As shown in Fig. 11.8C and Fig. 11.10A, Zone A corresponds to the core of TLV, and there is only low-intensity turbulent dissipation in the core region of TLV, which indicates that the fluid rotation is not the essential cause of the loss. The high loss

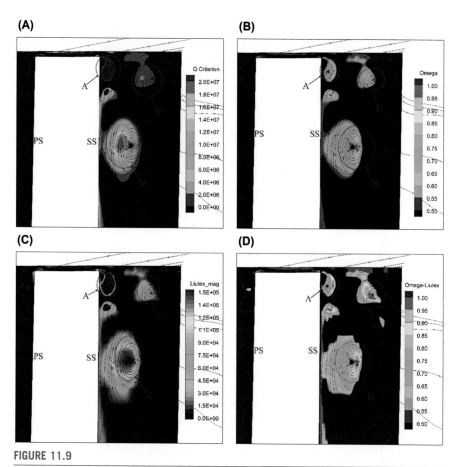

FIGURE 11.9

The distribution of identification parameters by four methods and surface streamline at 50% of rotor chord (*A*, tip leakage vortex; *B*, upper passage vortex; *C*, scraping vortex; *PS*, pressure side; *SS*, suction side).

Reproduced from Wang, Y., et al., The applicability of vortex identification methods for complex vortex structures in axial turbine rotor passages, Journal of Hydrodynamics 31 (4), 2019, with the permission of Journal of Hydrodynamics.

region near tip clearance is mainly caused by the shear layer on the casing and the shear between the tip leakage flow and the mainstream flow.

11.2.4 Summary

There are complex vortex structures in turbine rotor passages, which may be weak or strong, large or small, interacting with each other and generating most of aerodynamic loss in turbomachines. Therefore, it is significant to identify the vortices

(A)   (B)

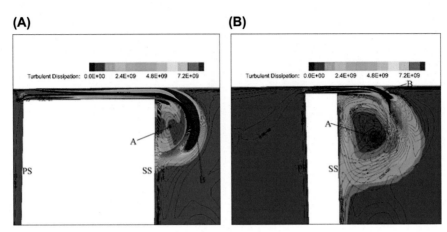

FIGURE 11.10

Turbulent dissipation and vorticity distribution (*A*, tip leakage vortex core; *PS*, pressure side; *SS*, suction side).

Reproduced from Wang, Y., et al., The applicability of vortex identification methods for complex vortex structures in axial turbine rotor passages, Journal of Hydrodynamics 31 (4), 2019, with the permission of Journal of Hydrodynamics.

accurately for flow field analysis and aerodynamic performance optimization in turbomachines. In this section, four vortex identification methods are applied to the turbine rotor passages, and the applicability and robustness of each method are summarized. The Liutex method is also applied to the vortices analysis of two cases of a turbine rotor passage with different angles of attack. The following conclusions are drawn:

The Omega method and the Liutex-Omega method can capture strong and weak vortices simultaneously in turbine rotor passages and have high robustness. The Q criterion and the Liutex method are greatly affected by the threshold. When the threshold is small, the weak vortices can be identified by the Q criterion and the Liutex method, however, whose structures are not clear enough. Therefore, the Omega method and the Liutex-Omega method are very helpful to capture and analyze the vortex structures in turbine rotor passage comprehensively.

A vortex is intuitively recognized as the rotational motion of the fluids, and streamlines are often used as an intuitive and rough method for judging the location of vortices. There is high consistency between vortices visualized by the Liutex method and vortices visualized by streamlines in identification of strong vortices, and weak vortices visualized by the Liutex-Omega method and streamlines have high consistency.

The Omega method, the Liutex method, and the Liutex-Omega method have obvious advantages in distinguishing vortices from high shear regions, compared with the Q criterion, which may be due to less shear contamination in these three methods. By using the Liutex method, it is found that the high loss regions in turbine rotor passages are concentrated in the high shear regions, while the losses within the core of the vortex are small.

With the application of the Liutex method, the difference in the structure and evolution of vortices in the turbine rotor passages with different attack angles can be clearly observed. At negative angle of attack, TLV and UPV are effectively suppressed, and the interaction between the two vortices is enhanced, which reduces the mixing between the vortices and the mainstream. It is shown that Liutex is an effective method, which can be applied to the analysis of vortex evolution in turbine rotor passages.

11.3 Vortex identification methods in marine hydrodynamics (Zhao et al., 2020)

11.3.1 Short review

Marine hydrodynamic flows are turbulent and disordered, not only because the flow is usually at high Reynolds number, but also due to the complexity of wall-bounded flows especially for those involving complex geometries. These two main factors lead to the ubiquitous vortex structures. For example, the tip and hub vortex of a rotating propeller (Felli and Falchi, 2018; Wang L. et al., 2018), the vortices shed from ship hull when performing large drift angle maneuvering (Xing et al., 2012; Shen et al., 2014), the vortices shed from appendages (such as bilge, fin) of a fully appended ship (Sakamoto et al., 2012; Wang J. et al., 2016), vortex shedding of deep-draft column stabilized floaters (Liang and Tao, 2017; Liu M. et al., 2017; Zhao et al., 2018), and so on. In the past, the research of marine hydrodynamics was focused on macroscopic quantities such as forces, moments, and motions due to the limitation of experimental facilities and potential theory methods. However, the advancement of particle imagine velocimetry (PIV) and computational fluid dynamic (CFD) for viscous flow makes it possible to provide detailed flow field information. Researchers are paying more and more attentions on the analysis of vortex structures for hydrodynamics. By extracting and studying the vortex dynamics, researchers can have a deep insight of the detailed flow field such as pressure fluctuation, loads, vibrations, and fatigue on structures.

Typically, vortex can be visualized by regions or lines (Epps, 2017). The region method takes vortex as a connected region of a specific quantity and uses iso-surfaces to identify vortex. The iso-surfaces could describe the extended distance from the vortex core. The line method tracks the trajectory of fluid particle through extracting the cores of swirl motion in given area (Zhang et al., 2018a). These two classes of visualization methods are complementary to each other for different purposes. Vortex identification methods can be categorized into either Eulerian or Lagrangian. The first type uses Eulerian quantities such as velocity, pressure to help identifying vortex. The second type is usually based on the trajectory of fluid particle motion.

Several Eulerian vortex identification methods are discussed with the relative merits of these methods. Then selective numerical examples of marine hydrodynamic problems are processed by the Q, λ_2 criterion and the modified normalized Liutex $\widetilde{\Omega}_R$ method. The strengths and weaknesses of different methods are discussed based on the extracted and visualized vortex structures.

11.3.2 Vortex identification methods

There are many existing vortex identification methods. According to Liu et al. (2019), vorticity is the first generation of vortex identification methods.

11.3.2.1 Vorticity

Vorticity is defined as the curl of velocity. It can be written in the following formula

$$\vec{\omega} = \nabla \times \vec{u} \tag{11.13}$$

It has no doubt that the magnitude of vorticity is the most widely used quantity to represent vortex cores for free shear flows, especially for 2-D case. However, vorticity magnitude is not necessarily an appropriate vortex identifier for wall-bounded flows because vorticity includes both the near-wall shear motion and swirling motion.

The swirling part cannot be extracted from vorticity. For example, in 2-D wall-bounded flow, the maxima and minima of vorticity magnitude occur at the wall, where a vortex obviously does not exist. The large-scale vorticity is due to the strong shear effect of boundary layer in near-wall region. In addition, vorticity cannot identify vortex cores in shear flows if the background shear is comparable to the vorticity magnitude (Jeong et al., 1995). Vorticity is a necessary but not sufficient condition for vortical motion. Thus vorticity is not suitable nor accurate to identify vortex.

11.3.2.2 Second generation of vortex identification methods

All vortex identification methods developed in the eighties and nineties such as $Q, \Delta, \lambda_2, \lambda_{ci}$ are scalar and completely determined by the eigenvalues of velocity gradient tensor.

11.3.2.3 Ω method

According to Liu et al. (2016), the Ω method defines a normalized scalar representing the ratio of rotational and shear motion, given by

$$\Omega = \frac{||\mathbf{B}||_F^2}{||\mathbf{A}||_F^2 + ||\mathbf{B}||_F^2} = \frac{b}{a+b} \tag{11.14}$$

In application, one picks:

$$\Omega = \frac{b}{a+b+\varepsilon} \tag{11.15}$$

where $a = \text{trace}(A^{\mathrm{T}}A) = \sum_{i=1}^{3}\sum_{j=1}^{3}\left(A_{ij}^2\right)$, $b = \text{trace}(B^{\mathrm{T}}B) = \sum_{i=1}^{3}\sum_{j=1}^{3}\left(B_{ij}^2\right)$, A is the symmetric part of velocity gradient tensor ∇V and B is the antisymmetric part, and $\|\cdot\|_F$ is the Frobenius norm.

The normalized Ω has a clear physical meaning. When the rotation strength is larger than deformation (i.e., $\Omega > 0.5$), vortex exists in this region. In practice, $\Omega = 0.52$ is recommended. This method can capture both the strong and weak vortices due to the definition that is a relative ratio rather than an absolute value.

11.3.2.4 Liutex method

According to Liu et al (Liu et al., 2018a; Gao et al., 2018; Wang Y. et al., 2019a), Liutex is defined as $\vec{R} = R\vec{r}$ and $\vec{\omega}\cdot\vec{r} > 0$ where:

$$R = \vec{\omega}\cdot\vec{r} - \sqrt{\left(\vec{\omega}\cdot\vec{r}\right)^2 - 4\lambda_{ci}^2}, \tag{11.16}$$

$\vec{\omega}$ is the vorticity, λ_{ci} is the imaginary part of the eigenvalue, and $\vec{r}$ is the real eigenvector of $\nabla\vec{v}$.

11.3.2.5 Liutex-Ω method

By combining the advantages of Ω and the Liutex methods, Dong et al. (2019b) proposed the normalized Liutex method, which defines a scalar Ω_R with the following formula

$$\Omega_R = \frac{\beta^2}{\alpha^2 + \beta^2 + \varepsilon} \tag{11.17}$$

$$\beta = \frac{1}{2}\vec{\omega}\cdot\vec{r} \tag{11.18}$$

$$\alpha = \frac{1}{2}\sqrt{\left(\vec{\omega}\cdot\vec{r}\right)^2 - 4\lambda_{ci}^2} \tag{11.19}$$

$\varepsilon = b_0\text{max}(\beta^2 - \alpha^2)$ is a small parameter to avoid division by zero and b_0 is a small positive number around $0.001-0.002$.

The combined method shows promising results. It can not only measure the relative rotation strength on the plane perpendicular to the local rotation axis, but also separate rotational vortices from shear layers, discontinuity structures, and other nonphysical structures. However, Liu and Liu (2019d) found that the extracted iso-surfaces formed by Ω_R are not smooth enough and contain many bulges. They improved it and proposed the modified normalized $\widetilde{\Omega}_R$ method, which is given by

$$\widetilde{\Omega}_R = \frac{\beta^2}{\alpha^2 + \beta^2 + \lambda_{cr}^2 + 0.5\lambda_r^2 + \varepsilon} \tag{11.20}$$

where λ_{cr} and λ_r are real part of conjugate complex eigenvalues and real eigenvalues of velocity gradient tensor, respectively.

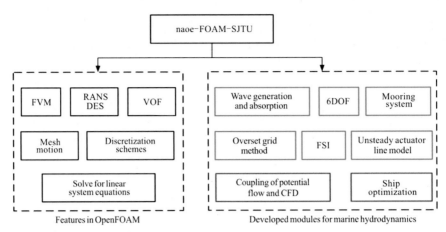

FIGURE 11.11

Main framework of naoe-FOAM-SJTU solver.

Reproduced from Zhao, W., et al., Vortex identification methods in marine hydrodynamics, Journal of Hydro-dynamics 32 (2), 2020, with the permission of Journal of Hydrodynamics.

With the help of explicit Liutex formula, the aforementioned equation can be rewritten as follows

$$\tilde{\Omega}_R = \frac{(\omega \cdot r)^2}{2\left[(\omega \cdot r)^2 - 2\lambda_{ci}^2 + 2\lambda_{cr}^2 + \lambda_r^2\right] + \varepsilon} \tag{11.21}$$

11.3.3 Numerical approaches

In this section, several typical flows in marine hydrodynamics are selected to demonstrate the characteristics of several vortex identification methods. All the selected examples are numerically computed using the naoe-FOAM-SJTU solver (Shen et al., 2015; Wang J. et al., 2017, 2018), which is developed on top of the OpenFOAM framework. It utilizes the data structures and low-level infrastructures of Open-FOAM and consists of several specialized features for marine hydrodynamics. Fig. 11.11 shows the functional modules of the solver, where the blue frame and red frame represent developed and developing modules, respectively.

Several vortex identification methods can be applied to arbitrary unstructured polyhedral computational mesh with complex geometries.

11.3.4 Applications of vortex identification methods

11.3.4.1 Propeller open water

In this subsection, a propeller open-water test case is used for demonstration. The model is a four-bladed propeller for the ONR Tumblehome (ONRT) model 5613, which is a modern surface combatant and publicly available for fundamental research (Sanada et al., 2013). Some studies of the hydrodynamic performance of the propeller have been published in previous work (Wang et al., 2019c). The propeller is simulated using dynamic overset grid and the open-water curves are obtained by a single-run procedure. Two mesh blocks were generated for background and propeller individually and assembled to a single overset grid system. Here, the condition of advance coefficient is set as $J = V_A/(nD_P) = 0.9$, where V_A is the advance speed, n and D_P are the rotational speed and diameter of propeller, respectively.

Fig. 11.12 shows iso-surfaces of different λ_2 values colored with vorticity magnitude. The tip vortices of the propeller are resolved clearly. However, using different λ_2 values as the threshold for contour, various lengths of tip vortices are extracted. For small λ_2 values shown in Fig. 11.12A, more detailed wake structures are identified and extracted. As the threshold of λ_2 increases, those weak vortices (fluid regions with smaller vorticity magnitude) cannot be identified. The structures of vortex filament are also distinct. The vortex tube extracted by larger λ_2 threshold value is thinner. Fig. 11.13 shows the propeller wake structures, which are identified and extracted by different values of modified normalized Liutex $\widetilde{\Omega}_R$ and colored by vorticity magnitude. The contour threshold value 0.52 is the recommended value in the original paper (Dong et al., 2019b). Besides this, three other threshold values are chosen for comparison. As depicted in the pictures, although the vortex lengths in wake region are different, the vortex tube structures extracted by different $\widetilde{\Omega}_R$ values and the corresponding vorticities remain almost the same.

11.3.4.2 Ship drag test

Japan Bulk Carrier (JBC) is another model of benchmark case for the Tokyo 2015 CFD workshop. It is a full-form ship with a block coefficient up to 0.858. The bare hull is drag in still water with a towing speed of 1.179 m/s, corresponding to $Fr = 0.142$ and $Re = 7.46 \times 10^7$. In this study, the turbulence is modeled by delayed detached-eddy simulations (DDES) and the total grid number is 7.48×10^6. Details about the computational setup can be referred to Wang and Wan (2019a).

Figs. 11.14 and 11.15 illustrate the vortical structures extracted by the Q criterion and the modified normalized Liutex method, respectively. The vortices above free surfaces are eliminated for the sake of simplicity. At first glance, the two figures show similar vortex structures. In the vicinity of stern, massively separation flow due to the sharp curvature change of the hull surface is observed. The captured structures are, however, very distinctive. In Fig. 11.14, the vortices identified by Q contain redundant shear motions near the hull surface. These deformations are excluded in the vortices extracted by $\widetilde{\Omega}_R$.

11.3.4.3 Ship propeller—rudder interaction

The propeller—rudder interaction of the ONRT surface combatant during a zigzag maneuver is visualized. The simulation starts from the starboard side. The model is scaled at a ratio of 1:48.935 and is self-propelled with free running, which means all 6 degrees of freedom for the model are released, at model point in calm water. The corresponding Froude and Reynolds numbers are $Fr = 0.200$ and $Re = 3.39 \times 10^6$, respectively. In the 20/20 zigzag maneuver test, the rudder is

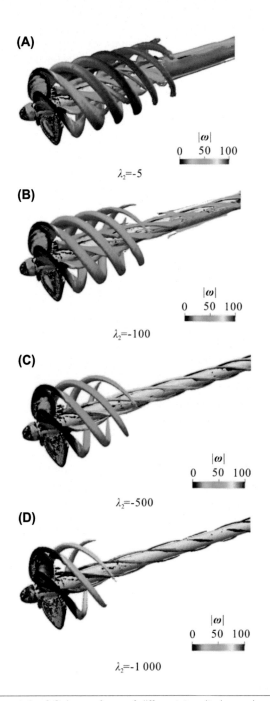

FIGURE 11.12

Propeller operating at $J = 0.9$. Iso-surfaces of different λ_2 criterion, colored by vorticity magnitude.

Reproduced from Zhao, W., et al., Vortex identification methods in marine hydrodynamics, Journal of Hydrodynamics 32 (2), 2020, with the permission of Journal of Hydrodynamics.

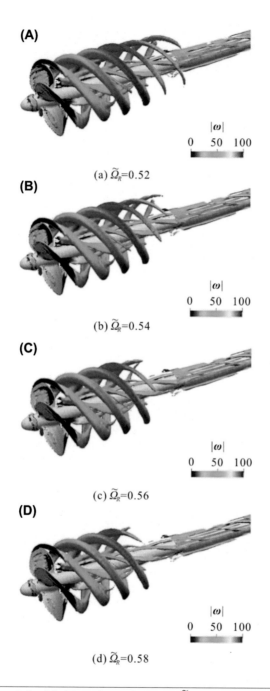

(A)

(a) $\widetilde{\Omega}_R = 0.52$

(B)

(b) $\widetilde{\Omega}_R = 0.54$

(C)

(c) $\widetilde{\Omega}_R = 0.56$

(D)

(d) $\widetilde{\Omega}_R = 0.58$

FIGURE 11.13

Propeller operating at $J = 0.9$. Iso-surfaces of different $\widetilde{\Omega}_R$, colored by vorticity magnitude.

Reproduced from Zhao, W., et al., Vortex identification methods in marine hydrodynamics, Journal of Hydrodynamics 32 (2), 2020, with the permission of Journal of Hydrodynamics.

FIGURE 11.14

Vortical structures in wake region presented by iso-surfaces with $Q = 5$ and colored by vorticity magnitude.

Reproduced from Zhao, W., et al., Vortex identification methods in marine hydrodynamics, Journal of Hydrodynamics 32 (2), 2020, with the permission of Journal of Hydrodynamics.

FIGURE 11.15

Vortical structures in wake region presented by iso-surfaces with $\widetilde{\Omega}_R = 0.52$ and colored by vorticity magnitude.

Reproduced from Zhao, W., et al., Vortex identification methods in marine hydrodynamics, Journal of Hydrodynamics 32 (2), 2020, with the permission of Journal of Hydrodynamics.

turned to the opposite direction at a rate of 35°/s to a rudder angle of 20° every time the ship heading reaches ±20°. The computational grids for moving components such as hull, propellers, and rudders are generated separately and are then assembled into a composite overset grid. All component grids can translate and rotate freely with respect to other grids.

Fig. 11.16 gives the instantaneous flow visualizations of hull–propeller–rudder interaction for ONRT during zigzag maneuvering. The vortical structures are identified by different criterions and colored by vorticity magnitude. Similar to the results in Section (1), the dominant wake structures are the clockwise and anticlockwise tip vortices generated by the contrarotating twin-screw propellers. Strong hub vortices are also observed. Owing to the numerical dissipation and coarse grid resolution in the rudder downstream wake regions, the vorticity of hub vortices becomes smaller. These weak vortices are not identified by the Q criterion. However, the $\widetilde{\Omega}_R$ method can capture both the weak and strong tip vortices simultaneously.

Fig. 11.17 depicts another view of the instantaneous vortical structures around the fully appended ONRT model during zigzag maneuver test. Although the simulations are performed using the two-phase VOF model, only vortices in the water are displayed for the sake of simplicity. The rudders have been executed and turned to the port side, which provide lift rudder force. A consequence turning moment

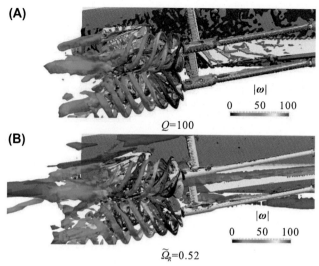

FIGURE 11.16

Hull—propeller—rudder interaction of the fully appended ONRT during 20/20 zigzag maneuver test. Iso-surfaces by (A), (B) and colored by vorticity magnitude.

Reproduced from Zhao, W., et al., Vortex identification methods in marine hydrodynamics, Journal of Hydrodynamics 32 (2), 2020, with the permission of Journal of Hydrodynamics.

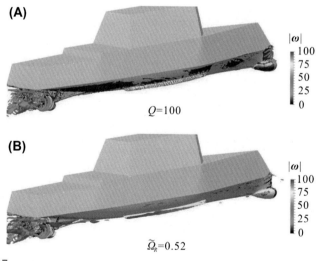

FIGURE 11.17

Vortical structures around the fully appended ONRT during 20/20 zigzag maneuver test. Iso-surfaces by (A), (B) and colored by vorticity magnitude.

Reproduced from Zhao, W., et al., Vortex identification methods in marine hydrodynamics, Journal of Hydrodynamics 32 (2), 2020, with the permission of Journal of Hydrodynamics.

turns the ship toward left. The "fake" vortices owing to the boundary layer shear motions near the hull surface such as bow, bilge keel, and skeg vortices are removed from the iso-surfaces of $\widetilde{\Omega}_R = 0.52$. Moreover, iso-surfaces of $Q = 100$ can only capture propeller and rudder vortices, with some redundant nonphysical vortices near hull, since these are strong vortices and have large vorticity. Other weak vortices generated by free surface ship waves, bilge keel, and skeg failed to be captured by iso-surfaces of $Q = 100$, but can be successfully observed in the iso-surfaces of $\widetilde{\Omega}_R = 0.52$.

11.3.4.4 VIV of a flexible riser

Vortex-induced vibration (VIV) is the main cause of marine risers' fatigue damage. It is very hard to predict VIV of marine risers due to the nonlinearity and instability of the flow. By taking the flexible structural deforming into account, the fluid–structure interaction problem becomes more complex. In this subsection, a riser under stepped flow is simulated. The diameter and length of the riser are 0.028 and 13.12 m, respectively. Lower part (45% of total length) of the riser is submerged in water with speed of 0.16 m/s. The top is tensioning with a pretension force of 405 N. This case setup is in accordance with the experiment that is performed by Chaplin et al. (2005a,b). The VIV of marine riser is simulated by 2.5-D strip method, i.e., the fluid is solved in a finite set of 2-D strip sections and the structure is obtained with 3-D model. In the current solver's implementation, the flow fields in 2-D domains are solved using finite volume method provided by OpenFOAM, and the dynamic structural response is solved using finite element method with Euler–Bernoulli beam model. The hydrodynamic force is obtained from fluid and passed to structure and the displacements are calculated in structure and passed to moving boundaries of fluid mesh in each section.

Fig. 11.8 depicts the 2-D vortices of VIV for a flexible marine riser in fluid sections, represented and colored by different methods. In both pictures, clear Kármán vortex streets are observed in each section. As elaborated in the previous section, the maxima of vorticity magnitude occurs at the near wall region (Fig. 11.18A). It is convinced that the disadvantages of vorticity magnitude in representing vortices are obvious. Different vortices in the vicinity of wall are identified as a single connected region, which is nonphysical. Furthermore, the vorticity magnitude is so small in the far wake region that it is hard to distinguish vortical structures. On the other hand, $\widetilde{\Omega}_R$, unlike vorticity magnitude, can clearly resolve the individual vortex cores in both near wake region and far wake region.

11.3.4.5 VIM of a spar platform

Spar platform is a kind of offshore floating product unit with relatively large aspect ratio (draft over diameter). Such kind of floating structures are subject to vortex-induced motions (VIM) when exposed in current with certain speed. The large amplitude oscillating motion is a critical issue for the fatigue failure and safety operations. To mitigate VIM, helical strakes are designed and installed on the surface of main spar hull. A truncated cylindrical hard tank with three-start helical

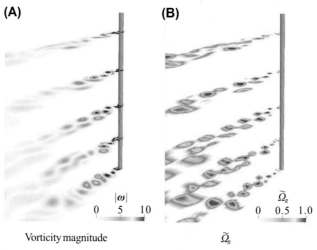

FIGURE 11.18

Vortices of VIV for a flexible marine riser in 2-D strip sections. The vortices are represented and colored by (A), (B).

Reproduced from Zhao, W., et al., Vortex identification methods in marine hydrodynamics, Journal of Hydro-dynamics 32 (2), 2020, with the permission of Journal of Hydrodynamics.

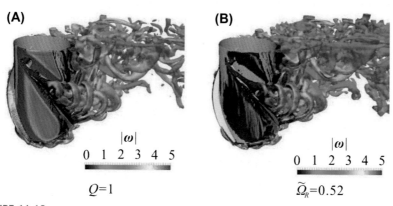

FIGURE 11.19

Vortical structures of VIM for a Spar with helical strakes. Iso-surfaces by (A), (B) and colored by vorticity magnitude.

Reproduced from Zhao, W., et al., Vortex identification methods in marine hydrodynamics, Journal of Hydro-dynamics 32 (2), 2020, with the permission of Journal of Hydrodynamics.

strakes of a truss spar scaled at 1:22.3 is selected for study (Finnigan and Roddier, 2007). Truss structure and keel tank are ignored in the simulation.

Fig. 11.19 shows the instantaneous vortical structures of VIM for the Spar at reduced velocity $U_r = UT_n/D = 6$, with U the current velocity, T_n the natural transverse period, and D the diameter of Spar. The vortical structures are represented by the iso-surfaces identified with $Q = 1$ and $\widetilde{\varOmega}_R = 0.52$, respectively. The picture vividly illustrates that the dominant vortices shed at the tip of helical strakes. After flow separation, the vortices form several horseshoe-like vortical structures. This reduces the formation of synchronized shedding of coherent structures along vertical direction and breaks the large-scale eddies into small structures. In such a way, the mitigation of VIM is achieved. Similar to previous results, the $\widetilde{\varOmega}_R$ method captures strong and weak vortices simultaneously.

11.3.5 Discussion on the sensitivity of b_0

According to the aforementioned numerical examples, $b_0 = 0.002$ is too large for typical marine hydrodynamic flows. This is due to the strong fluid rotations that can result in large ε, which will consequently affect the iso-surfaces by $\widetilde{\varOmega}_R$. If ε is large enough, even if the vortical rotation part strength is larger than nonvortical part, $\widetilde{\varOmega}_R = 0.52$ will extract wrong vortices, which contain nonvortical structures. According to the experience, $b_0 = 10^{-6}$ is suitable for most marine hydrodynamic problems. All the presented results demonstrated here are obtained with this value. However, in some special cases with very strong fluid rotations, $b_0 = 10^{-6}$ is still too large. In general, b_0 is still empirical and case-related.

An example of a self-propelled ONRT in still water is given to illustrate the sensitivity of b_0 in $\widetilde{\varOmega}_R$ method. The self-propelled ship is moving forward at model point with moving speed of 1.11 m/s corresponding to Froude number $Fr = 0.200$. The rotational speed is 8.75 RPS obtained with a PI controller, which makes sure the propeller thrust is equal to the ship resistance.

Fig. 11.20 shows the vortical structures of the self-propelled ONRT ship at model point. The iso-surfaces are extracted by $\widetilde{\varOmega}_R = 0.52$ with different b_0. For this case, $b_0 = 10^{-3}$ is so large that ε reaches the order of rotation strength magnitude and affects the identification of vortex. No vortices are identified with $b_0 = 10^{-3}$. Even with value $b_0 = 10^{-6}$, the vortices are emerged as thin vortex tube, which can be observed in Fig. 11.20(B). For the zigzag case with same Froude number in Section 11.3.4.3, $b_0 = 10^{-6}$ works smoothly. During free running ship zigzag maneuver, the speed loss and oblique inflow make the inflow condition different from self-propulsion. The absolute vortices strength for zigzag maneuver is smaller than self-propulsion condition. Thus large b_0 is reasonable to capture vortices for zigzag maneuver with weaker absolute vortices strength. b_0 should be set to a small value for flows with strong absolute vortices strength.

11.3.6 Summary

In this section, the most commonly used vortex identification methods for marine hydrodynamic flow problems are examined. The in-house solver, in Shanghai University of China, naoe-FOAM-SJTU is used and several numerical examples are demonstrated. Several classical examples are selected to demonstrate the advantages

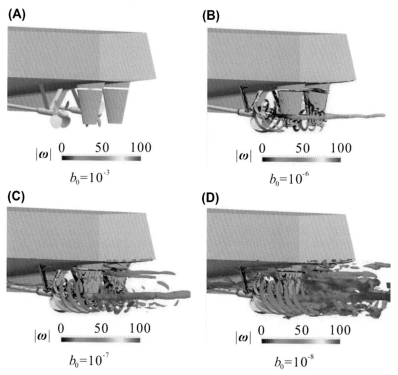

(A) $|\omega|$ 0   50   100 $b_0 = 10^{-3}$

(B) $|\omega|$ 0   50   100 $b_0 = 10^{-6}$

(C) $|\omega|$ 0   50   100 $b_0 = 10^{-7}$

(D) $|\omega|$ 0   50   100 $b_0 = 10^{-8}$

FIGURE 11.20

Hull–propeller–rudder interaction of the self-propelled ONRT at model point. Iso-surfaces by $\widetilde{\Omega}_R = 0.52$.

Reproduced from Zhao, W., et al., Vortex identification methods in marine hydrodynamics, Journal of Hydro-dynamics 32 (2), 2020, with the permission of Journal of Hydrodynamics.

and disadvantages of different vortex identification methods. After the discussions, the following conclusions can be made.

(1) Vortex identification methods such as Q and λ_2 criterion are quite sensitive to the threshold value of iso-surface. Hence, it is hard to extract the correct vortical structures for further vortex dynamics analysis.

(2) The Liutex-Omega (modified normalized Liutex) method shows promising results in vortex identification and visualization for marine hydrodynamics. It removes vorticity due to shear motion and only keeps the pure rotational motion part. Furthermore, it can identify strong and weak vortices simultaneously and is insensitive to iso-surface threshold value.

(3) For marine hydrodynamic flows, which involve strong absolute vortices strength, b_0 should be set to a small value. It is recommended that b_0 is set to 10^{-6}. However, b_0 is a empirical parameter and may need adjustment for different cases.

11.4 Prediction the precessing vortex core in the Francis-99 draft tube at off-design conditions with Liutex method (Tran et al., 2020)

11.4.1 Short review

The Francis turbine operating far from the best efficiency regime is characterized by abnormal flow in the draft tube and the appearance of a spiral vortex or columnar vortex, which is called the vortex rope. Arpe et al. (2009) found out the dominant frequency of a vortex rope lies between 0.2 and 0.4 times of the runner frequency. Understanding of the periodical precession of this vortex, as well as investigation of the vortex rope structure in the draft tube, is necessary for preventing structural vibrations and increasing the number of operation hours at off-design conditions. Nevertheless, detailed characteristics of the vortex structures are shown to be challenging to accurate visualizations.

Several attempts have been made to capture the vortex structures in the draft tube of the Francis turbine. Gavrilov et al. (2017) focused on detecting and analyzing the vortical structures and evolution of vortex core at deep partial-load points (flow rate of only 35%), using two URANS models and a hybrid LES/RANS method, where the vortex structures are visualized by λ_2 (Jeong et al., 1995) and Q criterions (Hunt et al., 1988). However, their results are strongly influenced by choice of threshold values, which will indicate different vortex structures by different threshold selections. In particular, using the λ_2 or Q criterion, there exist nonphysical vortex structures, and "vortex breakdown" could be observed for some large thresholds while no "vortex breakdown" can be found for some smaller thresholds (Dong et al., 2019b). These will easily lead to misunderstandings on the physics of the turbulent flow. Liu et al. (2016) proposed the Omega method (Ω method), which is not sensitive to the threshold selection and can successfully capture both strong and weak vortices simultaneously. However, the Ω method has some limitations like the introduction of an uncertain parameter of epsilon (ε) (Dong et al., 2018c).

The vortex structures are identified by several methods aiming at extracting a line feature called the vortex core line that is developed. For instance, the vorticity is a traditional and common indicator for the presence of vortices. However, this technique still has some limitations including: (i) sensitive to other nonlocal vector features; (ii) not producing contiguous lines (Haimes, 2000). Recently, a new vector called "Liutex" is introduced by Liu et al. (2018a, 2019) to describe the local rigid rotation of fluids. This method, which is not only Liutex iso-surfaces but also different strength along with the Liutex cores, can be used to analyze the process of the vortex generation and development (Gao et al., 2019b; Xu et al., 2019a). The Liutex method has not yet been applied to investigate the complex fluid flows such as cavitation flow in the Francis turbine in literature. Hence, to reduce the difficulty for more accurate visualization of vortex structures, a well-defined method such as the Liutex method is necessary to be applied for the vortex identification in the Francis turbine.

11.4.2 **Case setup and numerical results**

Inspired by the aforementioned work, this study mainly focused on identifying and clarifying the processing vortex rope in the draft tube of the Francis-99 turbine (NITU, 2016) under the off-design conditions. Therefore, by the shear—stress transport turbulence model (SST) (Menter, 1994) and Zwart—Gerber—Belamri (ZGB) cavitation model (Zwart, 2004), the cavitating flow in the draft tube is simulated, and the vortex structures on draft tube cone are visualized by the Liutex method. The periodical evolution of vortex rope on the draft tube is further revealed with the Liutex core line.

The numerical configurations were the Francis-99 model turbine, which consists of a runner with 15 long blades and 15 splitter blades, a spiral casing, 28 guide vanes, and a draft tube (as shown in Fig. 11.21). The three operating points: high load (HL), best efficiency point (BEP), and partial-load (PL) were created with ANSYS ICEM CFD using the ICEM files provided by the second workshop (NTNU, 2016). The mesh size for the complete model at the BEP was about 20 million elements. The quality of the mesh satisfies the common industrial standard, which was reported by Trivedi C. et al. (2013) and Goyal R. et al. (2018). In this study, the complete turbine was simulated and performed in two steps including steady and unsteady simulations. The unsteady simulations were conducted with the initial field obtained from the steady simulations.

The mass flow inlet boundary was set at the casing inlet, and static pressure was set at the draft tube outlet. The runner is considered rotating part while casing, stay vanes, guide vanes, and draft tubes are stationary parts. The rotational speed of the runner is set as 332.59 rpm and components are connected with others by the domain

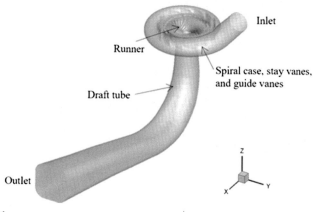

FIGURE 11.21

Simulation domain of Francis-99 turbine.

Reproduced from Tran, C.T., et al., Prediction of the precessing vortex core in the Francis-99 draft tube under off-design conditions by using Liutex/Rortex method, Journal of Hydrodynamics 32 (3), 2020, with the permission of Journal of Hydrodynamics.

interface. The General Grid Interfaces (GGI) connects stationary domains to the rotating domain.

The SST turbulent model is the widely used turbulence model for turbo machinery (Tran et al., 2019; Zhao et al., 2015). In this study, the SST turbulence model and ZGB cavitation model are adopted for simulation of the unsteady cavitating flow through a Francis-99 turbine. To investigate the cavitating vortex rope, the unsteady simulation was performed for 10 complete rotations of the runner where total computation time was 1.8s. The time step was set as t = 5E-4s (10 of the runner rotation per time step). The convergence criterion was set to a root-mean-square (RMS) value maximum 10E-5.

To validate the simulation method used in this study, Table 11.2 shows a comparison of hydraulic efficiency and torque by simulation with SST and experiment at three operating conditions. The maximum discrepancies between the simulated and experimental efficiencies are observed as 4.24% under the PL operating condition, while it is 3.05% for the BEP and 2.65% for HL. The numerical torque is 465 Nm (PL), 706 Nm (BEP), and 820 Nm (HL), which are higher than the experimental torque and the rate of numerical torque to experimental torque was 11.7%, 14.6%, and 10.7%, respectively. During the simulation, the numerical efficiency is higher than the experimental efficiency at all operating points because the flow leakage losses and other losses that occurred during the measurements were not considered. In the numerical simulation, the mesh quality, vortex, and flow separation may cause inaccuracy in the torque calculation. Taking into account the earlier comparison, the overall accuracy of simulation is acceptable.

11.4.3 Vortex structure

For further clarification about the reliability of Liutex method in the vortex definition for the turbulent flow, in the present study, the vortex rope frequency is investigated. The unsteady pressures at two levels DT5 and DT7 (see Fig. 11.22) are plotted in Fig. 11.23. The analysis of the vortex rope morphology is performed during a low-frequency period in order to examine its dynamics. As a result, a low-frequency cycle of 0.6 s (1.66 Hz) is observed. And to reveal the time evolution of the vortex rope, six snapshots with a time step of 0.1 s are plotted in Fig. 11.24 by Liutex iso-surface. The pressure amplitudes corresponded to a vortex rope frequency of 1.66 Hz (about 0.3 times of the runner frequency).

Here, the vortex rope is compressed during the first phase of the low-frequency cycle, after that stretching, breakdown, shedding, and moving downstream. The period of low frequency is responsible for the pressure fluctuations associated with the precession of the vortex rope.

On the other way, by using a fast Fourier transform (FFT) on the results, the dominant frequency of the pressure fluctuations can be obtained. The frequency spectrum obtained from present simulations under PL condition at DT5 pressure monitoring point is shown in Fig. 11.25. The vortex rope frequency is found to be about 0.3 times the runner frequency. The result is consistent with that numerically

Table 11.2 Comparison between numerical and experimental of turbine energy characteristics.

	PL			BEP			HL		
	Net head (m)	Hydraulic efficiency (%)	Torque (Nm)	Net head (m)	Hydraulic efficiency (%)	Torque (Nm)	Net head (m)	Hydraulic efficiency (%)	Torque (Nm)
SST in this study	11.87	94.37	465	11.94	95.44	706	11.88	94.36	820
Experiment (NTU, 2016)	11.87	90.13	416.39	11.94	92.39	616.13	11.88	91.71	740.54
Discrepancy (%)	0	4.24	11.7	0	3.05	14.6	0	2.65	0.7

Reproduced from Tran, C.T., et al., Prediction of the precessing vortex core in the Francis-99 draft tube under off-design conditions by using Liutex/Rortex method, Journal of Hydrodynamics 32 (3), 2020, with the permission of Journal of Hydrodynamics.

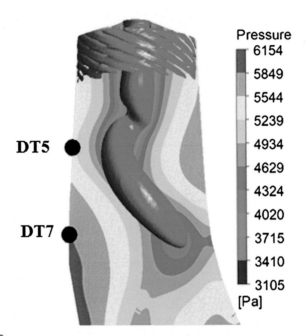

FIGURE 11.22

Side view of the Francis-99 draft tube cone.

Reproduced from Tran, C.T., et al., Prediction of the precessing vortex core in the Francis-99 draft tube under off-design conditions by using Liutex/Rortex method, Journal of Hydrodynamics 32 (3), 2020, with the permission of Journal of Hydrodynamics.

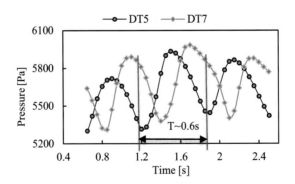

FIGURE 11.23

Unsteady pressure at two levels on the cone DT5 and DT7, based on 3D numerical simulation with SST model.

Reproduced from Tran, C.T., et al., Prediction of the precessing vortex core in the Francis-99 draft tube under off-design conditions by using Liutex/Rortex method, Journal of Hydrodynamics 32 (3), 2020, with the permission of Journal of Hydrodynamics.

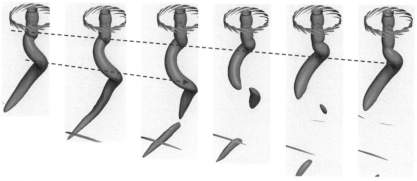

FIGURE 11.24

Vortex rope structures at several time steps visualized by Liutex method.

Reproduced from Tran, C.T., et al., Prediction of the precessing vortex core in the Francis-99 draft tube under off-design conditions by using Liutex/Rortex method, Journal of Hydrodynamics 32 (3), 2020, with the permission of Journal of Hydrodynamics.

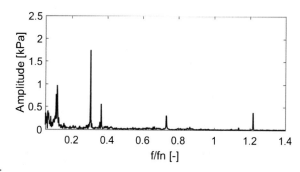

FIGURE 11.25

Pressure fluctuation frequency for the PL condition from cavitating flow analysis at DT5 monitor pressure point.

Reproduced from Tran, C.T., et al., Prediction of the precessing vortex core in the Francis-99 draft tube under off-design conditions by using Liutex/Rortex method, Journal of Hydrodynamics 32 (3), 2020, with the permission of Journal of Hydrodynamics.

studied by Arpe (2009) and has very good agreement with the value of 0.294 seen in the experimental studies (Bergan et al., 2016). The frequency of the vortex rope has been obtained by the pressure fluctuations and Liutex method, see Table 11.3, and compares well with the experimental result.

The vortex rope structure is composed of two different paths: the vortex core center line and the surrounding regime. The consideration of the dynamics of the vortex core line will further illuminate the mechanisms behind these observations. According to Gao et al. (2019b), the vortex core line is defined as a Liutex line that passes the points satisfying the condition of $\nabla R \times \vec{r} = 0$ and $R > 0$ where $\vec{r}$ represents

Table 11.3 Comparisons of the frequency of the vortex rope obtained by the pressure fluctuations and Liutex method.

	Pressure fluctuations	Liutex method	Experimental (Bergan et al. 2016)
f_{vortex}/f_{runner}	0.3	0.3	0.294

Reproduced from Tran, C.T., et al., Prediction of the precessing vortex core in the Francis-99 draft tube under off-design conditions by using Liutex/Rortex method, Journal of Hydrodynamics 32 (3), 2020, with the permission of Journal of Hydrodynamics.

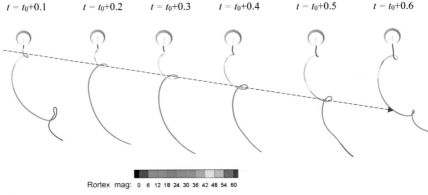

| $t = t_0+0.1$ | $t = t_0+0.2$ | $t = t_0+0.3$ | $t = t_0+0.4$ | $t = t_0+0.5$ | $t = t_0+0.6$ |

Rortex mag: 0 6 12 18 24 30 36 42 48 54 60

FIGURE 11.26

Vortex core structures evolution in one cycle at PL condition operation.

Reproduced from Tran, C.T., et al., Prediction of the precessing vortex core in the Francis-99 draft tube under off-design conditions by using Liutex/Rortex method, Journal of Hydrodynamics 32 (3), 2020, with the permission of Journal of Hydrodynamics.

the direction of the Liutex vector. This definition is used to find the Liutex (vortex) core lines in the flow field, which is uniquely defined without any threshold requirement (Gao et al., 2019b, Xu et al., 2019).

For PL condition, Fig. 11.26 visualizes serial snapshots of the vortex core lines by temporal evolutions in one cycle. The vortex core lines are colored by the Liutex magnitude, chosen as an indicator of the vortex strength. The picture shows significant motions in the vortex core as it rotates with the precession frequency and it is described by a stream line. The vortex core line is shown as a conical spring with a variable helix angle. Therefore, the vortex structure and precessing vortex core are represented as a unique morphology by the Liutex core lines.

For HL condition, Fig. 11.27 visualizes serial snapshots of the vortex core lines by temporal evolutions in one cycle. When the vortex rope occurs at HL condition, the core is centered in the draft tube cone. Fig. 11.27 shows that the Liutex vortex core line moving over time from runner outlet center to downstream with different vortex strength (red/blue color), which is described by a stream line. The movement

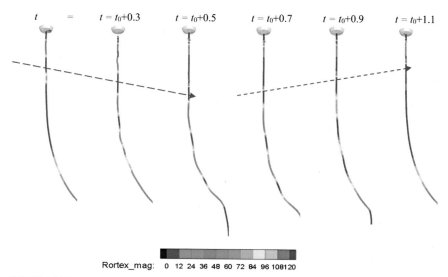

FIGURE 11.27

Vortex core structures evolution in one cycle at HL condition operation.

Reproduced from Tran, C.T., et al., Prediction of the precessing vortex core in the Francis-99 draft tube under off-design conditions by using Liutex/Rortex method, Journal of Hydrodynamics 32 (3), 2020, with the permission of Journal of Hydrodynamics.

of the vortex core segments with different strengths reflects the change in pressure distribution in the draft tube cone, which causes the pressure fluctuation with a smaller frequency. It is clearly shown in FFT analysis of the unsteady wall pressure signals measured at DT5, as shown in Fig. 11.28, which would not be observed if traditional vortex definition methods were applied.

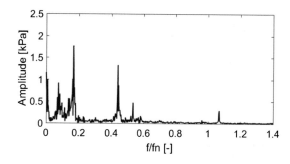

FIGURE 11.28

Pressure fluctuation frequency for the HL condition from cavitating flow analysis at DT5 monitor pressure point.

Reproduced from Tran, C.T., et al., Prediction of the precessing vortex core in the Francis-99 draft tube under off-design conditions by using Liutex/Rortex method, Journal of Hydrodynamics 32 (3), 2020, with the permission of Journal of Hydrodynamics.

In addition, since it seems to be very hard to define a sharp boundary surface for the whole vortex structure, focusing only on the vortex core line has the advantage that different vortex structures can be clearly distinguished. The Liutex method focusing on the vortex core lines is more stable, as the vortex core lines concentrate on the center of the vortex where it should be very evident.

11.4.4 Summary

Based on the earlier visualization illustrated by the Liutex core line for the turbulent flow in the draft tube of the Francis-99 turbine, the following conclusions can be made:

(1) Through this study, the Liutex method is verified to be able to successfully represent the structure and process of vortex rope on turbulent flow of Francis turbine. The process of vortex breakdown at off-design conditions operation is addressed with the use of the Liutex method.
(2) From the variation in time of the precessing Liutex core, the vortex core line in a draft tube at off-design conditions can be described by stream lines properly.
(3) By properly extracting the Liutex core line from the unsteady three-dimensional velocity field, it is revealed that a periodic sequence of vortex precessing appears to be responsible at the draft tube cone under off-design conditions.

Liutex methods for science and engineering applications

12

As a new physical quantity, Liutex can be applied to many theoretical research topics of fluid mechanics and practical engineering applications such as shock boundary layer interaction, flow control, turbulence generation mechanism, energy dissipation study, turbine machinery design, swirling jets, and so on. This chapter covers some examples of Liutex for science and engineering applications.

12.1 Omega method and spectrum analysis for shock boundary layer interaction (Dong et al., 2018d)

12.1.1 Short review

Shock wave and boundary layer interactions, commonly called SWBLIs, can significantly influence the performance of aircraft and decrease the quality of flow field by causing serious flow distortion and separation (Dolling, 2001). The shock-induced separation, with the low-frequency unsteadiness, is harmful to the environment and flying vehicle structures. These oscillations could be strong in aerodynamic loads and lead to heavy drag rise, acoustic noises, and even structural damage (Délery and Dussauge, 2009). After several decades of heavy research, mechanism on the low-frequency unsteadiness of SWBLI still remains mythical in the SWBLI research.

Many people believe that the unsteadiness of SWBLI is stochastic and unpredictable (Anderson, 2016). However, due to the advances in direct numerical simulation (DNS) and large eddy simulation (LES) for SWBLI (Garnier et al., 2002; Pirozzoli and Grasso, 2006; Morgan et al., 2010), some researchers have found that there is a correlation between the unsteadiness of SWBLI and separation bubble (Erengil and Dolling, 1991; Piponniau et al., 2009; Touber et al., 2009). However, the mechanism of low-frequency generation is still under debate. The relationship between the upstream boundary layer and the low-frequency and the large-scale unsteadiness of the separated flow in a compression ramp corner was investigated by Ganapathisubramani et al. (2007, 2009), which indicated that the incoming turbulent boundary layer could be the origin of the low-frequency unsteadiness. In recent years, a series of studies on the statistical relationship study between the low-frequency shock motion and the upstream/downstream flow fluctuation were conducted by Priebe and Martín (2012). They found that the shock motion is related to the separation bubble breathing and the separated shear layer flapping. In addition, the inherent instability in downstream separated flow is the physical origin of the low-frequency unsteadiness. The related experiment was carried out by Erengil and Dolling (1991). They found the shock frequency is directly related to the intensity of the SWBLI interaction under the same inflow conditions and obtained a typical dimensionless frequency with a Strouhal number of about 0.03, based on the interaction length L and on the velocity outside the separation bubble. Wu and Martin (2008) investigated the shock motion in a compression ramp flow with $Ma = 2.9$, $\alpha = 24°$. They found that the spanwise-averaged separation point undergoes a low-frequency motion and was highly correlated with the shock motion. This indicated that the low-frequency shock unsteadiness is affected by the downstream flow separation. The characteristic low frequency is in the range of $0.007U_\infty/\delta \sim 0.013U_\infty/\delta$ (δ is the boundary layer thickness). Touber and Sandham (2009) presented a LES investigation on the low-frequency unsteadiness of the interaction between an impinging oblique shock and a turbulent boundary layer at $Ma = 2.3$. They gave a similar conclusion that the energetic low frequencies, which were observed near the reflected shock, were not introduced by the inlet conditions and demonstrated the existence of energetic broadband low-frequency motions near the separation point with a peak near $St = fL_{sep}/U_\infty \approx 0.03$ (L_{sep} is the length of the separation bubble). Clemens and Narayanaswamy (2009) and Souverein et al. (2010) believe that both upstream and downstream effects existed in the low-frequency unsteadiness, downstream effects dominated for fully separated flow, whereas upstream effects became dominant for mild interactions. As turbulence is still a complex problem in fluid dynamics, it is difficult to study the unsteadiness of shock and turbulent boundary layer interaction. However, Li and Liu (2010) developed a new theory that the SWBLI is really shock and vortex interaction and the shock oscillation frequencies are determined by the vortex size and motion. Therefore, the low frequencies are really correlated with Liutex spectrum.

Micro-vortex generator (MVG), first proposed in the 1980s (Lin et al., 1989), which is widely used in SWBLI control to decrease the adverse effects of the separation (Bohannon, 2006; Anderson et al., 2006), is a kind of passive control device with a lower height (less than 50% of the boundary layer thickness), which is lower compared to the conventional vortex generator (VG). It has been receiving research interests because of its remarkable capacity of alleviating flow separation while carrying much lower drag penalty. Thus, it is encountered in many aerospace applications, such as supersonic inlets, propulsion wing, and so on. As a kind of miniature and a passive device, MVG has clear advantages in terms of the low profile drag, lack of intrusiveness, and the robustness (Lu et al., 2012a,b). Therefore, many efforts for experimental and computational investigations (Scarano, 2007; Sun et al., 2014b; Ligrani and Frendi, 2016) have been devoted to study both the shock/boundary layer flows and the MVG wake, in order to clarify the mechanism of the microramp in flow separation control.

Most researchers believe that a pair of streamwise vortices generated by MVG is the dominant element in flow separation control, which can generate strong energy and momentum exchange between the high-speed freestream and the lower-momentum boundary layer through the up-wash and down-wash motions and finally modify the pressure distribution of the boundary layer and finally make it less likely to separate. An experiment was performed by Holden and Babinsky (2007) in a supersonic wind tunnel to investigate the effect of the wedge-shaped and vane-type subboundary layer vortex generators (SBVGs) placed upstream of a normal shock/turbulent boundary layer interaction at $Ma = 1.5$ and $Re = 28 \times 10^6$. The results showed that both types of SBVGs could generate a pair of counterrotating vortices to reduce the scale of the shock-induced separation, especially, the vane-type SBVGs, which could eliminate the separation entirely due to the generation of more widely spaced primary vortices. Under the experimental conditions given by Babinsky, Ghosh et al. (2008) discussed the effects of MVGs in controlling oblique shock and turbulent boundary layer interactions by Reynolds-averaged Navier–Stokes (RANS) and RANS/LES models. They concluded that the major effect of the MVG array is to induce a pair of counterrotating longitudinal vortices, which force higher momentum fluid toward the wall surface and energize the lower momentum boundary layer.

Although numerous experimental and numerical studies of MVG for SWBLI control were carried out and showed that MVG is an efficient device to delay the flow separation induced by SWBLI, its control mechanism of vortices motions was still not clear until Li and Liu (2010) first discovered the large scale vortex rings behind the MVG in a supersonic compression ramp flow at $Ma = 2.5$, $Re_\theta = 1440$. A new mechanism on the SWBLI control by MVG was also discovered, associated with a train of vortex rings, which could strongly interact with the shock and play an important role in the separation zone reduction. This kind of ring-like vortices were confirmed by the experiments at the University of Texas at Arlington (Lu et al., 2010). Sun et al. (2014a, b, c) gave a conceptual model describing the evolution of vortical organization behind the microramp shown in Fig. 12.1. In this model, the streamwise vortices may generate in the immediate downstream of the micro-ramp. However, the curved free shear layer around the wake quickly becomes unstable and induces the arch-shaped K-H vortices, which eventually develop into the vortex rings further downstream. Accompanying the evolution of the K-H vortex, the streamwise vortices decay at a fast rate and are rather weak.

This train of vortex rings could be a dominant factor of the mechanism of MVG in the control of shock turbulent boundary layer interaction. Yan et al. (2013, 2014b) then showed more details about the vortex ring structures behind the microramp in supersonic flow at $Ma = 2.5$ and indicated that this train of vortex rings play dominant role in reduction of flow separation and did not even break down after passing through the strong oblique shock. Conversely, the shock was broken when interacting with vortex rings, and the scale of separation zone was significantly reduced due to the interaction of the shock and vortex rings. A corresponding experimental study on wake organization downstream of a microramp immersed in a low-noise supersonic wind tunnel at $Ma = 2.7$, $Re_\theta = 5845$ was also performed by Wang B. et al. (2012, 2014). They mentioned that the microramp should not be simply treated as

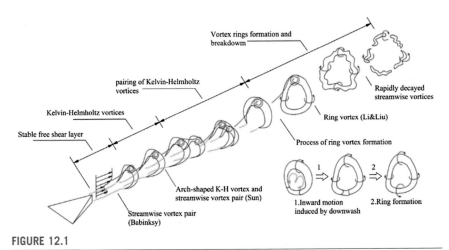

FIGURE 12.1

The conceptual model of the vortical organization in the microramp wake.

Reproduced from Z. Sun, et al., Decay of the supersonic turbulent wakes from micro-ramps, Phys. Fluids 26 (2), 2014, with the permission of AIP Publishing.

a traditional streamwise vortex-inducing device. The initial strong streamwise vortex pair begins to break down and its up-wash motions are weakened under the effect of the hairpin vortices. Then the initial small hairpin vortices merge into larger ones after the breaking down of the primary streamwise vortex pair. This indicates that the flow field downstream of the microramp would mainly be dominated by large-scale hairpin vortices. Detailed analyses on the mechanism of MVG and the vortex rings behind the MVG have been made in a supersonic ramp flow (Yan et al., 2013, 2014b; Yang et al., 2016). It is clear that the K-H vortices generated by MVG play a key role to reduce the flow separation.

12.1.2 Case description

The dimensions of the compression ramp flow domain with MVG are shown in Fig. 12.2, where h is the height of MVG and is set as 4 mm. The axes x, y, and z represent the streamwise, normal, and spanwise directions, respectively. The corner of MVG is located at $x = 0$, and the compression ramp angle α is 24°. The configuration of MVG follows the experimental study performed by Babinsky et al. (2009) with $\alpha = 24°$, $c/h = 7.2$, $s/h = 7.5$, where c is the chord length of MVG and s denotes the distance between MVGs. The trailing-edge declining angle behind MVG is set as 70° to alleviate the difficulty of grid generation.

The computational domain has a total number of $n_{\text{streamwise}} \times n_{\text{normal}} \times n_{\text{spanwise}} = 1600 \times 192 \times 137$ grid nodes. Wall grid clustering is applied for the whole domain and y^+ for the first grid point from the bottom wall equals to 1.36, which shows that the grid resolution is adequate to resolve the large vortex structures in the computational domain. Due to the complex geometry of the MVG, a

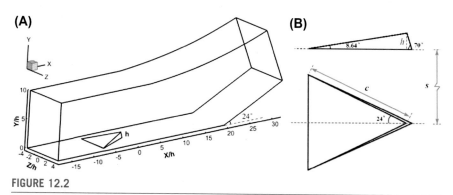

FIGURE 12.2

Configurations of (A) the computational domain with (B) MVG.

Reproduced from X. Dong, et al., Spectrum study on unsteadiness of shock wave—vortex ring interaction, Phys. Fluids 30 (056101), 2018, with the permission of AIP Publishing.

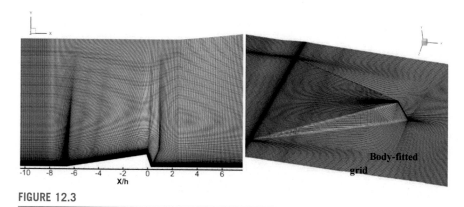

FIGURE 12.3

The grid system of the computational domain.

Reproduced from X. Dong, et al., Spectrum study on unsteadiness of shock wave—vortex ring interaction, Phys. Fluids 30 (056101), 2018, with the permission of AIP Publishing.

body-fitted grid system (Li and Liu, 2010) in Fig. 12.3 is generated to have sufficient orthogonality.

In addition, the initial parameters of the turbulent flow are listed in Table 12.1. Ma_∞, T_∞, and U_∞ are respectively the Mach number, temperature, and streamwise

Table 12.1 Initial parameters of the turbulent flow.

Ma_∞	Re_θ	T_∞	T_w	δ	U_∞
2.5	5760	288.15 K	300 K	9.44 mm	850 m/s

Reproduced from X. Dong, et al., Spectrum study on unsteadiness of shock wave—vortex ring interaction, Phys. Fluids 30 (056101), 2018, with the permission of AIP Publishing.

velocity of freestream; Re_θ is the Reynolds number based on the momentum thickness θ; T_w is the wall temperature; and δ is the undisturbed turbulent boundary layer thickness based on 99% of freestream velocity. The LES code has been well validated.

12.1.3 Vortex identification

A vortex identification method called Omega (Ω) method proposed by Liu et al. (2016) is widely used for capturing the vortex structures (Wang Y. et al., 2017; Dong X. et al., 2017, 2018a, b; Dong Y. et al., 2017). This method was also compared with other vortex identification methods in some research and review papers by Zhang et al. (2018a), Hu et al. (2017), Abdel-Raouf et al. (2017), and Epps (2017). The basic idea of this method is that the vortex can be described as the region where the vorticity overtakes the deformation. In application, Ω is calculated by

$$\Omega = \frac{b}{a + b + \varepsilon} \tag{12.1}$$

where $a = \text{trace}(A^T A) = \sum_{i=1}^{3}\sum_{j=1}^{3}\left(A_{ij}^2\right)$, $b = \text{trace}(B^T B) = \sum_{i=1}^{3}\sum_{j=1}^{3}\left(B_{ij}^2\right)$, A is the symmetric part of velocity gradient tensor $\nabla \vec{v}$, B the antisymmetric part, $\|\cdot\|_F$ the Frobenius norm, and $\varepsilon = 0.001 * (b - a)_{\max} = 0.002 * Q_{\max}$.

12.1.4 Results and discussions

According to previous studies, a train of vortex rings behind the MVG plays a dominant role in reducing separation zone. It can strongly interact with the shock and could be a factor of the mechanisms of the MVG in SWBLI control.

(1) Process of the shock and vortex ring interaction

Both iso-surfaces of the pressure gradient magnitude ($\nabla p = 0.5$) and of the Omega ($\Omega = 0.52$) colored by the streamwise velocity are shown in Fig. 12.4, which suggests the details on the shock and vortex ring interaction. A chain of vortex rings, recognized by $\Omega = 0.52$, has a high value of streamwise velocity near the ring head, due to the shear between the high-speed freestream and the wake generated by the MVG. As can be seen, the vortex ring near the shock can better keep its original shape while passing through the shock, which is recognized by the iso-surface of $\nabla p = 0.5$.

Fig. 12.5A displays a chain of vortex rings with several spanwise planes extracted from the whole domain. Here Lz is the nondimensional length of the domain in the spamwise direction. Moreover, the large-scale vortex rings can be found concentrating on the central plane $z = 50\% \ Lz$, which are zoomed in Fig. 12.5B.

(2) The signals of vortex rings in the time and frequency domain

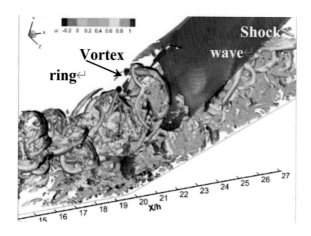

FIGURE 12.4

Iso-surface of the shock wave and the vortex ring.

Reproduced from X. Dong, et al., Spectrum study on unsteadiness of shock wave–vortex ring interaction, Phys. Fluids 30 (056101), 2018, with the permission of AIP Publishing.

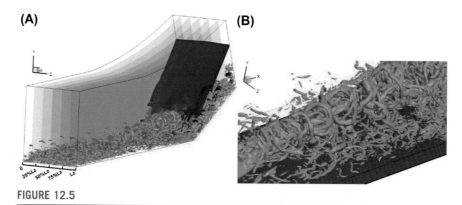

FIGURE 12.5

(A) Iso-surface of $\Omega = 0.52$ and spanwise planes in whole domain. (B) Zoom view of vortex rings.

Reproduced from X. Dong, et al., Spectrum study on unsteadiness of shock wave–vortex ring interaction, Phys. Fluids 30 (056101), 2018, with the permission of AIP Publishing.

According to the earlier discussions, the basic idea on the mechanism of shock and vortex ring interaction can be obtained. The current work is to make the analysis quantitative. As can be seen from Fig. 12.5, the large-scale vortex rings behind MVG concentrate in the central region along the spanwise direction; therefore, the interaction between vortex rings and the shock is the strongest on the central plane. Due to the distribution of vortex rings and the location of the interaction area, the plane $z = 50\% \; Lz$ is chosen to give the details, and 130,000 time steps from

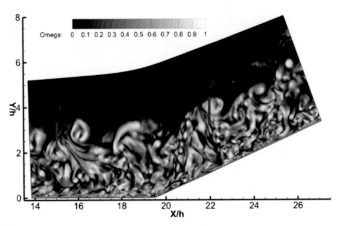

FIGURE 12.6

The positions upstream/downstream the shock for tracking vortex rings.

Reproduced from X. Dong, et al., Spectrum study on unsteadiness of shock wave–vortex ring interaction, Phys. Fluids 30 (056101), 2018, with the permission of AIP Publishing.

$t = 1990T$ to $t = 2510T$ (T is the characteristic time) are recorded to carry out the frequency analysis.

According to Liu et al. (2016), the vortex can be identified by $\Omega > 0.52$; therefore, the Omega flux is considered to be a vortex indicator signal, which is recorded during the vortex ring passing through the red lines 1, 2, which are shown in Fig. 12.6. The omega flux can be defined by using $\Omega > 0.52$ as the threshold,

$$\Omega_{Flux} = \frac{\sum_{n=1}^{100} r_0}{100}, r_0 = \begin{cases} 1, & \text{if } \Omega > 0.52 \\ 0, & \text{if } \Omega \le 0.52 \end{cases} \tag{12.2}$$

So it can be seen that 100 points were extracted from the red line, the node n goes from 1 to 100. Then the Omega value (Ω) of each point n was recorded. If the Ω is larger than 0.52, the rate of the vortex ring covering this point, r_0, can be treated as 1, otherwise 0. Therefore, the Ω_{Flux} represents the rate of $\Omega > 0.52$ over this red line. Two positions (red lines) respectively distributed upstream and downstream of the shock are chosen to detect the vortex ring motion.

Similarly, the density flux is defined by using $\rho < 0.7$ as the threshold of the vortex ring at the red line 1 (Eq. 12.2), thereby detecting the vortex rings by lower density. It can also be a characteristic of the vortex ring.

$$\rho_{Flux} = \frac{\sum_{n=1}^{100} r_0}{100}, r_0 = \begin{cases} 1, & \text{if } \rho < 0.7 \\ 0, & \text{if } \rho \ge 0.7 \end{cases} \tag{12.3}$$

It can also be seen that 100 points were extracted from the red line, the node n goes from 1 to 100. Then the density value (ρ) of each n point was recorded, if the ρ is smaller than 0.7, the rate of the vortex ring covering this point, r_0, can be

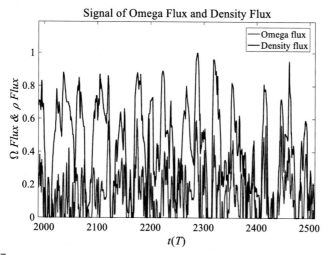

FIGURE 12.7

Signals of Omega flux and density flux over the red line 1 in Fig. 12.6.

Reproduced from X. Dong, et al., Spectrum study on unsteadiness of shock wave–vortex ring interaction, Phys. Fluids 30 (056101), 2018, with the permission of AIP Publishing.

treated as 1, otherwise 0. Fig. 12.7 compares the signals of Omega flux and density flux over the red line 1 (marked in Fig. 12.6) in the time domain. It can be seen that the Omega flux of the vortex rings over the red line violently changes with time, which means the frequency of the motion of the vortex rings passing through the shock is relatively high. However, due to the less distinctive feature of the vortex density, the value of density flux is not as fluctuated as the Omega flux signal. The curve of the Omega flux is totally under the density flux and more fluctuated. There is no doubt that the Omega flux is better as an effective and accurate signal of vortex rings, since both large vortex rings and small-scale vortex structures can be captured based on the signal of Omega flux.

Figs. 12.8 and 12.9 respectively give the Omega flux signal of the vortex rings passing through the red lines 1 and 2 in the time domain and their power spectrum in the frequency domain. Comparing the signals upstream and downstream the shock, it is obvious that the Omega flux is higher after the vortex rings break the shock, which indicates that the structures of the vortex rings are more complex and more discrete. Fig. 12.8B shows the power spectrum of the signal in Fig. 12.8A versus the nondimensional frequency, Strouhal number, St, which is defined as $St = f \cdot h/U_\infty$, where f denotes the frequency and h is the height of the MVG. As can be seen, the dominant frequency of the vortex rings upstream the shock is $St = 0.0379$ corresponding to a dimensional frequency f of around 8054 Hz, which is a relatively high frequency. However, in Fig. 12.9B, although the dominant frequency of the vortex ring motion is $St = 0.0552$ ($f = 11{,}687$ Hz), which is higher than the dominant frequency before interacting with the shock,

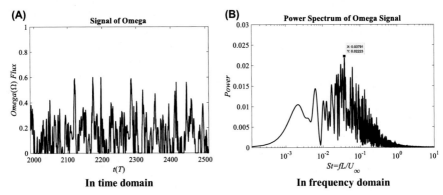

FIGURE 12.8

Omega flux signal of the vortex rings upstream the shock passing the red line 1.

Reproduced from X. Dong, et al., Spectrum study on unsteadiness of shock wave–vortex ring interaction, Phys. Fluids 30 (056101), 2018, with the permission of AIP Publishing.

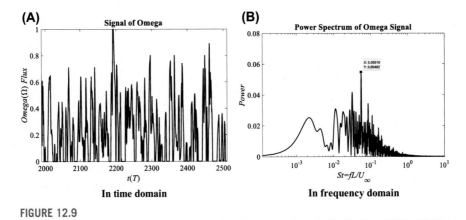

FIGURE 12.9

Omega flux signal of the vortex rings downstream the shock passing the red line 2.

Reproduced from X. Dong, et al., Spectrum study on unsteadiness of shock wave–vortex ring interaction, Phys. Fluids 30 (056101), 2018, with the permission of AIP Publishing.

the lower frequency components of the vortex rings after interacting with the shock have larger power than the one before the interaction. Therefore, it can be concluded that the vortex rings generated by the MVG in a supersonic ramp flow at $Ma = 2.5$ have a dominant frequency around $St = 0.0379$.

(3) The shock oscillation and its frequency characteristics

The shock oscillation in a compression ramp flow could generate locally large pressure fluctuation. This pressure fluctuation could cause the heavy drag rise and the acoustic noises as strong aerodynamic loads. Therefore, the frequency

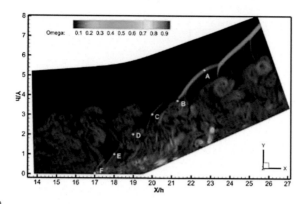

FIGURE 12.10

The positions of six probing points with the distribution of omega Ω and pressure gradient ∇p (*white schlieren*).

Reproduced from X. Dong, et al., Spectrum study on unsteadiness of shock wave–vortex ring interaction, Phys. Fluids 30 (056101), 2018, with the permission of AIP Publishing.

characteristics of the shock oscillation will be discussed in this section. The density gradient magnitude $\nabla\rho$ at the points from A to F is treated as signals for discussion and further frequency analysis. Point A is above the interaction area and Point B is located on the shock, which is far from the interaction area; Points C and D are located in the interaction area; Points E and F are located near the root of the shock. The locations of six points are shown in Fig. 12.10 and Table 12.2.

The density gradient magnitude signal $\nabla\rho$ at the Points A, B, D, and F in the time domain and in the frequency domain is shown in Figs. 12.11–12.14. In Fig. 12.11, the dominant low frequency $St = 0.002$ is also captured by the $\nabla\rho$ signal at Point (A) Comparing the $\nabla\rho$ signal in the time domain at these four locations, the range of $\nabla\rho$ is the smallest, which is from 0 to 5, but is most fluctuated at Point (D) As can be seen from Fig. 12.14B, the dominant frequency of the shock at Point D is $St = 0.0364$ ($f = 7735$ Hz), which is corresponding to the dominant frequency ($St = 0.0379$) of the vortex ring behind MVG.

Based on earlier discussion, a relative low frequency, which is around $St = 0.002$ ($f = 425$ Hz), can always be detected as the dominant frequency of the clean shock (undisturbed by the vortex rings) by these three signals, p, ∇p, and $\nabla\rho$. However, in the shock and vortex ring interaction region, a higher dominant frequency

Table 12.2 The locations of the points A, B, C, D, E, and F.

Point	A	B	C	D	E	F
x/h	22.7	21.2	20.0	19.0	18.0	17.0
y/h	5.1	3.8	3.0	2.0	1.0	0.0

Reproduced from X. Dong, et al., Spectrum study on unsteadiness of shock wave–vortex ring interaction, Phys. Fluids 30 (056101), 2018, with the permission of AIP Publishing.

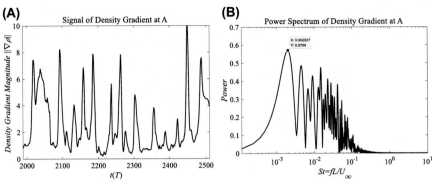

FIGURE 12.11

Density gradient signal $\nabla\rho$ (A) in the time domain and (B) in the frequency domain at Point A.

Reproduced from X. Dong, et al., Spectrum study on unsteadiness of shock wave–vortex ring interaction, Phys. Fluids 30 (056101), 2018, with the permission of AIP Publishing.

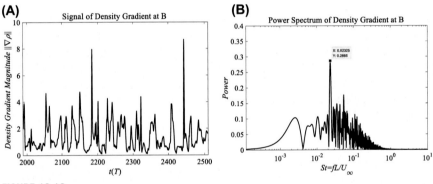

FIGURE 12.12

Density gradient signal $\nabla\rho$ (A) in the time domain and (B) in the frequency domain at Point B.

Reproduced from X. Dong, et al., Spectrum study on unsteadiness of shock wave–vortex ring interaction, Phys. Fluids 30 (056101), 2018, with the permission of AIP Publishing.

$St = 0.037 - 0.038$ can be detected, rather than the low frequency $St = 0.002$. This indicates that the vortex ring is stiff enough to break or weaken the shock.

12.1.5 Summary

Shock oscillation with low-frequency unsteadiness is commonly observed in supersonic flows and is a top priority for control of flow separation caused by shock wave and boundary layer interaction or SWBLI. The interaction of the shock caused by the compression ramp and the vortex rings generated by an MVG in a supersonic

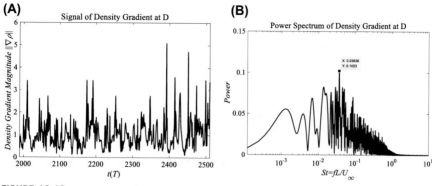

FIGURE 12.13

Density gradient signal $\nabla \rho$ (A) in the time domain and (B) in the frequency domain at Point D.

Reproduced from X. Dong, et al., Spectrum study on unsteadiness of shock wave–vortex ring interaction, Phys. Fluids 30 (056101), 2018, with the permission of AIP Publishing.

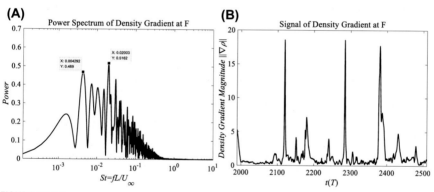

FIGURE 12.14

Density gradient signal $\nabla \rho$ (A) in the time domain and (B) in the frequency domain at Point F.

Reproduced from X. Dong, et al., Spectrum study on unsteadiness of shock wave–vortex ring interaction, Phys. Fluids 30 (056101), 2018, with the permission of AIP Publishing.

flow at $Ma = 2.5$ is simulated by the implicit large eddy simulation (ILES) method. The analysis on observation and the frequency of both the vortex ring motion and the shock oscillation is carried out. Several conclusions based on the power spectrum analysis of the vortex ring motion and the shock oscillation can be described as follows.

(1) The mechanism of the shock and vortex ring interaction, which aims at the reduction of the shock-induced flow separation, can be illustrated by the fact that the shock wave can be distorted and weakened by the vortex rings behind MVG, through decreasing the pressure gradient between upstream and downstream of the shock.

(2) The vortex rings generated by the MVG in a supersonic ramp flow at $Ma = 2.5$ have a dominant high frequency around $St = 0.0379$ ($f = 8054$ Hz).

(3) Based on the power spectrum analysis of the signals, the shock produced by a compression ramp flow at $Ma = 2.5$ is confirmed to have a dominant low nondimensional frequency, which is around $St = 0.002$, with the corresponding dimensional frequency $f = 425$ Hz.

(4) The dominant low frequency of the shock can be weakened or removed by the vortex rings through the shock and vortex ring interaction due to the higher frequency of the vortex ring motion. In the shock and vortex ring interaction region, a higher dominant frequency $St = 0.037 - 0.038$ can be detected, rather than the low frequency $St = 0.002$, which indicates that the vortex ring is strong and stiff enough to break or weaken the shock.

(5) This analysis could provide an effective tool, the MVG optimization through the high-frequency vortex generation, to weaken or remove the low-frequency pressure fluctuation of the SWBLI below 500 Hz, which is harmful to the flight vehicle structures and the environmental protection.

Again, this example shows that the SWBLI is really shock-vortex interaction and the shock oscillation frequencies are dependent to the vortex size and moving speed. Although the study is carried out by using the Omega method, the basic conclusions should be similar to the correlation analysis between density fluctuation and Liutex spectrum.

12.2 Energy dissipation analysis based on Liutex and velocity gradient tensor decomposition (Wu et al., 2020)

12.2.1 Short review

The study of the velocity gradient tensor has great significance in understanding of flow complex motion. It contains geometric information about the orientation of vorticity and strain-rate eigenvectors, so it can identify the areas of the flow in which either strain rate or vorticity prevails (Martins and Meneveau, 2010). The local statistics and geometric structure of three-dimensional turbulence can be described by the properties of velocity gradient tensor (Chevillard and Meneveau, 2006), and many studies on turbulent flows have been based on the velocity gradient tensor. Chong et al. (1990) classified the local structure of velocity fields in terms of the three invariants of the velocity gradient tensor. Contours of joint distributions of these invariants obtained from turbulent flows have a tear-drop shape. In homogeneous, isotropic, and incompressible turbulent flow, Ooi et al. (1999) observed that successive positions of a fluid particle occupy different parts of this tear drop, representing flow in sheet-like regions followed by stretching, then compression, in a repetitive cycle. These studies mainly focus on the evolution of the velocity gradient tensor and the full tensor for incompressible flow. Studies of the velocity gradient tensor in compressible flows have received far less attention, but essential features from a homogeneous, isotropic, decaying turbulence field were presented by Lee et al. (2009). Suman and Girimaji (2009, 2010) closed the equations for velocity gradient tensor evolution assuming a polytropic process to obtain the Homogenized Euler Equation (HEE) model, which reproduced the deformation caused by compressibility as observed in attenuated turbulence data. Mathew et al. (2016) studied the relationship between the flow structure and compressibility at the turbulent and nonturbulent interface of temporal plane mixing layers by analyzing the invariants of the velocity gradient tensor. Processes in turbulent flows may be understood from the view of the dynamics of coherent structures and their influence on surrounding flow. By analyzing the velocity gradient tensor, the local flow topologies can be revealed, which is helpful to understand the flow mechanism. The analysis based on the velocity gradient tensor of fluid element has made the complex motion of fluid element understood to a certain extent (Li et al., 2014; Gao et al., 2019a).

It is a great challenge to understand how and where energy is dissipated in turbulent flow, which is of great significance in many fields such as fundamental research, aeronautics, industry, and so on. In the classical three-dimensional turbulence phenomenology, energy is injected at large scales by the forcing mechanism, transferred downscale at a constant rate following a self-similar cascade, and then dissipated into heat at the Kolmogorov length scale, where viscous effects become dominant. On the basis of the properties of the velocity gradient tensor, the feature of the fine-scale motion can be studied properly (Atkinson et al., 2012). Wu et al. (2015) studied the evolution of the dynamics and the geometry of the planar jets along with the flow transition based on the characteristics of the invariants of the velocity gradient tensor. Their results show that the flow transition is accompanied by a severe rotation and straining of the flow elements, where the vortex structure evolves faster than the fluid element deformation, as well as the initial flow near the jet exit is strongly predominated by the dissipation over the entropy. Kuzzay et al. (2015) investigate the relations between global and local energy transfers in a turbulent von Kármán flow by using particle image velocimetry (PIV) measurements and a new method based on the work of Duchon and Robert (2000). The results evidence a stationary energy cycle within the flow where energy is injected at the top and the bottom impellers and dissipated within the shear layer. However, there are few studies on the proportion of shear in deformation and the effect of shear on energy dissipation with quantitative analyses.

Based on the principal coordinate, the principal decomposition of the velocity gradient tensor is conducted. The physical meaning of each tensor term is discussed and the effects on energy dissipation are analyzed. A separated boundary layer flow with pressure gradient is taken as an example for the analysis.

12.2.2 Two-dimensional velocity gradient tensor decomposition and its physical meaning

According to the definition of Liutex, the two-dimensional velocity gradient tensor should have a pair of conjugate complex eigenvalues, $\lambda_{cr} \pm i\lambda_{ci}$. If the fluid has rotation, which can be expressed as:

$$\nabla \vec{V} = \begin{bmatrix} \lambda_{cr} & -R/2 \\ R/2 + \varepsilon & \lambda_{ci} \end{bmatrix} \tag{12.4}$$

Then, $\nabla \vec{V}$ can be decomposed into normal tensor R and NR (*UTA R-NR Decomposition*).

$$\nabla \vec{V} = R + NR \tag{12.5}$$

$$R = \begin{bmatrix} 0 & -R/2 \\ R/2 & 0 \end{bmatrix}$$

$$NR = \begin{bmatrix} \lambda_{cr} & 0 \\ \varepsilon & \lambda_{cr} \end{bmatrix} = C + S \tag{12.6}$$

$$C = \begin{bmatrix} \lambda_{cr} & 0 \\ 0 & \lambda_{cr} \end{bmatrix}$$

$$S = \begin{bmatrix} 0 & 0 \\ \varepsilon & 0 \end{bmatrix}$$

Thus, the physical meaning of tensor C is the compression–stretching deformation rate tensor, tensor R is the rotation tensor, and tensor S is the shear tensor in the two-dimensional plane under the principal frame.

12.2.3 Three-dimensional velocity gradient tensor decomposition and its physical significance

In the three-dimensional flow field, the motion of fluid element is more complicated. Chong et al. (1990) proposed that the discriminant of the root of characteristic polynomial can be used to determine whether the fluid is rotating or not.

If $\Delta > 0$, $\nabla \vec{V}$ has a real eigenvalue λ_r and a pair of complex conjugate eigenvalues $\lambda_{cr} \pm \lambda_{ci}i$ under the principal frame, which can be written as:

$$\nabla \vec{V} = \begin{bmatrix} \lambda_{cr} & -R/2 & 0 \\ R/2 + \varepsilon & \lambda_{cr} & 0 \\ \xi & \eta & \lambda_r \end{bmatrix} \tag{12.7}$$

Then, $\nabla \vec{V}$ can be further decomposed into three parts:

$$\nabla \vec{V} = \begin{bmatrix} \lambda_{cr} & -R/2 & 0 \\ R/2 + \varepsilon & \lambda_{cr} & 0 \\ \xi & \eta & \lambda_r \end{bmatrix}$$

$$= \begin{bmatrix} 0 & -R/2 & 0 \\ R/2 & 0 & 0 \\ 0 & 0 & 0 \end{bmatrix} + \begin{bmatrix} 0 & 0 & 0 \\ \varepsilon & 0 & 0 \\ \xi & \eta & 0 \end{bmatrix} + \begin{bmatrix} \lambda_{cr} & 0 & 0 \\ 0 & \lambda_{cr} & 0 \\ 0 & 0 & \lambda_r \end{bmatrix}$$

$$= R + S + C = R + NR, \quad \text{and} \quad NR = S + C$$

$$R = \begin{bmatrix} 0 & -\dfrac{R}{2} & 0 \\ \dfrac{R}{2} & 0 & 0 \\ 0 & 0 & 0 \end{bmatrix} \quad S = \begin{bmatrix} 0 & 0 & 0 \\ \varepsilon & 0 & 0 \\ \xi & \eta & 0 \end{bmatrix} \quad C = \begin{bmatrix} \lambda_{cr} & 0 & 0 \\ 0 & \lambda_{cr} & 0 \\ 0 & 0 & \lambda_r \end{bmatrix} \qquad (12.8)$$

where tensor R represents the Liutex or rotation, S for the shear, and C for the compression-stretching deformation rate tensor, which is consistent with the local rigid rotation of fluid defined by Liutex vector proposed by Liu et al (2018a, 2019), Gao et al. (2018), and Wang Y. et al. (2018). This is a Liutex-based decomposition in a principal coordinate, which is called **UTA R-NR** decomposition of the velocity gradient tensor.

If $\Delta \leq 0$, the eigenvalues of $\nabla \vec{V}$ are real numbers $\lambda_1, \lambda_2, \lambda_3$, and $\nabla \vec{V}$ can be transformed as a triangular matrix under the normal frame, i.e.,

$$\nabla \vec{V} = \begin{bmatrix} \lambda_1 & 0 & 0 \\ \varepsilon & \lambda_2 & 0 \\ \xi & \eta & \lambda_3 \end{bmatrix} \qquad (12.9)$$

Then, $\nabla \vec{V}$ can be written as the sum of normal tensor C and shear tensor S.

$$\nabla \vec{V} = C + S \qquad (12.10)$$

$$C = \begin{bmatrix} \lambda_1 & 0 & 0 \\ 0 & \lambda_2 & 0 \\ 0 & 0 & \lambda_3 \end{bmatrix}$$

$$S = \begin{bmatrix} 0 & 0 & 0 \\ \varepsilon & 0 & 0 \\ \xi & \eta & 0 \end{bmatrix}$$

By the Cauchy–Stokes decomposition, $\nabla \vec{V}$ can be decomposed to a symmetric and an antisymmetric part:

$$\nabla \vec{V} = \begin{bmatrix} \lambda_{cr} & -R/2 & 0 \\ R/2 + \varepsilon & \lambda_{cr} & 0 \\ \xi & \eta & \lambda_r \end{bmatrix} = A + B \tag{12.11}$$

$$A = \begin{bmatrix} \lambda_{cr} & \varepsilon/2 & \xi/2 \\ \varepsilon/2 & \lambda_{cr} & \eta/2 \\ \xi/2 & \eta/2 & \lambda_r \end{bmatrix}$$

$$B = \begin{bmatrix} 0 & -R/2 - \varepsilon/2 & -\xi/2 \\ R/2 + \varepsilon/2 & 0 & -\eta/2 \\ \xi/2 & \eta/2 & 0 \end{bmatrix}$$

The symmetric tensor A contains the compression-stretching tensor and a part of the shear tensor, and the antisymmetric tensor B contains the pure rotation tensor and another part of the shear tensor. The total pure shear tensor is a new tensor, which combines the shear part in symmetric tensor with the shear part in the antisymmetric tensor. It is incomplete to regard antisymmetric deformation tensor as shear or vorticity tensor as rotation. The strain-rate tensor and the average rotation tensor both contain a part of the shear tensor.

12.2.4 Composition of energy dissipation

It was generally believed that shear would cause energy dissipation, and energy dissipation can also be caused by small-scale vortices, which are transformed by large-scale vortices. Here, energy dissipation is analyzed based on the physical meaning of the velocity gradient tensor decomposition described earlier.

The kinetic energy equation of viscous fluid can be written in the following form (Chen, 2002):

$$\frac{D}{Dt}\left(\frac{1}{2}u_i u_i\right) = u_i F_{x_i} + \frac{1}{\rho}\frac{\partial(m_{ji}u_i)}{\partial x} - \frac{1}{\rho}\frac{\partial(pu_i)}{\partial x_j} + \frac{p}{\rho}\frac{\partial u_i}{\partial x_j} - \frac{m_{ji}}{\rho}\frac{\partial u_i}{\partial x_j} \tag{12.12}$$

The first item on the right-hand side of this equation is the work done by the body force per unit mass of fluid per unit time. The second item is the mechanical energy transported by convection. The energy transported by convection does not change the total energy in volume but only changes the energy distribution, so that this term is a transport term. The third item on the right-hand side of this equation is the work done by the pressure per unit mass of fluid per unit time. The fourth item $\frac{\partial u_i}{\partial x_i}$ is the volume expansion rate, whose product with pressure P represents the expansion work. The fifth item is the deformation work done by the viscous force, which is the work resulted from the fluid viscous force resistance to deformation. It converts the mechanical energy of fluid motion into heat irreversibly and is dissipated, so that it is called dissipation term. Then, for compressible fluids, the dissipation function is defined as:

$$\Phi = \frac{\mu}{2}\left(\frac{\partial u_i}{\partial x_j} + \frac{\partial u_j}{\partial x_i}\right)^2 - \frac{2}{3}\mu\left(\frac{\partial u_i}{\partial x_i}\right)^2 \qquad (12.13)$$

And for incompressible fluids, the dissipation function is:

$$\Phi = \frac{\mu}{2}\left(\frac{\partial u_i}{\partial x_j} + \frac{\partial u_j}{\partial x_i}\right)^2 \qquad (12.14)$$

It can be concluded that energy dissipation is caused by shear deformation and expansion (contraction) deformation of fluid. Pure rotation is not the source of energy dissipation. For incompressible fluids, the energy dissipation is only caused by the shear deformation.

12.2.5 Energy dissipation for separated boundary layer

(1) Case setup

In the research of the flow transition mechanism, the boundary layer flow is often a proper choice of a model to explore the typical instantaneous vortex structure and the general statistical flow characteristics. The separated boundary layer of a channel flow with pressure gradient is simulated by using an in-house multiblock parallel LES code (Beihang University of China). The accuracy and robustness of the LES code have been verified by a series of test cases (Ye and Zou, 2007; Zhang et al., 2012, 2015), which has a satisfactory ability for unsteady complex flow field and can be applied to the study of time-space multiscale flow mechanism for unsteady complex flow.

The computational domain consists of four structured hexahedral meshes matched by interfaces, as shown in Fig. 12.15. Grid independency analysis is performed and the total number of grids that meet the requirements of grid independency is about 2.4 million, where the numbers of the cells in normal and spanwise directions are 80 and 48, respectively, and the number of the cells in the streamwise direction is 384. The wall-normal maximum wall distance of the first cell center to the flat plate wall is $\Delta y^+ < 1$. The wall-normal grid expansion ratio near the wall is less than 1.05. The inflow is uniform and Mach number is $Ma = 0.309$, the dimensionless stagnation temperature value is $T^* = 1.0325$, and the dimensionless stagnation pressure is $p^* = 4.9405$. The outlet boundary condition is the outlet static pressure $p = 4.4191$. The symmetrical boundary condition is enforced in the spanwise direction. The heat transfer wall is the down wall in Block 2, which is the no-slip boundary and the wall temperature is set at a uniform temperature of $0.8T^*$. Other walls are specified as inviscid Euler walls.

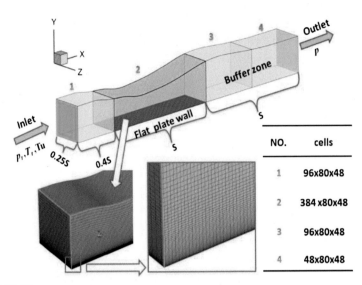

NO.	cells
1	96x80x48
2	384 x80x48
3	96x80x48
4	48x80x48

FIGURE 12.15

Computational domain and mesh of the converging—diverging channel.

Reproduced from Y. Wu, et al., Energy dissipation analysis based on velocity gradient tensor decomposition, Phys. Fluids 32 (035114), 2020, with the permission of AIP Publishing.

(2) Energy dissipation analysis in a separated boundary layer

Fig. 12.16A and B show multiscale vortex structures in the boundary layer of a transient flow field captured by iso-surface Liutex = 10 and colored by vorticity (omega-z) and by rotation strength (Liutex-z), respectively. The two-dimensional projections of vorticity and rotation intensity are given in Fig. 12.16C and D. The separation streamlines, reattachment region boundary lines, and displacement thickness lines of the time—space mean flow field are also shown in these figures. It can be seen from the vorticity distribution that the most of the vorticity is shear, because the velocity gradient $\frac{\partial u}{\partial y}$ in this case is larger than $\frac{\partial v}{\partial x}$. Therefore, the vorticity distribution reflects the distribution of energy loss rather than that of the momentum transfer. In the Liutex-z distribution, there is a region with very large rotational intensity near the position of $x/S = 0.8$ where different quality fluids are mixing and the momentum transfer is very large. The large vorticity in the upstream boundary layer is due to the large shear rather than the rotation of the fluid, so that the vortex structure is correctly captured by the Liutex method. After the momentum transfer between inside and outside the boundary layer, the fluid reattaches to the wall. Compared with vorticity, Liutex method is more suitable to describe the rotational strength of the vortex structure in the flow field. Fig. 12.16E and F are pure shear and divergence on the XY plane, respectively. It can be seen from the figures that shear is the same order of magnitude as vorticity and one order larger than the pure rotation and divergence. In the separated boundary layer flow field studied here, shear is

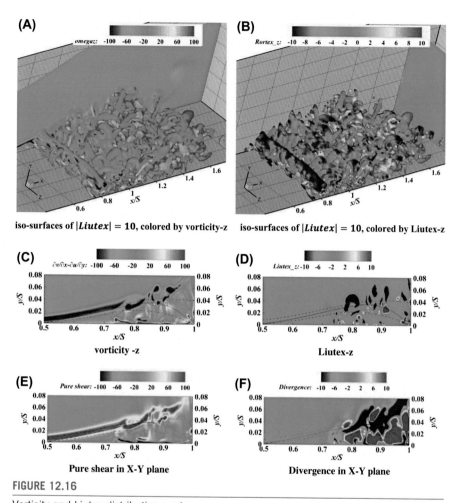

FIGURE 12.16

Vorticity and Liutex distribution contours.

Reproduced from Y. Wu, et al., Energy dissipation analysis based on velocity gradient tensor decomposition, Phys. Fluids 32 (035114), 2020, with the permission of AIP Publishing.

absolutely dominant, and pure rotation and compression-stretching of fluid are much smaller.

In this study, the velocity gradient tensor is quantitatively analyzed by calculating the surface integral of the reorganized velocity gradient component in the time–space statistic mean flow field. The contribution of shear and volume expansion to energy dissipation in a separated boundary layer flow is shown in Table 12.3. It can be seen that the energy dissipation caused by shear accounts for a large proportion of the total energy dissipation, approximately 99%, while the energy dissipation caused by expansion and contraction is around 1%.

Table 12.3 Contribution of shear deformation and expansion deformation to energy dissipation.

Energy dissipation	Shear square	Expansion and contraction
$\Phi = \mu\left(\frac{\partial v}{\partial x}+\frac{\partial u}{\partial y}\right)^2 +\frac{2}{3}\mu\left[\left(\frac{\partial u}{\partial x}\right)^2 + \left(\frac{\partial v}{\partial y}\right)^2 + \left(\frac{\partial u}{\partial x}-\frac{\partial v}{\partial y}\right)^2\right]$	$\mu(\partial v/\partial x+\partial u/\partial y)^2$	$\frac{2}{3}\left[\left(\frac{\partial u}{\partial x}\right)^2 + \left(\frac{\partial v}{\partial y}\right)^2 + \left(\frac{\partial u}{\partial x}-\frac{\partial v}{\partial y}\right)^2\right]$
100%	99%	1%

12.2.6 Summary

A principal decomposition of velocity gradient tensor is conducted in the principal coordinate. Based on this decomposition and the physical meaning of each tensor term, the energy dissipation of a separated boundary layer flow is analyzed. Several conclusions are made as follows:

1. The velocity gradient tensor can be decomposed into a compression-stretching tensor, a pure rotation tensor, and a pure shear tensor in the principal coordinate. The pure shear tensor is obtained by combining the shear part in the symmetrical deformation tensor and in the antisymmetric deformation tensor. Both strain-rate tensor and average rotation tensor contain shear tensor components.
2. For incompressible fluids, energy dissipation is caused by the shear deformation. For compressible fluids, energy dissipation is caused by both shear deformation and expansion and contraction deformation of fluid. Pure rotation does not cause any energy dissipation.
3. It is not appropriate to use vorticity to express the rotational intensity of vortex structure in the boundary layer flow. Most of its identification is shear, which reflects the energy loss rather than the momentum transfer. The region with the strongest momentum transfer can be identified by the Liutex method, so it is more appropriate to use the Liutex method to express the rotational intensity of the large-scale flow structure in the boundary layer.
4. In the separated boundary layer flow studied in this example, the energy dissipation caused by shear is 99%, and the energy dissipation caused by bulk expansion and contraction is 1%.

12.3 Liutex (vortex) cores in transitional boundary layer with spanwise-wall oscillation (Wang and Liu, 2019c)

12.3.1 Short review

In general, vortices are presented in the literature as iso-surfaces of a selected threshold, which implicitly refers to the rotational or swirling strength. However, the appropriateness of this approach is questionable especially as a large threshold may omit weak vortices while a small threshold may smear the vortical structures. A better strategy would be representing the vortices by vortex core line colored by the vortex strength. The efforts in this regard had not been very successful until the introduction of vortex rotation axis line definition based on Liutex vector (Gao et al., 2019b) and its automatic version (Xu et al., 2019). Unlike vorticity lines, which always penetrate vortex surfaces, a special Liutex line where $\nabla R \times \vec{r} = 0$ can well capture the vortex core. The idea is that on the iso-surfaces of R, ∇R is perpendicular to any small line element $d\vec{l}$ that lays down on the iso-surface. As R goes up until the iso-surface reaches the vortex core, ∇R would be in the same direction as $\vec{R}$ or $\vec{r}$. The method to reduce Liutex core has been proved efficient and accurate and will be used here to investigate the influence on the vortical structures of spanwise-wall oscillation on a boundary layer transition.

Numerical investigations on the capacity of turbulent drag reduction by spanwise-wall oscillation can be dated back to 1992 when Jung et al. (1992) simulated a planar channel flow subjected to spanwise oscillatory motion of a channel wall. It is found that oscillating at nondimensional period of $T^+ = \frac{Tu_\tau^2}{\nu} = 100$ for the fixed maximum oscillation speed, where T, u_τ, and ν represent the oscillation period, the friction velocity, and the kinematic viscosity, respectively, can achieve friction drag reduction up to 40%. Baron and Quadrio (1995) then conducted DNS of turbulent channel flow with spanwise-wall movement and showed that the overall energy saving is achievable even taking into account the power necessary to drive the oscillation of the wall. Quadrio and Ricco (2003) and Xu and Huang (2005) further investigated the initial response of the first few cycles of wall oscillation and found that the spanwise velocity profile is basically identical to Stokes layer, which forms rather quickly during the first oscillation period, while it takes much longer for the longitudinal flow to reach its long-term developed state. Parameter studies have also been performed experimentally (Trujillo et al., 1997; Choi, 2002; Karniadakis and Choi, 2003) and numerically (Dhanak and Si, 1999; Quadrio et al., 2004), and it is found that the drag reduction rate depends on both the peak wall speed and the frequency of wall oscillation under the Reynolds numbers considered. Choi and Clayton (2001) used hot-wire measurements to study the mechanism of turbulent boundary layer drag reduction by spanwise-wall movement. They suggested that the interaction between the viscous sublayer and the periodic Stokes layer leads to spanwise vorticity generation, which reduces the mean velocity gradient and thus reduces the drag. Quadrio et al. (2009) further extended the idea of cyclic spanwise-wall oscillation to the so-called streamwise sinusoidal traveling waves of spanwise-wall velocity, in which the effect of wavenumber, temporal frequency is studied with a fixed forcing amplitude. A region of drag reduction is identified in the frequency and wavenumber plane, and it is shown that the flow structures are notably tilted when the drag is increased, while they remain aligned when drag reduction occurs.

In most numerical studies regarding wall oscillation, channel flows are by no doubt the favorite choice since the existence of one more statistical homogeneous direction, while boundary layer flows are preferred in experimental studies. Yudhistira and Skote (2011) reported the first DNS of a turbulent boundary layer with an oscillating wall and gave detailed turbulent statistics. Skote (2012) further investigated the transient behavior of boundary layer flow immediately after the oscillation starts. These results are qualitatively in agreement with the results from DNS of channel flows. Although many numerical and experimental investigations have been conducted as mentioned earlier, focuses have been placed on the fully developed turbulent region. The effect on transition from laminar to turbulence of spanwise oscillation with drag reduction parameters in turbulent region has not been investigated from the DNS perspective. To justify the study on the influence of wall spanwise oscillation on the transition region, one can refer to Wallace

(2013), who pointed out that study on transitional boundary layer flows in which the vortical structures are more organized might be helpful in understanding the physics of turbulence generation and sustenance.

The spanwise-wall oscillation is generally enforced as a boundary condition of spanwise velocity (V_m) in the form of

$$V_m = A \sin(\omega t) \tag{12.15}$$

where A is the maximum spanwise wall velocity, t is time, and ω is the oscillation frequency, which is related to the period T though $\omega = \frac{2\pi}{T}$.

12.3.2 Details of numerical simulations

(1) Case setup

A high-order finite difference is used to simulate a natural boundary layer transition with and without spanwise-wall oscillation. The three-dimensional compressible Navier–Stokes equations in generalized curvilinear coordinates (ξ, η, ζ) are solved and can be written in conservative form as:

$$\frac{\partial Q}{\partial t} + \frac{\partial(E - E_V)}{\partial \xi} + \frac{\partial(F - F_V)}{\partial \xi} + \frac{\partial(G - G_V)}{\partial \xi} = 0 \tag{12.16}$$

where Q is the vector of conserved quantities, while E, F, G are inviscid flux vectors, and E_V, F_V, G_V are viscous flux vector. A six-order compact scheme is used for spatial discretization in the streamwise and normal direction while the Fourier discretization is employed in the spanwise direction. A spatial filter is adopted to eliminate numerical oscillations caused by central difference scheme and a third-order TVD (total variation diminishing) Runge–Kutta method is employed for time marching. Nonreflecting boundary conditions are applied at both the outflow and far-field boundaries. Periodic condition is naturally used at the spanwise boundaries.

The computational domain is shown in Fig. 12.17, with x, y, and z representing the streamwise, spanwise, and normal direction coordinates. The flow conditions, including Reynolds number, Mach number, and so on are listed in Table 12.4. The quantities used in the simulation are nondimensionalized by freestream velocity U_∞ and the inlet displacement thickness δ_{in}. The inlet boundary of the computational domain is placed 300.79 δ_{in} downstream the leading edge of the flat plate. The Reynold number $\mathrm{Re}_{\delta_{in}} = \frac{\rho_\infty U_\infty \delta_{in}}{\mu_\infty}$ is set to 1000 in the simulations with and without spanwise-wall oscillation. The simulations include $1280 \times 256 \times 241$ grid points in the streamwise (x), spanwise (y), and wall normal (z) directions, respectively. Uniform grids are employed in both streamwise and spanwise directions, while a stretching grid with clustering points near the wall is used in the normal direction. The normal grid distribution is further scaled downstream according to $z = z_{in}\sqrt{\frac{x}{x_{in}}}$ to accommodate the development of boundary layer.

A two dimensional (2D) and a pair of three dimensional (3D) Tollmien–Schlichting (T-S) waves obtained by linear stability analysis are enforced at the

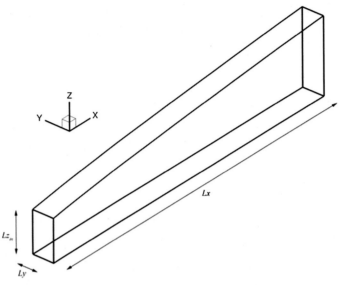

FIGURE 12.17

The computational domain.

Reproduced from Wang, Y., Liu, C., Liutex (vorex) cores in transitional boundary layer with spanwise-wall oscillation, Journal of Hydrodynamics 31 (6), 2019, with the permission of Journal of Hydrodynamics.

Table 12.4 Flow parameters.

M_∞	$Re_{\delta_{in}}$	x_{in}	Lx
0.5	1000	300.79 δ_{in}	672.02 δ_{in}
Ly	Lz_{in}	T_W	T_∞
22 δ_{in}	40 δ_{in}	273.15K	273.15K

Reproduced from Wang, Y., Liu, C., Liutex (vorex) cores in transitional boundary layer with spanwise-wall oscillation, Journal of Hydrodynamics 31 (6), 2019, with the permission of Journal of Hydrodynamics.

inflow boundary to trigger the flow transition. Thus, a primitive variable q at the inlet is gained according to

$$q = q_{lam} + A_{2d}\widehat{q}_{2d}e^{i(\alpha_{2d}x - \widetilde{\omega}t)} + A_{3d}\left(\widehat{q}_{3d+}e^{i(\alpha_{3d}x + \beta y - \widetilde{\omega}t)} + \widehat{q}_{3d-}e^{i(\alpha_{3d}x - \beta y - \widetilde{\omega}t)}\right)$$
$$+ c.c.$$

(12.17)

in which q_{lam} is the Blasius solution, A is the magnitude of T-S waves, α and β are the streamwise and spanwise wave number, $\widetilde{\omega}$ is the time frequency, and *c.c.* represents complex conjugates. In this study, $\widetilde{\omega}$ is set to 0.114027 and β is chosen to be

0.5712 to include two spanwise periods. α_{2d} is then solved via the compressible linear stability equation and found to be $0.2992 - 5.0954 \times 10^{-3}i$.

(2) Numerical results

The skin friction coefficient C_f is defined by:

$$C_f = \frac{\tau_w}{\frac{1}{2}\rho_\infty U_\infty^2} \tag{12.18}$$

where τ_w represents the local wall shear stress, which is calculated from the gradient of mean streamwise velocity $\bar{u}$ at the wall, i.e.,

$$\tau_w = \mu \frac{\partial \bar{u}}{\partial z}\bigg|_{z=0} \tag{12.19}$$

And the friction velocity and viscous length scale are defined as:

$$u_\tau = \sqrt{\frac{\tau_w}{\rho}} \tag{12.20}$$

$$z_\tau = \frac{\mu}{\rho u_\tau} \tag{12.21}$$

The turbulence statistics are collected from $t = 21.5\tilde{T}$ ($\tilde{T} = \frac{2\pi}{\tilde{\omega}}$ is the period of T-S waves), when the flow with freestream speed U_∞ has convected through the whole computational domain. Over the following time span of $5\tilde{T}$, time-independent statistics are collected and will be presented for the unmanipulated case later. For the spanwise oscillation case, $t = 21.5\tilde{T}$ marks the start of the wall oscillation, and the statistics are then collected every period of the wall oscillation. These statistics will be carefully presented later to exclude the transient effects of the cyclic motion start-up of the wall. Also note that the chosen parameter of $\omega = 10\tilde{\omega}$ leads to that the T-S wave period $\tilde{T}$ equals 10 times of the oscillation period T.

(3) The skin friction coefficient

The time and spanwise-averaged friction coefficient along streamwise direction for the baseline case is shown in Fig. 12.18A with that of laminar flow and an empirical formula proposed by Durcros et al. (1996). This comparison shows the present simulation can reliably capture the development of friction coefficient in flow transition. In addition, the velocity profile in local wall units at $x = 563$ where the flow can be considered fully turbulent from the friction coefficient is shown in Fig. 12.18B. This profile closely follows the linear profile $u^+ = z^+$ near the wall and is also in good agreement with the log law in the log region. The streamwise and spanwise resolutions in viscous length units are $\Delta x^+ = 25.5$ and $\Delta y^+ = 4.2$, based on z_τ at $x = 530.7$ where the maximum wall shear stress locates. At this same point, the first grid interval in the normal direction is $\Delta z^+ = 0.39$. Note that unless otherwise indicated, the superscript $+$ represents quantities normalized by

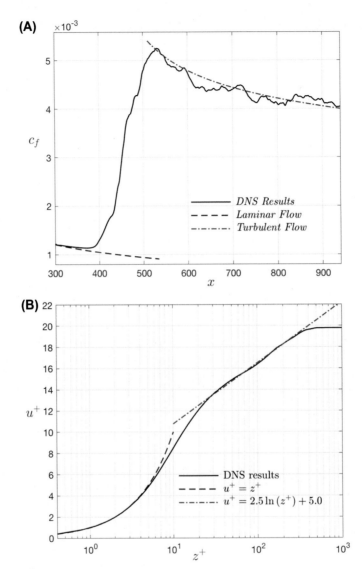

FIGURE 12.18

(A) Time and spanwise-averaged skin friction coefficient and (B) velocity profile at $x = 563$ for the baseline case.

Reproduced from Wang, Y., Liu, C., Liutex (vorex) cores in transitional boundary layer with spanwise-wall oscillation, Journal of Hydrodynamics 31 (6), 2019, with the permission of Journal of Hydrodynamics.

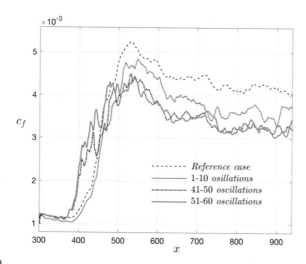

FIGURE 12.19

Skin friction coefficient averaged over $(- - -)$ 1–10 oscillations, $(\cdot \cdot \cdot \cdot)$ 41–50 oscillations, $(\cdot - \cdot - \cdot)$ 51–60 oscillations.

Reproduced from Wang, Y., Liu, C., Liutex (vorex) cores in transitional boundary layer with spanwise-wall oscillation, Journal of Hydrodynamics 31 (6), 2019, with the permission of Journal of Hydrodynamics.

the friction velocity and kinematic viscosity of the uncontrolled boundary layer. The region that can be considered fully developed turbulent boundary layer and free from the upstream influence of the outlet is $x = 500$ to $x = 900$, where the Reynolds number based on the momentum thickness Re_θ ranges from 910 to 1140.

The oscillation parameters in Eq. (12.15) for the present simulation are $V_m = 0.6$ and $\omega = 1.14027$, which leads to $V_m^+ = 11.3$ and $T^+ = 14.4$ based on the wall units of the nonoscillating boundary layer flow. For the controlled case with cyclic wall movements, the friction coefficients averaged over 1–10, 31–50, and 51–60 oscillation periods are shown in Fig. 12.19 to verify that the transient response of wall oscillation start-up has been excluded before collecting the turbulence statistics. The drag in the turbulent region converges to a statistical stationary state in a relatively quick way whereas it takes longer time for the drag in the transition region to reach its long-term state. The friction coefficient of the baseline case is also shown in Fig. 12.19, which obviously shows a significant drag reduction has been achieved in the turbulent region. While the drag in the transition region decreases a little after the start-up of the wall oscillation, eventually the transitional skin friction is increased due to the fact that transition point is moved upstream.

Both the transition and turbulent region reached the long-term state after 60 oscillations, and the statistical information is then collected from the 61 to 110 oscillation period. The time and spanwise-averaged skin friction coefficient is shown in Fig. 12.20. The percentage of drag reduction (DR) is defined as:

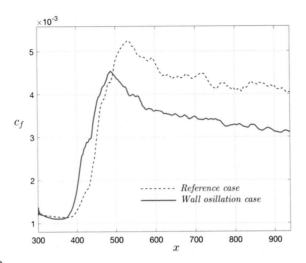

FIGURE 12.20

Time and spanwise-averaged skin friction coefficient comparison.

Reproduced from Wang, Y., Liu, C., Liutex (vorex) cores in transitional boundary layer with spanwise-wall oscillation, Journal of Hydrodynamics 31 (6), 2019, with the permission of Journal of Hydrodynamics.

$$DR(\%) = 100\,\frac{C_f^0 - C_f}{C_f^0} \qquad (12.22)$$

where C_f^0 is the skin friction coefficient of the unmanipulated case. Based on the definition, the drag is reduced up to 21.8% in the region from $x = 500$ to $x = 900$, which is comparable to the value of 23% reported by Ricco and Wu (2004) from their experiments on turbulent boundary layer with $V_m^+ = 11.3$ and $T^+ = 67$. The present numerical results show a fairly constant DR in the turbulent region, and it is expected that it will continue if the simulation domain is further extended in the streamwise direction.

(4) The alternation of vortical structures by spanwise wall oscillation

The Liutex (vortex) core line method is used to visualize vortices for boundary layer transition with and without spanwise wall oscillation. With the Liutex core line, the vortical structures for the baseline case at $t = 25.5\widetilde{T}$ and wall-oscillation case at $t = 32.5\widetilde{T}$ are shown in Fig. 12.22. These two time-steps are at the same T-S wave phase and the transient effects of wall oscillation have been avoided. The first observation is that the Liutex lines can well capture the transition process as T-S waves, Λ- and hairpin vortices, and entangled vortices of multiple scales, which are characteristics of the modal linear growth stage, transition section, and turbulent section are all well presented. In addition, the Liutex lines are colored by the magnitude of Liutex vector, which is the shear-contamination-free quantity representing the strength of rigid rotation part of the fluid motion. In agreement

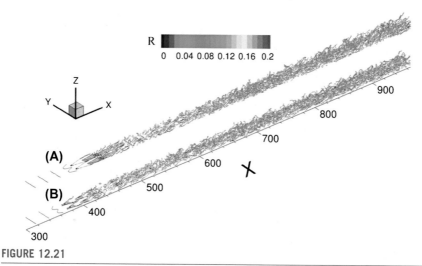

FIGURE 12.21

The vortical structures represented by Liutex core lines at (A) $t = 25.5\widetilde{T}$ for the baseline case; (B) $t = 32.5\widetilde{T}$ for the wall oscillation case.

Reproduced from Wang, Y., Liu, C., Liutex (vorex) cores in transitional boundary layer with spanwise-wall oscillation, Journal of Hydrodynamics 31 (6), 2019, with the permission of Journal of Hydrodynamics.

with Fig. 12.20, the laminar-turbulent transition point has been advanced upstream since the Λ- and hairpin vortices appear earlier in the wall-oscillation case than that in the baseline case.

(5) The development of T-S waves

An enlarged view of the linear growth and early transitional region is shown in Fig. 12.21 by Liutex core lines colored by Liutex magnitude. The distance between the first three vortex waves from both cases is about 21, which is the streamwise wave number of the 2D T-S wave. This stage is characterized by the linear growth of the T-S wave, whose interaction with the Stokes layer excited by the cyclic movement of the wall is small especially for the first two vortex waves. This phenomenon can be explained as the disturbance introduced by T-S waves is located further away from the wall surface with a distance about 1.2 while the Stokes layer is creeping on the wall below $z = 0.2$ in this region. For the third wave around $x = 365$, visual distinctions of the vortical structures can be observed, and the fourth wave from the oscillation case has developed into a chain of hairpin vortices, while it remains in a Λ-shape from the baseline case. Fig. 12.22 shows the spanwise velocity (v) distribution at three streamwise positions $x = 317, 338$, and 359, which centered at the first three T-S waves. It can be seen from Fig. 12.23 A and B that the maximum magnitude of spanwise velocity disturbances is about the same for the two cases, and the velocity profile in the Stokes layer is uniform in the spanwise direction. The wall oscillation enforces the spanwise velocity on the wall at this considered phase, which is $V_m = -0.6$. As the T-S waves develop, the disturbances grow

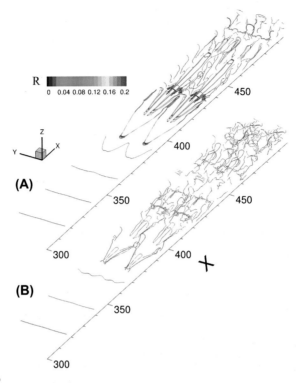

FIGURE 12.22

The vortical structures represented by Liutex core lines in the upstream section (A) the baseline case; (B) the wall oscillation case.

Reproduced from Wang, Y., Liu, C., Liutex (vorex) cores in transitional boundary layer with spanwise-wall oscillation, Journal of Hydrodynamics 31 (6), 2019, with the permission of Journal of Hydrodynamics.

and begin to interact with the Stokes layer. The spanwise velocity magnitude around $z = 1$ in Fig. 12.23D is much larger than that in Fig. 12.23C. In addition, the Stokes layer is altered and an uneven distribution is presented, especially just below $z = 0.2$. Further downstream, as the Λ-vortex begins to form, the vortex head begins to elevate and the tail further descends close to the wall, which leads to a stronger interaction with the Stokes layer. Obviously, this cyclic wall movement introduces additional disturbance to the flow and accelerates the development and formation of the Λ-vortex. As the Λ-vortex comes into being, the rotation induced by the pair of streamwise vortex legs brings low speed fluids upward and creates high shear layer away from the wall. Note that here low speed means low streamwise speed. This process is detailed by Wang Y. et al. (2016) and is similar for the case with spanwise wall oscillation.

To compare the Liutex core line method with iso-surface based methods, iso-surfaces of $R = 0.005, 0.02, 0.1$ are shown in Fig. 12.24A−C respectively for the

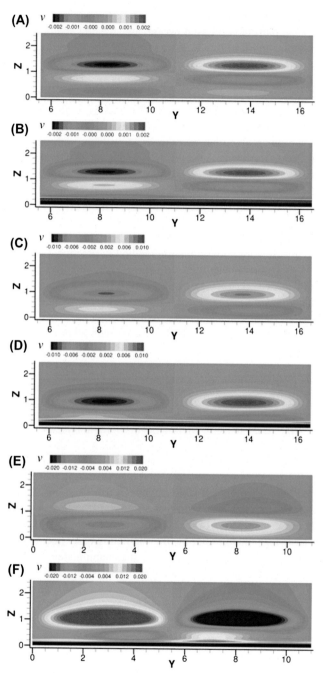

FIGURE 12.23

unmanipulated case. With a smaller threshold of $R = 0.005$, the T-S waves are captured. The Λ and hairpin vortices, however, are smeared and blurred in the transition region as shown in Fig. 12.24A. With $R = 0.02$, the Λ vortex around x = 385 is clearly visualized but the T-S waves are lost as in Fig. 12.24B. With an even larger threshold R = 0.1 in Fig. 12.24C the rings of hairpins are clearer. However, both the T-S waves and Λ vortices are lost. In addition, the legs of the hairpin chain around x = 430 are disconnected with hairpin rings. This inconsistency of threshold problem roots from the fact that vortices are of different rotational strengths and almost every iso-surface-based method has this threshold problem with one exception: the Ω method. Rather than capturing the swirling strength, the Ω parameter is a measure of local vorticity density, or fluid rigidity, which can be called relative vortex strength. It has been proved by many authors (Zhang et al., 2019; Gui et al., 2019a; Wang C. et al., 2019; Wang L. et al., 2019; Wang Y. et al., 2019) that the Ω method could well capture both strong and weak vortices in transient flows without adjustment of the threshold and thus could be viewed as the most reliable, robust, and easy-to-use vortex identification method in practice. On the other hand, Liutex vector is more precise without any shear contamination from a scientific perspective and Liutex core lines can well capture both weak and strong vortices at the same time as shown in Fig. 12.21. It should also be pointed out that the T-S wave obtained from stability analysis is actually very weak vortex with the precise definition of Liutex vector, which was quite obscure with other vortex identification methods especially when the physical meaning of the parameter is unclear. It is interesting that not only vortices are sinews and muscles of turbulence, but also the disturbances from stability analysis, which break laminar flow, are vortices.

(6) Early transition region

As shown in Fig. 12.20, the skin friction begins to rapid increase around $x = 400$, which is marked by Λ and hairpin vortex formation as shown in Fig. 12.22. Fig. 12.25 illustrates this early transition stage with Liutex core lines and iso-surface of $R = 0.1$ for the two cases. The first observation is that the longitudinal distance covered by one vortex package originated from a single T-S wave in this stage is larger for the baseline case than that of the wall-oscillation case. The cyclic wall movement makes generation of hairpin heads from high shear layer between the two counterrotating legs much easier for the oscillating case. Second, the near-wall quasi-streamwise vortices around $x = 415$ in the manipulated case as shown in Fig. 12.9B are disrupted while those in the baseline case are still ordered. In addition, the hairpin heads or hairpin rings are distorted for the oscillation case while

The spanwise velocity distribution at $x = 317, 338, 359$. (A, C, E) for the baseline case, (B, D, F) for the oscillation case.

Reproduced from Wang, Y., Liu, C., Liutex (vorex) cores in transitional boundary layer with spanwise-wall oscillation, Journal of Hydrodynamics 31 (6), 2019, with the permission of Journal of Hydrodynamics.

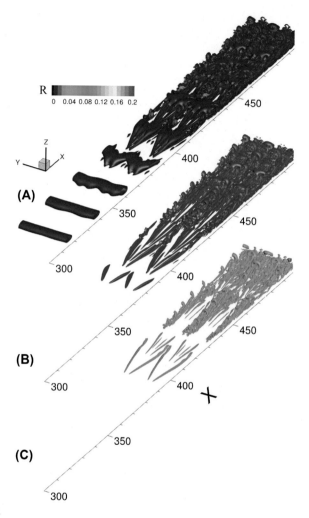

FIGURE 12.24

Iso-surfaces of R (Liutex magnitude) for the baseline case (A) $R = 0.005$, (B) $R = 0.02$,
(C) $R = 0.1$.

*Reproduced from Wang, Y., Liu, C., Liutex (vorex) cores in transitional boundary layer with spanwise-wall
oscillation, Journal of Hydrodynamics 31 (6), 2019, with the permission of Journal of Hydrodynamics.*

those in the baseline case are of typical hairpin shapes. This indicates an accelerated
transition process, which is in agreement with the skin friction development in
Fig. 12.20.

Packages of hairpin vortex chains are typically observed in boundary layer transition. Fig. 12.26 illustrates a chain of three hairpins for the oscillation case. It is

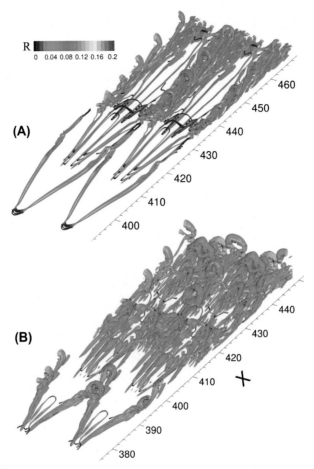

FIGURE 12.25

The Liutex core lines in the early transition region with transparent iso-surface of $R = 0.1$.
(A) The baseline case, (B) the wall oscillation case.

Reproduced from Wang, Y., Liu, C., Liutex (vorex) cores in transitional boundary layer with spanwise-wall oscillation, Journal of Hydrodynamics 31 (6), 2019, with the permission of Journal of Hydrodynamics.

shown that the legs of hairpin vortex are basically streamwise vortices, and the neck region below the hairpin heads is also rotating in the quasi-streamwise direction. Meanwhile, the three hairpin heads in the vortex package are mostly spanwise vortices. It has been reported (Liu et al., 2014; Wang et al., 2016) that the hairpin head and the hairpin leg are two different vortex structures generated separately by different mechanisms. The Λ vortex, which later becomes the hairpin legs, is generated first. As it becomes strong, a high shear layer is formed above the legs, which results from the ejection of Λ-vortex rotation that lifts low-speed fluids near the wall up, forming a low-speed zone. The instability of the high shear layer

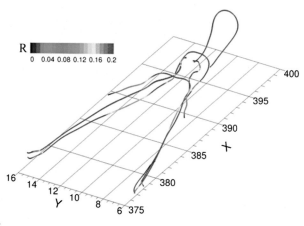

FIGURE 12.26

Hairpin chain in the oscillation case.

Reproduced from Wang, Y., Liu, C., Liutex (vorex) cores in transitional boundary layer with spanwise-wall
oscillation, Journal of Hydrodynamics 31 (6), 2019, with the permission of Journal of Hydrodynamics.

leads to the formation of vortex rings. With the Liutex core lines shown in Fig. 12.26, it is obvious that the Liutex core lines of hairpin rings are separated from that of the legs, which means they are actually not an integrated structure.

The instantaneous and time-averaged skin friction coefficients for the early transition stage are shown in Fig. 12.27. It is shown that the instantaneous drag is highly correlated with the vortical structures at this stage especially for the baseline case in Fig. 12.27A. The sweeps and ejections induced by vortex heads and legs are reflected as high and low friction regions on the wall. However, this trend is less obvious for the oscillation case in Fig. 12.27C, in which a large area with low skin friction is found under the vortex package considered. The high drag region is also enlarged and intensified as the tail part of the vortex leg has extended inside the Stokes layer. The time-averaged skin frictions are comparable for these two vortex packages as shown in Fig. 12.27B and D, and it can be inferred that the averaged drag is a result of the instantaneous drag moving through the confined region since the development of these vortical structures is limited in this region. In addition, it is shown in Fig. 12.28 that the drag is increased in the high friction region and decreased in the low friction region for the spanwise oscillation case, which leads to a slight drag reduction in the time-averaged sense compared to the baseline case in the corresponding region.

(7) Late transition region

As shown in Fig. 12.20, the skin friction profile quickly diverges from the laminar coefficient and overshoots the turbulent estimation in the corresponding transition region for both cases. The streamwise ranges covering the skin friction overshoots are shown in Fig. 12.29 respectively for the baseline and wall-

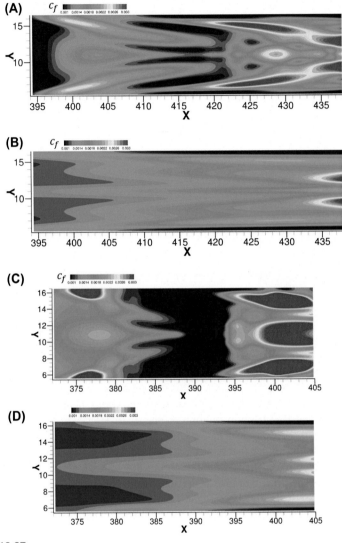

FIGURE 12.27

Instantaneous and time-averaged skin friction coefficients for the baseline flow (A) and (B) and for the spanwise oscillation case (C) and (D).

Reproduced from Wang, Y., Liu, C., Liutex (vorex) cores in transitional boundary layer with spanwise-wall oscillation, Journal of Hydrodynamics 31 (6), 2019, with the permission of Journal of Hydrodynamics.

oscillation case. Multiple-level hairpin vortices are easily recognized in this region as the characteristics of late transition region. Despite this organized pattern in this section, the statistical features are identical to that of turbulence. The drag reaches maximum at the locations of $x = 531$ and $x = 486$ respectively, which overshoots

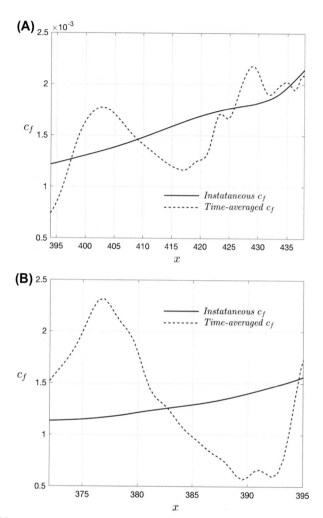

FIGURE 12.28

Spanwise-averaged skin friction coefficients on the wall surface (A) for the baseline case and (B) for the spanwise oscillation case.

Reproduced from Wang, Y., Liu, C., Liutex (vorex) cores in transitional boundary layer with spanwise-wall oscillation, Journal of Hydrodynamics 31 (6), 2019, with the permission of Journal of Hydrodynamics.

the turbulent estimation. For the baseline case, the inflow and spanwise boundary conditions are both periodic with period of 12.29 in the spanwise direction and symmetric about $y = 11$. The periodicity and symmetry are preserved in this region as shown in Fig. 12.29A and C with and without iso-surfaces of $R = 0.1$ respectively. For the oscillation case, however, the symmetry is lost in this region since the cyclic wall movement introduces an asymmetry disturbance into the flow field. The

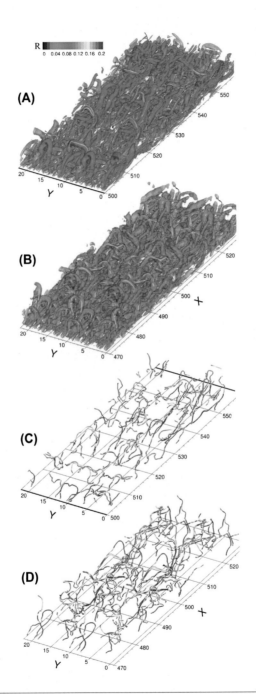

FIGURE 12.29

The vortical structures in the late transition region presented by Liutex core lines with iso-surface of $R = 0.1$ for (A) baseline case and (B) wall oscillation case, without iso-surfaces for (C) baseline case and (D) wall oscillation case.

Reproduced from Wang, Y., Liu, C., Liutex (vorex) cores in transitional boundary layer with spanwise-wall oscillation, Journal of Hydrodynamics 31 (6), 2019, with the permission of Journal of Hydrodynamics.

periodicity, on the other hand, can still be found for the wall oscillation case from Fig. 12.29B and D. It is also shown that the Liutex core lines can well capture the hairpin rings in this region and that the rotational strength of these hairpin rings is rather strong, which means in the late transition region the hairpin heads are very effective to bring high-speed fluids toward the wall and bring the low-speed fluids up. In comparison, the hairpins in the turbulent region are much weaker.

(8) Turbulent region

In the turbulent region, the vortical structures in the range between $x = 750$ and $x = 850$ are shown in Fig. 12.30. Extrusions of hairpin vortices at the edge of boundary layer are observed in both cases with and without spanwise wall oscillation. Tilted, twisted, and tangled vortices are also observed, which is a typical feature of turbulence. Compared with the late transition region, the rotational strength of vortices or the Liutex magnitude near the edge of boundary layer is rather weak so that the iso-surface based methods cannot capture these vortices correctly, while the Liutex core line method has the advantage of precisely capturing both strong and weak vortices. The periodicity and symmetry are completely lost in this section as the flow becomes fully developed turbulent flow.

The instantaneous skin friction coefficients C_f are shown in Fig. 12.31, with the legends centering at the mean C_f from $x = 750$ to $x = 850$ at 0.0043 for the baseline case and 0.0034 for the wall cyclic movement case. It shows the area where friction coefficient comparable to the mean values is much larger in the oscillation case than that of the baseline case. The cyclic wall movement might contribute to average of the drag in certain sense, and thus the high and low drag regions are not as intense as the baseline case.

12.3.3 Summary

The vortex core detection method based on the Liutex vector is utilized to investigate the alternation of vortical structures on a boundary layer transition subjected to spanwise-wall oscillation. Compared with iso-surface-based methods, the Liutex core line method is shown to be precise, free of threshold, and capable to capture both strong and weak vortices simultaneously. Tollmien-Schlichting (T-S) waves in the linear growth region, Λ- and hairpin vortices in the transition region, and twisted vortices in the turbulent region are all well captured by Liutex core lines. The cyclic wall movement accelerates the transition process while reducing the turbulent drag by 21.8% with selected parameters. For the wall oscillation case, the development from T-S wave to Λ-vortex is advanced about one T-S wave length in the streamwise direction. In the transition region, the Λ-vortex and legs of hairpin vortex are shortened in the wall oscillation case, and the symmetry of the vortical structures is lost in the late transition region since the introduction of asymmetry disturbances by the cyclic wall movement. Extrusions of weak vortices at the edge of boundary layer are found in the turbulent section, which is often omitted by iso-surface-based vortex identification method. Thus, it is demonstrated that for the

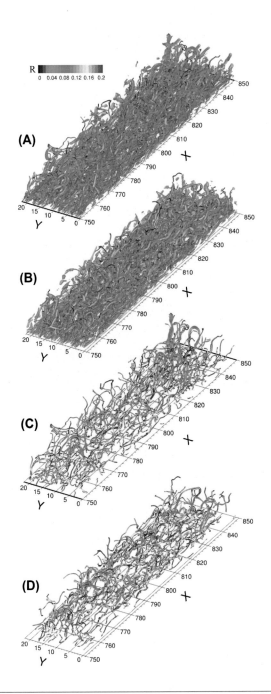

FIGURE 12.30

The vortical structures in the turbulent region presented by Liutex core lines with iso-surface of $R = 0.1$ for (A) baseline case and (B) wall oscillation case, without iso-surfaces for (C) baseline case and (D) wall oscillation case.

Reproduced from Wang, Y., Liu, C., Liutex (vorex) cores in transitional boundary layer with spanwise-wall oscillation, Journal of Hydrodynamics 31 (6), 2019, with the permission of Journal of Hydrodynamics.

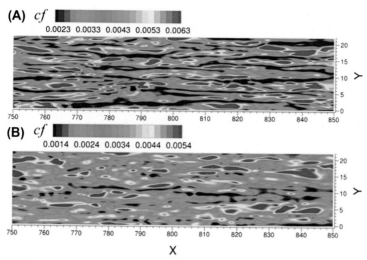

FIGURE 12.31

The instantaneous skin friction in the turbulent region (A) for the baseline case and (B) for the oscillation case.

Reproduced from Wang, Y., Liu, C., Liutex (vorex) cores in transitional boundary layer with spanwise-wall oscillation, Journal of Hydrodynamics 31 (6), 2019, with the permission of Journal of Hydrodynamics.

transitional boundary layer, the Liutex core line method provides a systematic and threshold-free vortex definition, which could serve as a powerful tool to understand and guide flow control.

In this study, the Liutex core line method is used to detect the vortices in transitional boundary layers with and without spanwise wall oscillation. First, the method is capable of capturing the T-S waves in the linear growth region, Lambda Λ and hairpin vortices in the transition region, and entangled twisted vortices in the turbulent region. Its superiority over iso-surface-based method including precise, threshold-free, capturing both weak and strong vortices simultaneously is clearly demonstrated. In addition, the alternation of vortical structures on boundary layer transition under spanwise-wall oscillation is investigated by comparing with the baseline case. It is shown that the additional introduced disturbance accelerates the transition process and the Λ and hairpin vortices appear earlier in the controlled case. In the transition region, the Λ vortex and hairpin legs are shortened and with the help of Liutex core lines, it is obvious the hairpin heads and legs are separate parts generated by different mechanisms. Weak vortical structures at the edge of boundary layer in the turbulent region are captured by the Liutex core lines, which are however often skipped by iso-surface-based vortex identification methods. In summary, the Liutex core line method provides a systematic approach and powerful tool to investigate vortex structures in various flows from a new perspective.

Liutex and experiments for turbulent flow of low Reynolds number

13.1 Introduction

Accurate and reliable experiments are important aspects of turbulence studies. A key to successful experimental research is the measuring method. In 2005, Ecke (2005) gave an experimentalist's perspective on the background, research progress, and future development of the turbulence problem, noting that "It highlights the application of modern supercomputers in simulating the multiscale velocity field of turbulence and the use of computerized data acquisition systems to follow the trajectories of individual fluid parcels in a turbulent flow. Finally, it suggests that these tools, combined with a resurgence in theoretical research, may lead to a 'solution' of the turbulence problem." This suggests that conventional scientific methods are still applicable if equipped with modern measuring techniques, and thus, numerous experimental measurement methods for capturing the fine structure of the dynamic turbulence process have been developed.

Various flow field measurements and visualization techniques have undergone considerable development in terms of experimental research and measurement methods. Several flow visualization methods, such as hydrogen bubble, smoke and dye methods, are widely applied in the flow visualization of turbulent boundary layers due to their minimal influence on the flow field. Kline et al. experimentally studied a turbulent boundary layer flow field by using the hydrogen bubble method (Kline, 1967). The results indicated the existence of low- and high-speed streaks of hydrogen bubbles in the near-wall region and suggested that bursting is an important factor for the energy generation of turbulence. Furthermore, abundant coherent structures were found in a turbulent boundary layer in an experimental investigation performed using the hydrogen bubble technique (Lian, 1990). This study also found that vortex rotation motion leads to the generation of downwash and upwash flow, and the streamwise vortex stretches along the flow direction and increases the rotation speed by absorbing energy from the surroundings. These methods mentioned earlier are economical and easy to implement to realize the qualitative observation of large-scale vortex structures in turbulence and obtain the macroscopic characteristics of turbulence; however, they are unable to obtain quantitative results or turbulent microstructural observations.

Liutex and Its Applications in Turbulence Research. https://doi.org/10.1016/B978-0-12-819023-4.00012-4
Copyright © 2021 Elsevier Inc. All rights reserved.

Flow field measuring methods include intrusive and nonintrusive methods. Earlier intrusive measurements included total and static pressure probes, hot-wire anemometry (HWA), and hot-film measurements (HFA). HWA/HFA is an essential method in turbulence fluctuation measurement. According to a recent review of near-wall turbulence experiments (Stanislas, 2017), turbulence intensity is anisotropic and takes the maximum value at the near-wall zone in a turbulent boundary layer. It is extremely difficult to apply HWA to the turbulence intensity measurement, even with the smallest probes or the largest wind tunnels.

Several nonintrusive measurements, such as laser Doppler velocimetry (LDV), laser-induced fluorescence (LIF), particle tracking velocimetry (PTV), and particle image velocimetry (PIV), have developed rapidly compared with intrusive measuring systems. Galmiche et al. (2014) performed a measurement on the turbulence statistics in a fan-stirred combustion vessel using complementary LDV and PIV and noted that LDV provides information on the mean velocity, higher-order moments, and frequency spectra with a high degree of accuracy; however, it fails to provide information on the spatial structure of the flow for single-point measurement.

Currently, PIV and PTV are the major experimental measurements because both employ very short double exposures at a certain time interval to obtain the distribution of tracer particles in the flow field at these two moments, thus obtaining the velocity field with a cross-correlation algorithm. PIV measures the velocity field of a particle group at a certain time in space, while PTV tracks the motion trajectory of a single particle to calculate its velocity. PIV and PTV belong to the Euler method and Lagrange method, respectively.

In recent years, several research groups have initiated more complex PIV setups in terms of dual-plane measurements and time-resolved measurements to obtain spatiotemporal information regarding turbulent boundary layers. The most straightforward approach is to increase the number of cameras; then, the velocity field can be reconstructed with a corresponding algorithm. Hutchins et al. (2005) utilized a 3D stereo-PIV measuring system that uses two cameras with separate viewing angles to extract the z-axis displacement to provide a unique quantitative view of the turbulent structure of planes inclined both with and against the principle vorticity axis of a proposed hairpin model (inclined at both 45° and 135° to the streamwise axis). Their detections of both swirl and low streamwise momentum suggested that multiple patches of counterrotating vortex pairs often occupy the vertically aligned shear layers between high- and low-speed fluids in the 135° plane. Dennis and Nickels (2011) performed high-speed stereo PIV in a turbulent boundary layer with $Re_\theta = 4700$. The light sheet was arranged to orientate perpendicular to the flow direction so that the coherent structures passing through the measurement plane could be effectively scanned.

Due to laser performance limitations, both 2D and 3D PIV can only measure the parameters of the flow field at a certain time; they cannot measure the parameters of a continuous process, such as the formation, evolution, and development of a vortex structure in a turbulent flow. In this situation, a time-resolved PIV (TR-PIV) is proposed that combines the conventional PIV with a high-speed camera and a high-energy continuous laser or a high-frequency pulsed laser; thus, continuous PIV images that have flow field information that changes over time can be obtained after image processing (Wang et al., 2014).

Tomographic PIV (Tomo-PIV) is enabled with high-power pulse light and many high-resolution cameras. Scattered light images are simultaneously recorded by these cameras, and the 3D spatial distribution of particles in the flow field can be obtained by the multiplicative algebraic reconstruction technique (MART). Experiments on studying the coherent structures related to negative Reynolds stress events (Q2 and Q4) and their time evolutions in the logarithmic region of a turbulent boundary layer have been performed (Schroder et al., 2011) using time-resolved tomographic PIV (TR-tomo-PIV) with six CMOS cameras, which provided an impetus to rethink the early stages of the classic hairpin development model in the turbulent boundary layer for high Reynolds numbers. A moving tomo-PIV was performed in a turbulent boundary layer at $Re_\theta = 2410$ (Gao et al., 2013). The measuring system traveled at the local flow velocity, and thus the evolution of coherent structures was obtained in time and space with a restricted field of view (FOV).

The basic principle of nanobased planar laser scattering (NPLS) measurement (Schroder et al., 2011) is to intersperse nanoscale tracer particles into the fluid. Since the distribution of particles in the field is inhomogeneous, the scattering intensity of nanoscale particles is related to the density of the particles. Assume that the concentration of nanoparticles does not change as they flow from one location to another at very short time intervals, and take the scattering intensity distribution of nanoparticles as the characteristic parameter; as a consequence, the velocity field can be obtained by matching the parameters of two frames. This method has been widely used in the measurement of supersonic and hypersonic flow fields. Moreover, molecular tagging velocimetry (MTV) (Gendrich and Koochesfahani, 1996; Hiroki et al., 2016) is another flow field measurement tool similar to NPLS. The difference is that the tracer particle is replaced by fluorescent water molecules.

Several flow field imaging measurements, such as single frame and long exposure (SFLE) (Yang et al., 2017), single frame and multiple exposure (SFME), and moving SFLE (M-SFLE), which have been proposed and widely applied in turbulent jet flows and plate turbulent boundary layers by Cai's group at University of Shanghai for Science and Technology (USST), will be introduced in detail further. Recently, a moving PIV/PTV (M-PIV/PTV) measuring method was implemented and used for quantitative research on the evolution process of vortex generation at high spatiotemporal resolution.

13.2 Imaging measurement techniques

Imaging methods employ tracer particles with good tracking performance to measure the velocity of a flow field. In a typical experiment, the flow field is illuminated by laser light. The scattered light information of the tracer particles can be received by charge-coupled device (CCD) or complementary metal oxide semiconductor (CMOS) image sensors in industrial cameras, which can convert these optical signals into orderly electrical signals to obtain images. In recent years, several imaging methods have been proposed and applied by Cai's group in the research of turbulence experiments, which is introduced as follows.

13.2.1 Basic knowledge of cameras and lenses

Cameras and lenses are important equipment used in imaging measuring systems that can determine the quality of an image. A brief introduction to the basic principles and main parameters of cameras and lenses is given in the following.

The most essential function of cameras is to convert optical signals into orderly electrical signals. Commonly used types include CCD cameras and CMOS cameras, categorized according to the type of image sensor. The major camera parameters include pixel size, resolution, frame rate, and pixel depth. The pixel size is the actual physical size of the smallest unit of the image sensor, such as 4.8 μm $\times$ 4.8 μm. Resolution denotes the number of pixels of the image collected by the camera and can be expressed as the number of pixels in the horizontal direction multiplied by the number of pixels in the vertical direction, such as 1280 $\times$ 1024 pixels or 1.3 megapixels. Both pixel size and resolution determine the sensor plane size of the camera. The frame rate represents the rate at which a camera collects and transmits images, and the typical frame rate unit is frames per second (fps). The pixel depth is the number of bits of signal per pixel, such as 8-bit, 10-bit, and 12-bit depths. For an 8-bit camera, the maximum of the image gray value is 2^8; thus, the gray value of the picture ranges from 0 to 2^8. The typical camera exposure methods include global shutters and rolling shutters. A global shutter provides simultaneous exposure to an entire image. A rolling shutter provides exposure from the top of an image to the bottom in turn. It should be noted that rolling shutters are not suitable for the measurement of moving objects.

Lenses are used to image an object on the sensor plane of the camera. For a general lens, the image becomes larger when moving the lens relatively close to the object; otherwise, it decreases. The parallax of the lens can be corrected, and then the real size of the object can be obtained by a telecentric lens due to a parallel optical path design, such as the trajectory length of particles in a flow field. The main parameters of a lens include the focal length, working distance, magnification, FOV, aperture, and depth of field (DOF). Focal length refers to the distance from the center of the lens to the focus. The focal length is the distance between the center of the lens and the camera sensor plane in the imaging measuring system. Working distance (WD) refers to the distance from the bottom mechanical surface of the lens to the object. Magnification is the ratio of the imaging size of an object to the actual size, and the FOV refers to the region of the space that can be measured by the lens. The aperture, also called the F number, controls the amount of light entering the lens. The aperture becomes larger as the F number decreases. The DOF refers to a certain distance within the area of the focal plane where the image is clear and the aperture grows larger as the DOF decreases. In the measurement, the area of the sensor plane of the lens should be equal to or larger than that of the camera to avoid corner shadows on the image.

13.2.2 Single frame and long exposure

For most imaging measurements, the camera is usually fixed without moving, such as in SFLE and SFME imaging methods. The principle of SFLE can be outlined as follows. First, tracer particles with good tracking properties are interspersed in the flow field; then, these tracer particles are illuminated by a light sheet from the laser; finally, the scattered light of the particles can be received by the camera. SFLE can record the trajectory of the tracer particles clearly by selecting a proper exposure time. Fig. 13.1 shows a typical trajectory image of a moving particle with a long exposure time. One can obtain the velocity v of a moving particle in the flow field by the following formula:

$$v = \frac{S - D}{M \Delta t_e} \tag{13.1}$$

where S is the total length of the trajectory, which represents the movement distance of the tracer particle during the exposure time Δt_e, D represents the diameter of the particle, and M is the magnification factor of the lens. Moreover, the direction of the particle velocity can be determined by two consecutive frames.

Both SFLE and PIV are Eulerian-type measurements, and their schematic diagrams are shown in Fig. 13.2. Compared with PIV measurement, SFLE has the following advantages:

(1) Compared with PIV and PTV, which obtain the starting and ending positions of a particle motion by using two exposures and two frames, SFLE can display the particle motion trajectory in a single frame with a single exposure. The flow direction, which is approximately treated as the direction of the tracer particle

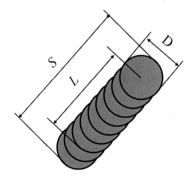

FIGURE 13.1

Typical trajectory image of a moving particle and calculation model.

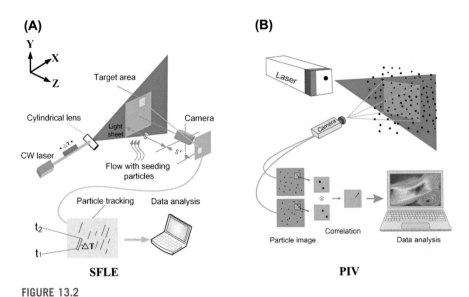

FIGURE 13.2

Schematic diagram of SFLE and PIV. (A) SFLE (B) PIV.

trajectory, can be obtained from two consecutive image frames; meanwhile, the velocity can be obtained from the magnification of the lens, the length of the trajectory in the image, and the exposure time. Thus, SFLE effectively avoids employing a complex cross-correlation algorithm in PIV.

(2) The SFLE method is cost-effective and easy to operate, as it only requires a simple measuring apparatus such as a continuous laser with low power and a common industrial camera. However, PIV requires a high-speed camera and a high-energy pulse laser, and the measuring apparatus is complex and expensive.

(3) SFLE can obtain different resolutions by selecting different cameras and lenses, which is suitable for the measurement of different flow fields, especially jet flow fields with micron vortex structures.

(4) If the size distribution of particles in a flow field is not uniform, employing the PIV cross-correlation algorithm becomes more difficult, and the flow field measurement can become inaccurate. However, the image obtained by the SFLE method can clearly record the trajectory of a moving particle and obtain its velocity.

However, there are also some disadvantages of the SFLE imaging method, which are listed as follows:

(1) The velocity of a particle obtained by the SFLE method is actually the average velocity during a specific exposure time; the particles in an actual flow field do not move at a constant speed, thus leading to a velocity measurement error.

(2) Ambiguity of the flow direction exists in the particle trajectory measured by SFLE.

(3) Once the flow field has a high concentration of tracer particles, the particle trajectories obtained by SFLE will overlap, which creates difficulties in the analysis of particle velocity and thus results in inaccurate experimental results.

13.2.3 Single frame and multiple exposure

The SFME imaging method proposed by Cai's group aims to obtain the trajectory of the tracer particle connected by continuous discontinuity points during multiple exposures in the same frame image. There are two ways to realize the multiple exposure of a single frame image. One way is to use multiple camera exposures with a continuous laser; the other is to use a multiple stroboflash of a pulsed laser during a long exposure time. Both methods can eventually obtain discrete particle trajectories in a single frame image. By setting multiple appropriate exposure times, the SFME method can identify the flow direction of the particles and measure the flow field parameters such as transient velocity and acceleration fields.

Fig. 13.3A shows the sketch of a particle trajectory during a relatively long exposure time Δt obtained by the SFLE method, and Fig. 13.3B shows the trajectory with three exposure moments t_1, t_2, and t_3 by the SFME imaging method, where $t_1 < t_2 < t_3$ and the time interval between two exposures is t_0. Fig. 13.3B shows that $L_1 < L_2 < L_3$; thus, it is known that the particle flows from left to right. Similar to the SFLE method, SFME can obtain the corresponding velocity of each small section of the entire particle trajectory by Eq. (13.1). Thus, in the figure, five different velocities can be calculated during three exposure times, including three for each small trace L_1, L_2, L_3 and two for each time interval t_0. Furthermore, the acceleration of the fluid can be calculated through the differences between the two continuous velocities and the exposure time. Similar to the velocity, in this way, the continuous

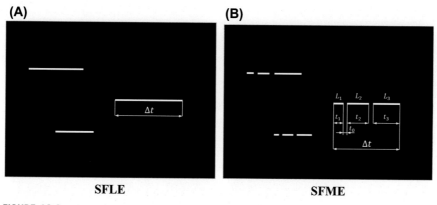

SFLE   **SFME**

FIGURE 13.3

Schematic diagram of SFLE and SFME. (A) SFLE (B) SFME.

accelerations of a particle can be obtained, and then other parameters, such as turbulence stress and turbulence intensity, can be calculated. Therefore, the SFME method can approximatively obtain the transient velocity of the fluid during a period of time as well as the acceleration of the fluid based on these continuous transient velocities. However, due to particle matching and other problems, SFME has difficulties with image processing, which will be a focus of further research.

13.2.4 Moving single frame and long exposure

13.2.4.1 Measuring system

Eulerian-type measurements, such as SFLE and PIV, fail to capture the process of a fast-moving vortex structure evolution if the camera is fixed. For instance, the coherent structure in a turbulent boundary layer has a migration velocity of flow. In addition, the FOV of cameras and the image resolution are contradictory for imaging measurements. When a high-resolution camera is applied, a fast-moving vortex structure may not be captured due to the small FOV. To address the aforementioned issues, the M-SFLE imaging method, which was developed from the SFLE method, was proposed by Cai's group to show both the temporal and spatial development of the vortex structure in a single frame without using vortex identification criteria or Galilean invariant. The M-SFLE is a Lagrangian-type measurement that can easily and intuitively observe the evolution process of the vortex structure.

Compared to the SFLE method, the advantage of the M-SFLE is that the camera can move along the flow direction so that vortex structures with nearly the same speed as the measuring system can be captured. A comparison of the imaging results by SFLE and M-SFLE imaging methods is shown in Fig. 13.4. Since a moving measuring system is applied in Fig. 13.4B, the numerous vortices that are marked in the figure move at the same speed as the measuring system. It can be seen that the heads of hairpin vortices can be entirely observed by the relatively longer particle

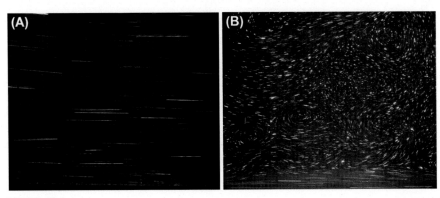

FIGURE 13.4

Comparison of frames by SFLE and M-SFLE.

path. M-SFLE can capture the vortex structure by using the particle lines with the moving camera with the similar speed as the vortex speed while SFLE cannot.

13.2.4.2 Image processing

The image processing of the velocity information of the flow field obtained from M-SFLE imaging mainly includes two tasks: particle trajectory identification and velocity calculation. The basic steps consist of (1) denoising and sharpening, (2) adaptive thresholding (OTSU), (3) removal of small particles, and (4) skeleton image processing, which are shown in Fig. 13.5 (Guo et al., 2020). Denoising can reduce the imaging error, and sharpening can clarify the edge of the particle trajectory. Adaptive thresholding means there is no need to set the threshold artificially. The threshold of the pixel is determined by the grayscale characteristics surrounding this pixel so that the image can keep more information after threshold segmentation. Based on adaptive thresholding, noises with smaller areas can be easily removed. For the last step, the particle trajectory has a certain width with a single pixel after the skeleton extraction, which makes the trajectory length and the velocity direction more accurate, especially for the curved path line. Therefore, in this study, this processed path line is named the pixelwise path line.

Fig. 13.6 shows an experimental image for a jet flow field and the corresponding result solved by the image processing. It can be seen in the comparison that the pixelwise path line distribution has good agreement with the distribution displayed in

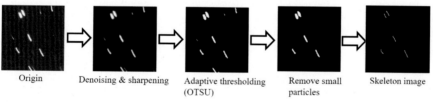

| Origin | Denoising & sharpening | Adaptive thresholding (OTSU) | Remove small particles | Skeleton image |

FIGURE 13.5

Image processing of M-SFLE.

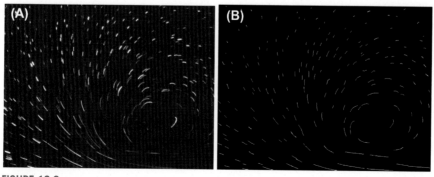

FIGURE 13.6

(A) Experimental image for a jet flow field and (B) result of image processing.

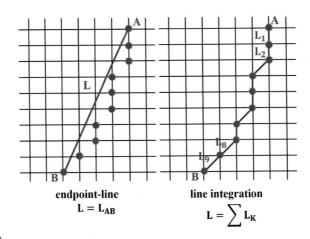

endpoint-line
$$L = L_{AB}$$

line integration
$$L = \sum L_K$$

FIGURE 13.7

The comparison of the total length of a pixelwise path line determined by endpoint-line and line integration functions.

the original image. All pixel points on a trajectory can be identified by the segmentation function. Endpoint-line and line integration are two segmentation functions, and the comparison of the results applying these two functions is shown in Fig. 13.7. It is clear that the endpoint-line function may lead to a relatively large error when solving the curved path line. However, for the line integration function, all the pixels on the pixelwise path line are arranged in order from the starting point to the ending point; then, each distance between two adjacent pixels can be calculated and summed up to the total length of a pixelwise path line.

13.2.5 Moving particle image velocimetry/particle tracing velocimetry

As outlined earlier, the M-SFLE method can intuitively obtain the evolution process of a vortex structure during a set time period, but the transient parameters cannot be obtained because of the long exposure time. Therefore, a nonintrusive optical velocimetry measuring method, PIV, is considered to measure the instantaneous velocity field for further quantitative analysis. For most 2D or 3D flow field measurements using the conventional PIV technique, several cameras should be arranged along the flow direction to spatially capture the motions of a vortex structure with a high velocity. Therefore, a moving PIV/PTV method is proposed. A continuous laser light and measuring system move synchronously at the same speed as the migratory vortex. Compared to M-SFLE, M-PIV/PTV employs a shorter exposure time, a higher tracer particle concentration, and the PIV cross-correlation algorithm and/or PTV tracking algorithm to achieve quantitative analysis. Consequently, M-PIV/PTV is a Lagrangian–Eulerian-type measurement that temporally and spatially captures the whole process of vortex generation and evolution in a turbulent boundary layer.

13.3 Submerged water jet experiments

Jet flow widely occurs in the fields of aerospace engineering, water conservancy and hydropower engineering, environmental engineering, chemical metallurgy, and energy machinery. Turbulent jet free flows, such as the fuel injection process in an internal combustion engine and the gas injection in a jet engine, commonly exist in nature and engineering technology.

The jet flow fluid injects into static surroundings with an initial velocity, forms shear layers, and exchanges momentum with the static fluid. This action makes the surrounding static fluid move with the jet fluid, which is called entrainment of the jet flow.

13.3.1 Experimental measuring system

An experimental measuring system of a submerged water jet, consisting of a water circulation system, a rectangular container, and an optical measuring system, is shown in Fig. 13.8. The rectangular container has a cross-sectional size of 40×50 mm and a length of 200 mm. In the middle of one end of the container, a jet nozzle is placed with an outer diameter of 2 mm and an inner diameter of 1.6 mm, and the other end is equipped with a water outlet. The size of the jet nozzle is extremely small compared to the size of the experimental container; thus, it can be considered to be a jet incident in an infinite space. In an experiment with a small Reynolds number, a syringe pump is used for the water supply, while a peristaltic pump is used for a larger Reynolds number. The light sheet is arranged perpendicularly at the central position in the spanwise direction. The camera is fixed on one

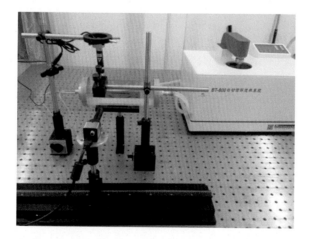

FIGURE 13.8

Measuring system of the submerged jet flow.

side of the container, which is normal to the sheet light. The tracer particle is PS (1 μm), and the particle density is 1.05 g/cm^3.

A 4x telecentric lens is used to measure the entrained structure, while a 1x lens is used to capture the core area of the jet. The camera has a resolution of 1600 $\times$ 1200, and a pixel size of 4.5 μm is selected for this study. The spatial resolution of the image is 4.5 $\times$ 4.5 μm when measured with a 1x lens, while it is 1.125 $\times$ 1.125 μm using a 4x magnification lens.

13.3.2 Measuring results by SFLE

The flow rate of the jet is 230 ml/min with a velocity of 2.07 m/s; thus, the Reynolds number is approximately 3294. The exposure time is set to 3 ms, and a 4x magnification lens is used. The FOV of the image is 1.80 $\times$ 1.35 mm. The fine structure of the vortex near the jet outlet is shown in Fig. 13.9. Due to the entrainment of the main jet flow, many small vortices with a size of several microns to tens of microns are generated once a large entrained vortex forms.

Fig. 13.10 shows the entrainment of the static surrounding fluid by jet flow. During the process, the fluid starts to rotate and increases the moving speed. As shown in the figure, the fluid accelerates from a linear motion to a spiral vortex motion of approximately 800 − 1000r/s with a distance of approximately 120 μm, and the velocity increases by 76 % from 33 mm/s to 58 mm/s within approximately 3 ms. The increase in rotation speed is much higher than that in flow rate. As the rotation speed increases, the rotation diameter of the vortex continues to increase. As shown in Fig. 13.11, the rotation diameter of the vortex increases by 150 % from approximately 16 μm to approximately 40 μm within a distance of approximately 50 μm.

Under the same experimental conditions, several singular vortex structures are observed by the SFLE measurement method. As shown in Fig. 13.12, the shapes of these vortices are similar to tennis rackets and headphones; thus, they can be named the "tennis racket" vortex and "headphone" vortex. It was found in the

FIGURE 13.9

Micron vortex structures near the jet exit.

FIGURE 13.10

Entrainment of the static surrounding fluid by jet flow.

FIGURE 13.11

Spiral vortex acceleration to the mainstream.

(A)　　　　　　　　　　**(B)**

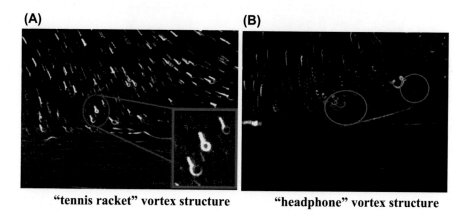

"tennis racket" vortex structure　　**"headphone" vortex structure**

FIGURE 13.12

Singular vortex structures in jet flow. (A) "tennis racket" vortex structure, (B) "headphone" vortex structure.

experiment that the "tennis racket" vortex was caused by fluid retardation of the entrained boundary layer. The "headphone vortex" was caused by the stretching and shearing of the jet stream along the mainstream direction.

A 1x lens is used to observe the structure of the jet core area, and the FOV is 7.2 mm × 5.4 mm. The Reynolds number is 292, and the exposure time is 5 ms. Thus, the structures in the jet flow field in the range of 25.00 mm − 34.69 mm

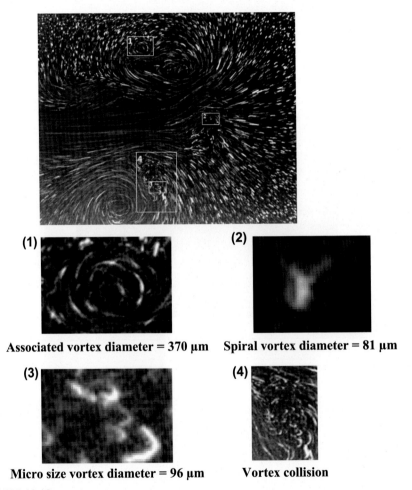

(1)

Associated vortex diameter = 370 μm

(2)

Spiral vortex diameter = 81 μm

(3)

Micro size vortex diameter = 96 μm

(4)

Vortex collision

FIGURE 13.13

Vortex structures in the core area of the jet flow with a distance of 25.00 mm − 34.69 mm from the nozzle. (1) Associated vortex diameter = 370 μm (2) Spiral vortex diameter = 81 μm. (3) Micro-size vortex diameter = 96 μm (4) Vortex collision.

from the nozzle outlet are measured. Fig. 13.13 shows the large vortex ring caused by the jet head. There are also three vortices with a medium size beside the large vortex ring and many small-scale vortices with turbulent characteristics.

13.3.3 Measuring results by SFME
13.3.3.1 Discrimination of flow direction
Fig. 13.14 shows the result obtained by SFME with four consecutive exposures. The exposure time parameters are 0.5 ms + 1 ms + 2 ms + 3 ms, and the exposure time interval is 0.3 ms. Fig. 13.14 shows the particle trajectories during entrainment. According to the exposure time and the tracer trace length in Fig. 13.14, the flow direction moves from the upper left to the lower right. It should be noted that the fluid is not linear or rotational but fluctuates while flowing diagonally downward.

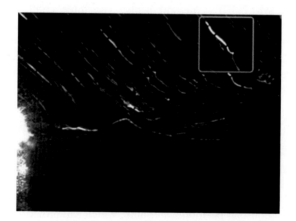

FIGURE 13.14

Image of jet entrainment obtained by SFME with four exposures in a single frame.

FIGURE 13.15

Image of jet entrainment obtained by SFME with three exposures in a single frame.

13.3.3.2 Flow acceleration measurement

Fig. 13.15 shows the entrainment obtained by the SFME method with three exposures. The velocity and acceleration can be obtained by processing the trajectories of the tracer particles marked in Fig. 13.15. The velocity of the particle marked No. 10 obtained from three trajectories with three exposures is $v_1 = 0.012\text{m/s}$, $v_2 = 0.027\text{m/s}$, $v_3 = 0.035\text{ m/s}$, and their accelerations are $a_1 = 2.612 \text{ m/s}^2$ and $a_2 = 1.737 \text{ m/s}^2$. Since the second-stage acceleration is less than the first-stage acceleration, it can be seen that the fluid in this area flows with reduced acceleration.

13.4 Plate turbulent boundary layer experiments

In this section, the coherent structure characteristics and the spatiotemporal evolution process of the turbulent boundary layer are experimentally studied by M-SFLE as well as M-PIV/PTV measurements. The three configurations of the measuring system include the measurement in (1) the streamwise-normal plane $(x - y)$, (2) at an angle of 53° between the light sheet and flow direction, and (3) at an angle of 135° between the light sheet and flow direction. Different measuring configurations can capture various structural characteristics of the turbulent boundary layer, such as an intact hairpin vortex, multilayer vortices, and vortex merging.

13.4.1 Measuring system and experimental validation

13.4.1.1 Measuring system

A schematic sketch of the measuring system and the experimental equipment is shown in Figs. 13.16 and 13.17, respectively. The experiment is conducted in a low-speed circulating water tunnel. The water pump can transport the water from the water tank to a constant position water tank, which can provide a constant inlet

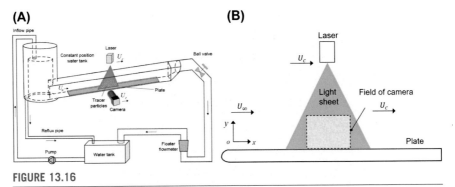

FIGURE 13.16

Schematic diagram of the measuring system.

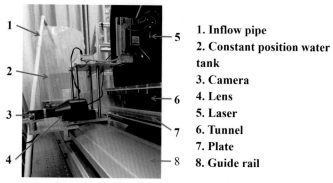

1. Inflow pipe
2. Constant position water tank
3. Camera
4. Lens
5. Laser
6. Tunnel
7. Plate
8. Guide rail

FIGURE 13.17

Measuring equipment.

pressure. Then, the water can be recycled after flowing through the ball valves and the floater flowmeter and finally back to the water tank.

The length of the water tunnel is 2200 mm, and the cross-sectional area of the tunnel is 80×80 mm^2. A plexiglass plate with a size of 1500 mm $\times$ 78 mm $\times$ 4.5 mm is positioned at the bottom of the water tunnel. The leading edge of the plate is set as an ellipse, with a ratio of 4 : 1 from the long axis to the short axis, to alleviate the disturbance of the incoming flow. In addition, a wire with a diameter of 4 mm is placed at a distance of 50 mm from the leading edge of the plate to accelerate the transition of the boundary layer. The tracer particles are polystyrene (PS) with an average diameter of 5 μm, density of 1.05 g/ml, and refractive index of 1.59. The light is a 450 nm wavelength continuous laser diode, which produces a light sheet with 1 mm thickness. Images with a FOV of 44×35 mm^2 are captured by a CMOS camera that has a resolution of 1280×1024 pixels and a pixel size of 4.8 μm, and the lens has a magnification factor of 0.14. The velocity of the main flow U_∞ can be calculated by Eq. (13.1) and should be approximately 100 mm/s when the flow flux is 1.6 m^3/h. The front view of the measuring field in the $x - y$ plane with detailed parameters is shown in Fig. 13.16B. The x and y axes indicate coordinates in the streamwise and normal directions, respectively. The speed of the camera system is U_c, and the dimensionless parameter of U_c is U^*, $U^* = U_c/U_\infty$. Using extensive tests, the most abundant vortex structures can be captured from the flow field by setting $U^* \in [0.8, 0.9]$. The measuring range in the flow direction is from $x = 400$ mm to $x = 1300$ mm, and the corresponding boundary layer thickness δ is 10 mm and 20 mm. The momentum thickness of the boundary layer θ is 0.97 mm and 1.94 mm at these two positions, and thus, the Reynolds numbers $Re_\theta = \theta U_\infty/\nu$ are 97 and 194, respectively, where ν is the coefficient of kinematic viscosity.

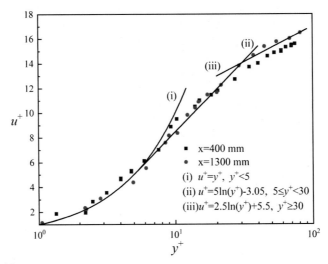

FIGURE 13.18

Dimensionless averaged-velocity profile of the experimental result at $x = 400$ mm and $x = 1300$ mm.

13.4.1.2 Experimental validation

To verify that the test region displays fully developed turbulence, the velocity profile of the boundary layer is measured by the SFLE method at $x = 400$ mm and $x = 1300$ mm. The normalized average velocity u^+ and the normalized normal distance from the wall y^+ are respectively calculated by $y^+ = yu_\tau/\nu$ and $u^+ = \bar{u}/u_\tau$, where the friction velocity u_τ is obtained by the wall shear stress τ_ω, which is calculated by the velocity gradient of the viscous sublayer:

$$u_\tau = \sqrt{\frac{\tau_\omega}{\rho}}, \ \tau_\omega = \mu \frac{\partial \bar{u}}{\partial y} \tag{13.2}$$

where ρ is the density and μ is the dynamic viscosity coefficient. Thus, u_τ is equal to 6.10 mm/s at $x = 400$ mm and equals 5.48 mm/s at $x = 1300$ mm. The results are compared in Fig. 13.18. The experimental result is in good agreement with the theory of log law, which means that the experimental method in this study is relatively accurate. The results of two different locations are similar to each other, which demonstrates that the turbulent boundary layer is about fully developed in the test region.

13.4.2 Hairpin vortex structure

To capture an intact hairpin vortex in the turbulent boundary layer, the light sheet is placed in the position shown in Fig. 13.19. The intact hairpin vortex was measured when the angle between the light sheet and flow direction was 53° and the camera

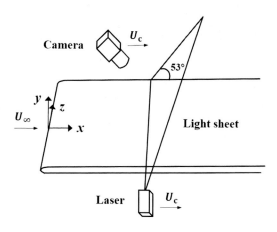

FIGURE 13.19

Measuring system with a light sheet placed with 53° inclination to the flow direction.

was placed normal to the light sheet. The moving speed of the measuring system is $U^* = 0.8$, the exposure time is 700 ms, the frame rate is 1.42 fps, and the laser power is 1.6 W.

Fig. 13.20A shows the coherent structures in the low Reynolds number but fully developed turbulent boundary layer obtained by the M-SFLE measurement, and Fig. 13.20B shows the particle path lines calculated by Liu's DNS data (2014). Nearly 30,000 seed points are placed, and the points are selected from one plane, which has the same direction as the light sheet. In Fig. 13.20A, the bright dots indicate the tracer particles moving mainly in the streamwise direction, and the long bright lines denote the tracer particles moving in the spanwise or wall-normal direction since a dot means the particle moving direction is orthogonal to the light sheet. The structure obtained in Fig. 13.20A is a hairpin structure. To demonstrate this result, the hairpin vortex is divided into five different sections: sections A and B represent the left and right legs, sections C and D represent the left and right necks, and section E represents the center of the hairpin vortex.

Fig. 13.20A shows that the bright lines in section A are nearly horizontal, and the tracer particles in section A moving to the left are obtained by the two consecutive frames. In the same way, the particles in section B move to the right. The bright lines in section C move toward the upper left, which means particles in section C move to the upper left. Similarly, the particles in section D move to the upper right. In section E, there are a quantity of bright dots instead of long bright lines, so that the particles in section E move opposite of the streamwise direction. The trajectories of tracer particles of five different sections obtained by the experiment have a similar trend as analyzed earlier. Apparently, the structure obtained in Fig. 13.20A is a hairpin structure. In addition, two vortex cores at the bottom of sections A and B are discovered, since the light sheet cuts through the left and right legs of the hairpin. The hairpin-like structure is also obtained by Liu's DNS result,

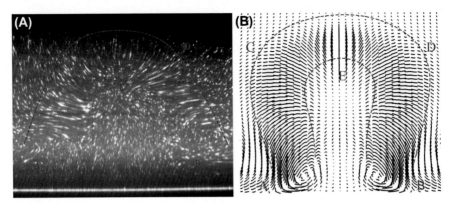

FIGURE 13.20

(A) Hairpin vortex structure obtained by the M-SFLE method. (B) Hairpin vortex structure obtained by DNS data.

as shown in Fig. 13.20B. The result is similar to that obtained by the M-SFLE measurement.

13.4.3 Multilayer vortices

When the angle between the light sheet and flow direction is changed to 135°, the light sheet is perpendicular to the rotation axis of the hairpin vortex leg; see the measuring system in Fig. 13.21. Therefore, in this section, the multilayer vortex structure with hairpin legs at different normal heights is discussed in detail.

Fig. 13.22A shows the M-SFLE experimental result with the light sheet placed at a 135° inclination to the flow direction, as illustrated in Fig. 13.21. Fig. 13.22B shows the particle trace calculated by Liu's DNS data with seed points selected from a plane that has the same direction as the light sheet. Fig. 13.22B is colored

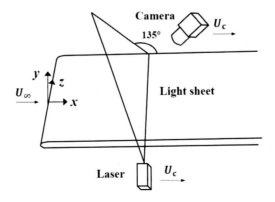

FIGURE 13.21

Measuring system with a light sheet placed with 135° inclination to the flow direction.

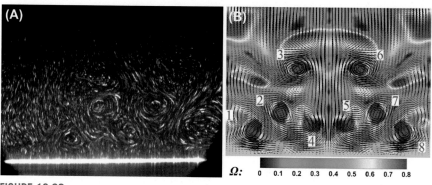

FIGURE 13.22

(A) Multilayer hairpin vortex legs obtained by the M-SFLE method. (B) Multilayer hairpin vortex legs obtained by DNS data colored by Ω.

by Omega method (Liu et al., 2016), and the red area represents the large value of Ω, which most likely can capture the vortex cores. The vortices denoted by the particle trace lines are in accordance with the vortex cores represented by the Ω contour. This shows that both the Ω method and particle trace lines are appropriate for identifying vortex structures. As shown in Fig. 13.22A, seven streamwise vortices are captured by the M-SFLE experiment, and these vortices are believed to be the legs of the hairpin vortices. One can see that the vortices are located in different layers. A total of three layers of vortices can be observed: vortices 1, 4, and 7 lie in the first layer near the wall, vortices 2 and 5 are observed in the middle layer, and vortices 3 and 6 are found on the top layer. This phenomenon is also observed by Liu's DNS result, as shown in Fig. 13.22B. Similar multilayer vortex structures are obtained by calculating particle trace lines. In total, 8 vortices are captured. Vortices 1 and 8 lie in the first layer near the wall, vortices 2, 4, 5, and 7 appear in the middle layer, and vortices 3 and 6 are on the top layer.

It is clear that the M-SFLE method is able to capture the coherent vortex structures of turbulent flow. The hairpin vortex and hairpin vortex legs are obtained successfully. There are multilayer hairpin vortices inside the low Reynolds number turbulent boundary layer, and Liu's DNS results qualitatively agree with the experimental result.

13.4.4 Single vortex generation

To further investigate the physical mechanisms of single vortex generation and evolution in a plate turbulent boundary layer, a quantitative analysis by M-PIV/PTV is provided in this section. Under the M-PIV/PTV measurement, the corresponding single-frame exposure time is 5 ms and the frame rate is 172.1 fps. The time interval between two frames is approximately 810 μs. Fig. 13.23 shows the contours of the 2D streamwise velocity (u^*) field with streamlines, $u^* = u/U_\infty$, the vorticity field with streamlines, and the Liutex field with streamlines at four typical moments. In the figures, the fluid flows along the x direction, x/δ_0 and y/δ_0 are employed to achieve dimensionless of the horizontal and vertical coordinates, δ_0 is the boundary layer thickness at $x = 400$ mm, which is 10 mm. Moreover, the dimensionless time $t^* = U_\infty t/\delta_0$ of the first frame is assumed to be $t^* = 0$ since it shows the primary stage of vortex generation; thus, the subsequent moments can be extracted as $t^* = 10.2$, $t^* = 16.1$, and $t^* = 28.6$. According to u^* field on the left-hand side of Fig. 13.23A, the upper high-speed fluid and the low-speed fluid at the bottom can be clearly observed to generate a shear layer by the Kelvin–Helmholtz (K-H) instability. This shear layer can also be identified as a connected fluid region with a relatively high concentration of vorticity $|\omega|$ from the contour of the vorticity magnitude $|\omega|$ in Fig. 13.23A.

Schlichting and Gersten (2010) noted in their study that the shear part is unstable in a boundary layer and isotropic flow when the Reynolds number is large enough, while the rigid rotation is linearly stable based on Taylor's stability (Liu et al., 2014). Therefore, the flow tends to transform from shear to rotation once it moves away from the wall. It thus can be discerned in Fig. 13.23B that a vortex with rotation motion is basically formed due to the synergistic action between the downward sweep of the high-speed streaks and the upward ejection of the low-speed fluids shown by the streamlines. As noted by Robinson, the association between regions of strong vorticity and actual vortices can be rather weak in the turbulent boundary layer, especially in the near-wall region (Robinson, 1989; Robinson et al., 1990). This means that some nonrotational areas can also be polluted by higher vorticity; see the vorticity magnitude contour in Fig. 13.23C. With respect to this issue, Liutex, which was proposed by Liu et al. (2018), is applied to accurately capture the pure rotational region in terms of the size and intensity of a vortex and to predict the evolution, including the generation and dissipation process, of a single vortex (Gao et al., 2018; Liu et al., 2018; Dong et al., 2019; Wang et al., 2019). The Liutex magnitude (R) field is thus shown on the right-hand side of Fig. 13.23.

In Fig. 13.23C at $t^* = 16.1$, under a continuous roll-up of the surrounding shear and momentum transfer, the vortex rotates clockwise in an approximate circle and provides the strongest intensity of the rotational motion with the highest Liutex magnitude. It should be noted that two vorticity concentrated regions marked 1 and 2 are observed in the middle figure; however, there is actually only one vortex that can be detected by a spiral of streamlines, which indicates that the area where the vorticity is congregated may not be a vortex if the deformation is also very large. Therefore, compared with the distribution of $|\omega|$, a vortex can be well determined by the region of $R > 0$, and the vortex core can be exactly defined as the concentration of Liutex, which can be physically regarded as the rigid rotation region of the rotational fluid.

In Fig. 13.23D at $t^* = 28.6$, since the near-wall shear weakens, the vortex may slow down when moving downstream and may even leave the field of view. $|\omega|$ and R correspondingly decay in the core region but without a distinct size reduction. Actually, it can only become weaker in terms of the intensity due to the dissipation

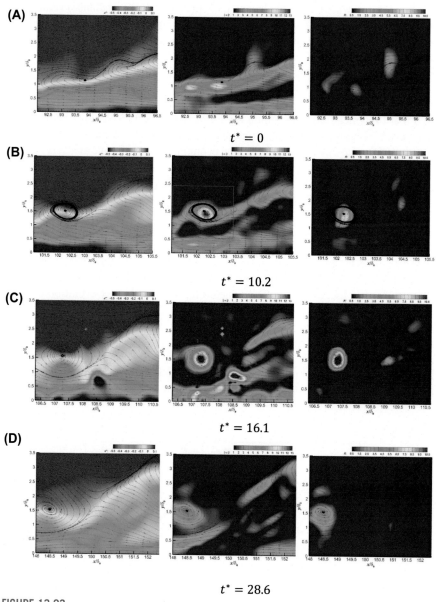

FIGURE 13.23

Process of vortex structure generation and evolution by streamwise velocity u^* (left), vorticity magnitude $|\omega|$ (center), and Liutex magnitude R (right). (A) $t^* = 0$, (B) $t^* = 10.2$, (C) $t^* = 16.1$, (D).$t^* = 28.6$

when traveling downstream. This means that small vortices (turbulence) cannot be generated by a vortex breakdown, and a vortex with any (large or small) scale can only be generated by the shear layer, without exception. This confirms that "shear layer instability is the mother of turbulence" according to Liu et al. (2014).

It is mentioned earlier that a vortex can be well determined by the area of $R > 0$, and the vortex core can be precisely defined as the concentration of Liutex instead of vorticity. From the earlier experimental measurement and observation, it is clearly found that Liutex can always capture the vortices, but vorticity could mistreat the shear as vortex although it can sometimes capture the vortices. To further understand the process of vortex generation as well as evolution, a quantitative analysis is carried out by collecting the Liutex magnitude integration ($R_{Int} = \int R dS$) over the vortex areaΔS, where the vortex region is defined as $R > 0$. Fig. 13.24 presents the Liutex magnitude integral R_{Int} as well as the 2D scaling parameter ΔS changing with time t during the entire vortex evolution process. It is clearly seen in Fig. 13.24 that the variation of the Liutex magnitude integral R_{Int} basically occurs in three stages: the ascent stage $(0 < t^* < 16.1)$, equilibrium stage $(16.1 < t^* < 23.1)$, and descent stage $(23.1 < t^* < 39.8)$. It should be noted that the vortex starts to move left out of the FOV at $t^* = 32.1$ in the descent stage, and a steeper slope is found at $t^* = 32.1$. Therefore, only the vortex evolution during $0 \sim 32.1$ is considered for the following discussion.

It seems to be more reasonable to quantitatively divide a vortex evolution into four states, which are marked as I, II, III, and IV in Fig. 13.24, by considering both the rotation strength and the core size of the vortex. During State I $(0 < t^* < 6.2)$, a rotation motion of a shear layer initially forms with the approximate linear increase of both the rotation strength and the core size; thus, State I is

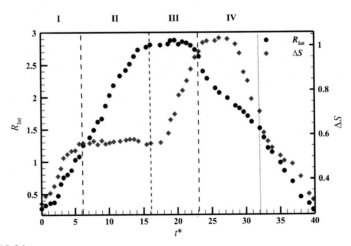

FIGURE 13.24

Temporal distributions of R_{Int} and ΔS during the whole process of vortex generation and evolution.

denoted as the "synchronous linear segment (SL)." Then, the vortex continues to increase its rotation strength but remains constant and stable in its core size in State II ($6.2 < t^* < 16.1$), which is thus treated as the "absolute enhancement segment (AE)." In contrast to State II, the rotation intensity of the vortex stops increasing and remains steady with the core size rapidly enlarging in State III ($16.1 < t^* < 23.1$). In this state, the local Liutex magnitude is diffused instead of dissipated; thus, State III is classified as an "absolute diffusion segment (AD)." It is observed that the rotation strength is sharply weakened and that the core size is reduced after a temporary equilibrium during State IV ($23.1 < t^* < 32.1$). However, the Liutex magnitude takes the characteristics of dissipation in this state, which is different from State III. Therefore, the final state is characterized as a "skewing dissipation segment (SD)." It can be concluded that the physical mechanism of vortex generation and evolution in a turbulent boundary layer can be summarized as a four-dominant-state course: SL-AE-AD-SD.

13.4.5 Vortex merging

In this section, M-SFLE measuring is utilized to study vortex merging in a plate turbulent boundary layer. Fig. 13.25 (Guo et al., 2020) shows the experimental image (left), streamwise velocity (u^*) contour with streamlines (center), and R contour with streamlines (right) of the vortex merging at $U^* = 0.9$. The dimensionless time $t^* = U_\infty t/\delta_0$ of the first frame in Fig. 13.25A is assumed to be $t^* = 0$; thus, the subsequent moments can be extracted as $t^* = 3$ and $t^* = 5$. It is clear that the concentration of Liutex (R) has good agreement with the rotation motion in the experimental image as well as the spiral streamlines in the middle of Fig. 13.25, which indicates that Liutex can capture the pure rotation of fluid, although there are some small errors observed in the nonrotational region due to the interpolation error of the gradient parameters caused by experimental data. The vortex merging progress can be revealed by the results at three typical moments.

In Fig. 13.25, three vortices with clockwise rotation are marked by A, B, and C in the figure. It appears that vortex C is fixed along the flow direction, which indicates that the camera has the same moving speed as vortex C. Meanwhile, vortex merging occurs on vortices A and B. In Fig. 13.25B, vortex A tends to be close to vortex B due to the velocity difference in the streamwise direction, and these two vortices undergo clockwise revolution since both the streamwise and normal directions have velocity differences. In this situation, a new vortex D with clockwise rotation forms; see Fig. 13.25C.

The Liutex magnitude integration, which is defined as $R_{Int} = \int R dS$ in Section 13.4.4, is used in this section to track the vortex core size as well as the rotation intensity during the merging process of vortices A and B in the turbulent boundary layer (see Fig. 13.26). In the figure, three stages marked I ($0 < t^* < 2$), II ($2 < t^* < 4$), and III ($4 < t^* < 5$) can be observed. In the first stage, vortex A increases its R_{Int} by getting close to vortex B and absorbing its energy, which thus leads to a

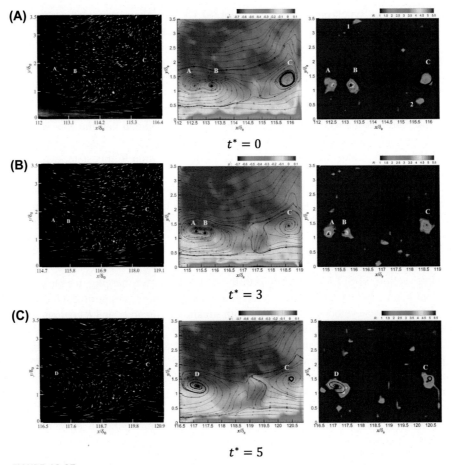

$t^* = 0$

$t^* = 3$

$t^* = 5$

FIGURE 13.25

Vortex merging at $U^* = 0.9$ shown by the experimental image (left), streamwise velocity (u^*) contour with streamlines (center), and R contour with streamlines (right). (A) $t^* = 0$, (B) $t^* = 3$, (C).$t^* = 5$

Reproduced from Guo, Y.A., Dong, X.R., Cai, X.S., Zhou, W., 2020. Experimental studies on vortices merging based on MSFLE and Liutex. Acta Aerodynamica Sinica 38 (3), 432–440. with the permission of Acta Aerodynamica Sinica.

decrease in the R_{Int} of vortex B. However, both A and B dissipate their energy due to the shear and rotation with each other during stage II, and this causes an increasing R_{Int} of vortex B and a decreasing R_{Int} of vortex A. At the initial moment of the first stage, vortex A has a rotation intensity of $R_{Int} = 0.43$ and a size of $\Delta S = 0.26$, while vortex B has a rotation intensity of $R_{Int} = 0.48$ and a size of $\Delta S = 0.27$. They finally merge into a new vortex D that has a higher rotation intensity ($R_{Int} = 0.85$) and larger size ($\Delta S = 0.47$). This indicates that the sizes of these original vortices are

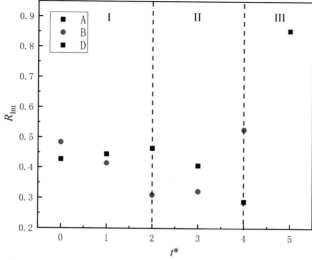

FIGURE 13.26

Distribution of Liutex magnitude integration with time during the vortex merging process.

Reproduced from Guo, Y.A., Dong, X.R., Cai, X.S., Zhou, W., 2020. Experimental studies on vortices merging based on MSFLE and Liutex. Acta Aerodynamica Sinica 38 (3), 432–440. with the permission of Acta Aerodynamica Sinica.

basically the same, while the size and rotation intensity of the newly formed vortex due to merging are approximately the sum of these two vortices at the initial time. The merge of two corotational vortices to a stronger vortex is well recognized by experiment.

13.5 Summary

The four image measuring methods proposed by Cai's group are introduced in this chapter, including SFLE, SFME, M-SFLE, and moving PIV (M-PIV/PTV). All of the methods are cost-effective, and the measurement apparatus is very simple. The SFLE measuring method can clearly show the moving trajectory of particles in a single frame with a long exposure; the highest spatial resolution of the SFLE can reach 1.125 μm in the submerged water jet experiment of this chapter. Moreover, the flow direction can be obtained from two consecutive frames of images; meanwhile, the velocity can be obtained from the magnification of the lens, the length of the trajectory in the image, and the exposure time. Thus, SFLE effectively avoids engaging the complex cross-correlation algorithm of PIV. Compared with SFLE, the SFME measuring method can identify the flow direction of the particles and measure the flow field parameters such as transient velocity and acceleration fields only in a single frame by employing multiple exposures. Both SFLE and

SFME are Eulerian-type measurements, and the camera is fixed in the experiment. However, a moving measuring system is applied in MSFLE and M-PIV/PTV measurements so that the vortex structure characteristics and the spatiotemporal evolution of the flow field can be measured more intuitively from Lagrangian and Lagrangian/Euler perspectives.

These four image measuring methods were applied to the experimental study on the vortex structures in the jet entrainment boundary layer and a plate turbulent boundary layer. The microscale structures within the entrained boundary layer and some singular vortices, named the "tennis racket" vortex and "headphone" vortex in this chapter, are captured in a submerged water jet flow by using SFLE and SFME. The physical mechanism of these singular vortices will be further analyzed. The experimental results of jet flow show that the physical mechanism of the turbulent fluctuation is actually the random movement of the vortex. The physical mechanism of turbulence fluctuation may be that the vortices of different scales randomly pass through the tiny space with different rotation velocities, mainstream moving velocities, and different positions and directions. Turbulence fluctuation frequency, fluctuation intensity, fluctuation velocity, and other parameters actually describe the rotation velocities, rotation radius, rotation phase, moving velocities, and random motion of vortices.

The spatiotemporal evolution of vortex structures in a plate turbulent boundary layer is experimentally studied by using the MSFLE and M-PIV/PTV methods. The vortex structure and its evolution process, such as the generation of a single hairpin vortex, multilayer vortices near the wall, and vortex merging in a turbulent boundary layer, were obtained by three measuring approaches: the streamwise-normal plane, an angle set as 53°, and an angle set as 135° between the light sheet and flow direction. Moreover, the Liutex theory was adopted to characterize the critical vortex core boundary and the accurate rotation strength of the local fluid rotation. M-SFLE and M-PIV/PTV measuring methods combined with Liutex can be applied to the visualization and quantification of vortex structure in a turbulent boundary layer, which provides a new concept for experimental turbulence studies.

References

Abdel-Raouf, E., Sharif, M.A., Baker, J., 2017. Impulsively started, steady and pulsated annular inflows. Fluid Dynam. Res. 49 (2), 025511.

Adrian, R.J., 2007. Hairpin vortex organization in wall turbulence. Phys. Fluids 19, 041301.

Anderson, B., 2016. The Statistical Nature of Unsteady Flows. 9th Annual Shock Wave/ Boundary Layer Interaction (SWBLI) Technical Interchange Meeting.

Anderson, B., Tinapple, J., Surber, L., 2006. Optimal Control of Shock Wave Turbulent Boundary Layer Interactions Using Micro-array Actuation. AIAA paper, AIAA 3197-2006.

Arpe, J., Nicolet, C., Avellan, F., 2009. Experimental evidence of hydroacoustic pressure waves in a Francis turbine elbow draft tube for low discharge conditions. J. Fluid Eng. 131 (8), 081102−081109.

Atkinson, C., Chumakov, S., Bermejomoreno, I., Soria, J., 2012. Lagrangian evolution of the invariants of the velocity gradient tensor in a turbulent boundary layer. Phys. Fluids 24, 65−75.

Babiano, C., Basdevant, B., Legras, B., Sadourny, R., 1987. Vorticity and passive-scalar dynamics in two-dimensional turbulence. J. Fluid Mech. 183, 379−397.

Babinsky, H., Li, Y., Pitt Ford, C.W., 2009. Microramp control of supersonic oblique shock-wave/boundary-layer interactions. AIAA J. 47, 668−675.

Bake, S., Meyer, D., Rist, U., 2002. Turbulence mechanism in Klebanoff transition: a quantitative comparison of experiment and direct numerical simulation. J. Fluid Mech. 459, 217−243.

Banks, D.C., Singer, B.A., 1994. Vortex tubes in turbulent flows: identication, representation, reconstruction. In: Proceedings of the Conference on Visualization '94, Los Alamitos, CA, USA.

Baron, A., Quadrio, M., 1995. Turbulent drag reduction by spanwise wall oscillations. Appl. Sci. Res. 55 (4), 311−326.

Batchelor, G.K., 2000. Introduction to Fluid Mechanics. Cambridge University Press, Cambridge, England.

Behara, S., Mittal, S., 2010. Wake transition in flow past a circular cylinder. Phys. Fluids 22, 114104.

Bergan, C., Goyal, R., Cervantes, M.J., Dahlhaug, O.G., 2016. Experimental investigation of a high head model Francis turbine during steady-state operation at off-design conditions. IOP Conf. Ser. Earth Environ. Sci. 49, 062018.

Bergmann, M., Cordier, L., 2008. Optimal control of the cylinder wake in the laminar regime by trust-region methods and POD reduced-order models. J. Comput. Phys. 227 (16), 7813−7840.

Berkooz, G., Holmes, P., Lumley, J.L., 1993. The proper orthogonal decomposition in the analysis of turbulent flows. Annu. Rev. Fluid Mech. 25 (1), 539−575.

Berry, M.G., Magstadt, A.S., Glauser, M.N., 2017. Application of POD on time-resolved schlieren in supersonic multi-stream rectangular jets. Phys. Fluids 29 (2), 020706.

Billant, P., Brancher, P., Chomaz, J.M., 1999. Three-dimensional stability of a vortex pair. Phys. Fluids 11 (8), 2069−2077.

Bohannon, K.S., 2006. Passive Flow Control on Civil Aircraft Flaps Using Sub-boundary Layer Vortex Generators in the AWIATOR Programme. AIAA Paper, AIAA 2006-2858.

Brachet, M.E., Meneguzzi, M., Politano, H., Sulem, P.L., 1988. The dynamics of freely decaying two-dimensional turbulence. J. Fluid Mech. 194, 333−349.

415

Braza, M., Faghani, D., Persillon, H., 2001. Successive stages and the role of natural vortex dislocations in three-dimensional wake transition. J. Fluid Mech. 439, 1–41.

Brooke, J.W., Hanratty, T.J., 1993. Origin of turbulence-producing eddies in a channel flow. Phys. Fluid. Fluid Dynam. 5, 1011–1022.

Buzica, A., Breitsamter, C., 2019. Turbulent and transitional flow around the AVT-183 diamond wing. Aero. Sci. Technol. 92, 520–535.

Camarri, S., Giannetti, F., 2010. Effect of confinement on three-dimensional stability in the wake of a circular cylinder. J. Fluid Mech. 642, 477–487.

Cao, B., Xu, H., 2018. Re-understanding the law-of-the-wall for wall-bounded turbulence based on in-depth investigation of DNS data. Acta Mech. Sin. 34 (5), 793–911.

Carmo, B.S., Meneghini, J.R., 2006. Numerical investigation of the flow around two circular cylinders in tandem. J. Fluid Struct. 22, 979–988.

Carmo, B.S., Meneghini, J.R., 2010. Secondary instabilities in the flow around two circular cylinders in tandem. J. Fluid Mech. 644, 395–431.

Cavar, D., Meyer, K.E., 2012. LES of turbulent jet in cross flow: part 2-POD analysis and identification of coherent structures. Int. J. Heat Fluid Flow 36, 35–46.

Chakraborty, P., Balachandar, S., Adrian, R.J., 2005. On the relationships between local vortex identification schemes. J. Fluid Mech. 535, 189–214.

Chaplin, J.R., Bearman, P.W., Cheng, Y., Fontaine, E., Graham, J.M.R., Herfjord, K., Huera Huarte, F.J., Isherwood, M., Lambrakos, K., Larsen, C.M., Meneghini, J.R., Moe, G., Pattenden, R.J., Triantafyllou, M.S., Willden, R.H.J., 2005a. Blind predictions of laboratory measurements of vortex-induced vibrations of a tension riser. J. Fluid Struct. 21 (1), 25–40.

Chaplin, J.R., Bearman, P.W., Huera Huarte, F.J., Pattendena, R.J., 2005b. Laboratory measurements of vortex-induced vibrations of a vertical tension riser in a stepped current. J. Fluid Struct. 21 (1), 3–24.

Charkrit, S., Dong, X., Liu, C., 2019. POD Analysis of Losing Symmetry in Late Flow Transition. AIAA paper 2019-1870.

Chashechkin, Y.D., 1993. Visualization and identification of vortex structures in stratified wakes. In: Bonnet, J.P., Glauser, M.N. (Eds.), Eddy Structure Identification in Free Turbulent Shear Flows, Fluid Mechanics and Its Applications, vol. 21. Springer, Dordrecht.

Chen, M., 2002. Fundamentals of Viscous Fluid Dynamics. Higher Education Press, Beijing (in Chinese).

Chen, J., Cai, X., 2015. Research on the single frame imaging method for measuring multi-parameter fields in flow field. J. Exp. Fluid Mechan. 29 (6), 67–73 (in Chinese).

Chen, Z., Kong, X.P., Li, T.J., Yang, K., 2019. MVG control on the supersonic compression ramp flow. In: 31st International Symposium on Shock Waves, 1, pp. 1033–1040.

Chen, C., Sun, D.J., 2010. Numerical investigation of a swirling flow under the optimal perturbation. J. Hydrodyn. 22 (5Suppl. l), 237–241.

Chevillard, L., Meneveau, C., 2006. Lagrangian dynamics and statistical geometric structure of turbulence. Phys. Rev. Lett. 97, 174501.

Choi, K.S., 2002. Near-wall structure of turbulent boundary layer with spanwise-wall oscillation. Phys. Fluids 14, 2530.

Choi, K.S., Clayton, B.R., 2001. The mechanism of turbulent drag reduction with wall oscillation. Int. J. Heat Fluid Flow 22 (1), 1–9.

Chong, M.S., Perry, A.E., Cantwell, B.J., 1990. A general classification of three-dimensional flow fields. Phys. Fluids 2 (5), 765–777.

Clemens, N.T., Narayanaswamy, V., 2009. Shock/Turbulent Boundary Layer Interactions: Review of Recent Work on Sources of Unsteadiness. AIAA paper, AIAA 2009-3710.

Dallmann, U., 1983. Topological structures of three-dimensional vortex flow separation. In: 16th AIAA Fluid and Plasma Dynamics Conference, Danvers, MA.

Deich, M.E., Zaryakin, A.E., Fillipov, G.A., Zatsepin, M.F., 1960. Method of increasing the efficiency of turbine stages and shon blades. Teploenergetika 2, 240–254.

Délery, J., Dussauge, J.P., 2009. Some physical aspects of shock wave/boundary layer interactions. Shock Waves 19 (6), 453–468.

Deng, J., Ren, A.L., Zou, J.F., Shao, X.M., 2006. Three-dimensional flow around two circular cylinders in tandem arrangement. Fluid Dynam. Res. 38, 386–404.

Dennis, D.J.C., Nickels, T.B., 2011. Experimental measurement of large-scale three-dimensional structures in a turbulent boundary layer. Part 1. Vortex packets. J. Fluid Mech. 673, 180–217.

Denton, J.D., 1993. Loss mechanisms in turbomachines. J. Turbomach. 115, 621–656.

Dhanak, M.R., Si, C., 1999. On reduction of turbulent wall friction through spanwise wall oscillations. J. Fluid Mech. 383, 175–195.

Dolling, D.S., 2001. Fifty years of shock-wave/boundary-layer interaction research. AIAA J. 39 (8), 1517–1531.

Dong, X., Chen, Y., Dong, G., Liu, Y., 2017a. Study on wake structure characteristics of a slotted micro-ramp with large-eddy simulation. Fluid Dynam. Res. 49 (3), 035507.

Dong, Y., Yang, Y., Liu, C., 2017b. DNS Study on Three Vortex Identification Methods, 55th AIAA Aerospace Sciences Meeting. AIAA paper 2017-0137.

Dong, X., Dong, G., Liu, C., 2018a. Study on vorticity structure in late flow transition. Phys. Fluids 30 (10), 104108.

Dong, X., Tian, S., Liu, C., 2018b. Correlation analysis on volume vorticity and vortex in late boundary layer transition. Phys. Fluids 30 (1), 014105.

Dong, X., Wang, Y., Chen, X., Dong, Y., Zhang, Y., Liu, C., 2018c. Determination of epsilon for Omega vortex identification method. J. Hydrodyn. 30, 541–548.

Dong, X., Yan, Y., Yang, Y., Dong, G., Liu, C., 2018d. Spectrum study on unsteadiness of shock wave-vortex ring interaction. Phys. Fluids 30, 056101.

Dong, X., Charkrit, S., Troung, X., Liu, C., 2019a. POD Study on Vortex Structures in MVG Wake. AIAA paper 2019-1136.

Dong, X., Gao, Y., Liu, C., 2019b. New normalized Rortex/vortex identification method. Phys. Fluids 31, 011701.

Dong, X., Cai, X., Dong, Y., Liu, C., 2020. POD analysis on vortical structures in MVG wake by Liutex core line identification. J. Hydrodyn. 32, 497–509.

Drouot, R., 1976. Définition D'un Transport Associé À Un Modèle de Fluide du Deuxième Ordre. In: Comparaison de diverses lois de comportement, Comptes rendus de l'Académie des Sciences, Série A, vol. 282, pp. 923–926.

Duchon, J., Robert, R., 2000. Inertial energy dissipation for weak solutions of incompressible Euler and Navier-Stokes equations. Nonlinearity 13, 249–255.

Duggleby, A., Ball, K.S., Paul, M.R., Ficher, P.F., 2007a. Dynamical eigenfunction decomposition of turbulent pipe flow. J. Turbul. 8, N43.

Duggleby, A., Ball, K.S., Paul, M.R., 2007b. The effect of spanwise wall oscillation on turbulent pipe flow structures resulting in drag reduction. Phys. Fluids 19, 125107.

Durcros, F., Comte, P., Lesieur, M., 1996. Large-eddy simulation of transition to turbulence in a boundary layer developing spatially over a flat plate. J. Fluid Mech. 326, 1–36.

Eitel-Amor, G., Örlü, R., Schlatter, P., Flores, O., 2015. Hairpin vortices in turbulent boundary layers. Phys. Fluids 27, 025108.

Epps, B., 2017. Review of Vortex Identification Methods. AIAA 2017-0989.

Erengil, M.E., Dolling, D.S., 1991. Correlation of separation shock motion with pressure fluctuations inthe incoming boundary layer. AIAA J. 29 (11), 1868–1877.

Falco, R.E., 1977. Coherent motions in the outer region of turbulent boundary layers. Phys. Fluid. 20, S124.

Felli, M., Falchi, M., 2018. Propeller wake evolution mechanisms in oblique flow conditions. J. Fluid Mech. 845, 520–559.

Feynman, R.F., 1955. The process. In: Gorter, C.J. (Ed.), Progress in Low Temperature Physics, vol. 1. North Holland Publishing Co, Amsterdam.

Filippov, G.A., Wang, Z.Q., 1963. The calculation of axial symmetric flow in an turbine stage with small ratio of diameter to blade length. J. Moscow Power Inst. 47, 63–78.

Finnigan, T., Roddier, D., 2007. Spar VIM model tests at super- critical Reynolds numbers. In: Proceedings of the 26th International Conference on Offshore Mechanics and Arctic Engineering, San Diego, California, USA, 3, pp. 731–740.

Fouladi, F., Henshaw, P., Ting, D.S.K., Ray, S., 2017. Flat plate convection heat transfer enhancement via a square rib. Int. J. Heat Mass Tran. 104, 1202–1216.

Frenander, H., Nordström, J., 2017. Constructing non-reflecting boundary conditions using summation-by-parts in time. J. Comput. Phys. 331, 38–48.

Freno, B.A., Cizmas, P., 2014. A proper orthogonal decomposition method for nonlinear flows with deforming meshes. Int. J. Heat Fluid Flow 50, 145–159.

Fuenteso, U.V., 2007. On the topology of vortex lines and tubes. J. Fluid Mech. 584, 147–156.

Galmiche, B., Mazellier, N., Halter, F., Foucher, F., 2014. Turbulence characterization of a high-pressure high-temperature fan-stirred combustion vessel using LDV, PIV and TR-PIV measurements. Exp. Fluid 55 (1), 1636.

Ganapathisubramani, B., Clemens, N.T., Dolling, D.S., 2007. Effects of upstream boundary layer on the unsteadiness of shock-induced separation. J. Fluid Mech. 585, 369–394.

Ganapathisubramani, B., Clemens, N.T., Dolling, D.S., 2009. Low-frequency dynamics of shock-induced separation in a compression ramp interaction. J. Fluid Mech. 636, 397–425.

Gao, Y., Liu, C., 2018. Rortex and comparison with eigenvalue-based vortex identification criteria. Phys. Fluids 30, 085107.

Gao, Y., Liu, C., 2019. Rortex based velocity gradient tensor decomposition. Phys. Fluids 31 (1), 011704.

Gao, Q., Ortiz-Dueñas, C., Longmire, E.K., 2011. Analysis of vortex populations in turbulent wall-bounded flows. J. Fluid Mech. 678, 87–123.

Gao, Q., Ortiz-Dueñas, C., Longmire, E.K., 2013. Evolution of coherent structures in turbulent boundary layers based on moving tomographic PIV. Exp. Fluid 54 (12), 1625.

Gao, Y., Liu, J., Yu, Y., Liu, C., 2019. A Liutex based definition and identification of vortex rotation axis line. J. Hydrodyn. 31, 445–454.

Garnier, E., Sagaut, P., Deville, M., 2002. Large eddy simulation of shock/boundary-layer interaction. AIAA J. 40 (10), 1935–1944.

Gavrilov, A.A., Sentyabov, A.V., Dekterev, A.A., Hanjali, K., 2017. Vortical structures and pressure pulsations in draft tube of a Francis-99 turbine at PL: RANS and hybrid RANS/LES analysis. Int. J. Heat Fluid Flow 63, 158–171.

Gendrich, C.P., Koochesfahani, M.M., 1996. A spatial correlation technique for estimating velocity fields using molecular tagging velocimetry (MTV). Exp. Fluid 22 (1), 67–77.

Ghosh, S., Choi, J.I., Edwards, J.R., 2008. RANS and Hybrid LES/RANS Simulations of the Effects of Micro Vortex Generators Using Immersed Boundary Methods. AIAA paper, AIAA 3728-2008.

Golub, G., Van Loan, C., 2012. Matrix Computations, fourth ed. Johns Hopkins University Press, Baltimore.

Goyal, R., Trivedi, C., Gandhi, B.K., Cervantes, M.J., 2018. Numerical simulation and validation of a high head model Francis turbine at PL operating condition. J. Inst. Eng. 99, 557–570.

Green, S.I., 1995. Fluid Vortices. Kluwer Academic Publishers, Dordrecht.

Gui, N., Fan, J.R., Chen, S., 2010a. Numerical study of particle-particle collision in swirling jets: a DEM-DNS coupling simulation. Chem. Eng. Sci. 65 (10), 3268–3278.

Gui, N., Fan, J., Chen, S., 2010b. Numerical study of particle-vortex interaction and turbulence modulation in swirling jets. Phys. Rev. 82 (2), 056323.

Gui, N., Fan, J.R., Zhou, Z., 2010c. Particle statistics in a gas-solid coaxial strongly swirling flow: a direct numerical simulation. Int. J. Multiphas. Flow 36 (3), 234–243.

Gui, N., Ge, L., Cheng, P., Yang, X., Tu, J., Jiang, S., 2019a. Comparative assessment and analysis of Rortex vortex in swirling jets. J. Hydrodyn. 31 (3), 495–503.

Gui, N., Qi, H., Ge, L., Cheng, P., Wu, H., Yang, X., Tu, J., Jiang, S., 2019b. Analysis and correlation of fluid acceleration with vorticity and Liutex (Rortex) in swirling jets. J. Hydrodyn. 31, 864–872.

Gunes, H., 2004. Proper orthogonal decomposition reconstruction of a transitional boundary layer with and without control. Phys. Fluids 16, 2763.

Guo, Y.A., Dong, X.R., Cai, X.S., Zhou, W., 2020. Experimental studies on vortices merging based on MSFLE and Liutex. Acta Aerodyn. Sin. 38 (3), 432–440 (in Chinese).

Haller, G., Hadjighasem, A., Farazmand, M., Huhn, F., 2016. Defining coherent vortices objectively from the vorticity. J. Fluid Mech. 795, 136–173.

Haller, G., 2005. An objective definition of a vortex. J. Fluid Mech. 525, 1–26.

Haller, G., 2015. Lagrangian coherent structures. Annu. Rev. Fluid Mech. 47, 137–162.

Hama, F.R., 1960. Boundary-layer transition induced by a vibrating ribbon on a flat plate. In: Proceedings of the 1960 Heat Transfer and Fluid Mechanics Institute. Stanford University Press, Palo Alto, CA, pp. 92–105.

Haimes, R.A.K.D., 2000. On the Velocity Gradient Tensor and Fluid Feature and Fluid Feature Extraction. AIAA Paper, No. 99-3288.

Hama, F.R., Nutant, J., 1963. Detailed flow-field observations in the transition process in a thick boundary layer. In: Proceedings of the 1963 Heat Transfer and Fluid Mechanics Institute. Stanford University Press, Palo Alto, CA, pp. 77–93.

Head, M., Bandyopadhyay, P., 1981. New aspects of turbulent boundary-layer structure. J. Fluid Mech. 107, 297–338.

Hellström, L., Ganapathisubramani, B., Smits, A.J., 2016. Coherent structures in transitional pipe flow. Phys. Rev. Fluids 1, 024403.

Hellström, L., Smits, A.J., 2017. Structure identification in pipe flow using proper orthogonal decomposition. Philos. Trans. R Soc. A 375, 20160086.

Helmholtz, H., 1858. Über Integrale der hydrodynamischen Gleichungen, welche den Wirbelbewegungen entsprechen. J. für die Reine Angewandte Math. (Crelle's J.) 55, 22–25 (in German).

Hiroki, Y.K.H., Yukihiro, I., Kensuke, T., Yu, M., Tomohide, N., 2016. Micro-molecular tagging velocimetry of internal gaseous flow. Microfluid. Nanofluid. 20 (2), 32.

Holden, H.A., Babinsky, H., 2007. Effect of microvortex generators on separated normal shock/boundary layer interactions. J. Aircraft 44 (1), 170–174.

Hosseinverdi, S., Fasel, H., 2019. Numerical investigation of laminar-turbulent transition in laminar separation bubbles: the effect of free-stream turbulence. J. Fluid Mech. 858, 714–759.

Hu, Y., Bi, W., Li, S., She, Z., 2017. β – Distribution for Reynolds stress and turbulent heat flux in relaxation turbulent boundary layer of compression ramp. Sci. China Phys. Mech. Astron. 60 (12), 124711.

Hunt, J.C.R., Wary, A.A., Moin, P., 1988. Eddies, streams, convergence zones in turbulent flows. In: Center for Turbulent Research Report CTR-S88, pp. 193–208.

Hussain, F., 1986. Coherent structures and turbulence. J. Fluid Mech. 173, 303–356.

Hutchins, N., Hambleton, W.T., Marusic, I., 2005. Inclined cross-stream stereo particle image velocimetry measurements in turbulent boundary layers. J. Fluid Mech. 541, 21–54.

Jeong, J., Hussain, F., 1995. On the identification of a vortex. J. Fluid Mech. 285, 69–94.

Jiménez, J., 2015. Direct detection of linearized bursts in turbulence. Phys. Fluids 27, 065102.

Jin, C., Ma, H., 2018. POD analysis of entropy generation in a laminar separation boundary layer. Energies 11, 3003.

Jung, J., Mangiavacchin, N., Akhavan, R., 1992. Suppression of turbulence in wall spanwise oscillations. Phys. Fluid. Fluid Dynam. 4, 1605.

Kanaris, N., Grigoriadis, D., Kassinos, S., 2011. Three-dimensional flow around a circular cylinder confined in a plane channel. Phys. Fluids 23, 064106.

Karniadakis, G.E., Choi, K.S., 2003. Mechanisms on transverse motions in turbulent wall flows. Annu. Rev. Fluid Mech. 35, 45–62.

Kida, S., Miura, H., 1998. Identification and analysis of vertical structures. Eur. J. Mech. B Fluid 17 (4), 471–488.

Kline, S.J., Reynolds, W.C., Schraub, F.A., Runstadler, P.W., 1967. The structure of turbulent boundary layers. J. Fluid Mech. 30 (4), 741–773.

Kolář, V., 2007. Vortex identification: new requirements and limitations. Int. J. Heat Fluid Flow 28 (4), 638–652.

Kolář, V., 2009. Compressibility effect in vortex identification. AIAA J. 47 (2), 473–475.

Kolář, V., Šístek, J., Cirak, F., Moses, P., 2013. Average corotation of line segments near a point and vortex identification. AIAA J. 51, 2678–2694.

Kolmogorov, A.N., 1941a. The local structure of turbulence in incompressible viscous fluid for very large Reynolds number. Proc. USSR Acad. Sci. 30, 299–303.

Kolmogorov, A.N., 1941b. On degeneration (decay) of isotropic turbulence in an incompressible viscous liquid. Proc. USSR Acad. Sci. 31, 538–540.

Kolmogorov, A.N., 1941c. Dissipation of energy in the locally isotropic turbulence. Proc. USSR Acad. Sci. 32, 16–18.

Knapp, C.F., Roache, P.J., 1968. A combined visual and hot-wire anemometer investigation of boundary-layer transition. AIAA J. 6, 29–36.

Kurz, A., Grundmann, S., Tropea, C., 2013. Boundary layer transition control using DBD plasma actuators. AerospaceLab 6, 1.

Kuzzay, D., Faranda, D., Dubrulle, B., 2015. Global vs local energy dissipation: the energy cycle of the turbulent von Kármán flow. Phys. Fluids 27, 075105.

Laizet, S., Lardeau, S., Lamballais, E., 2010. Direct numerical simulation of a mixing layer downstream a thick splitter plate. Phys. Fluids 22 (1), 1–15.

Lamb, H., 1932. Hydrodynamics. Cambridge university press, Cambridge.

Landau, L.D., Lifshitz, E.M., 1987. Fluid Mechanics, second ed. Pergamon Press, Oxford, UK.

Le, H., Moin, P., Kim, J., 1997. Direct numerical simulation of turbulent flow over a backward-facing step. J. Fluid Mech. 330, 349–374.

Lee, K., Girimaji, S.S., Kerimo, J., 2009. Effect of compressibility on turbulent velocity gradients and small-scale structure. J. Turbul. 10, N9.

Lele, S., 1992. Compact finite difference schemes with spectral- like resolution. J. Comput. Phys. 103, 16–42.

Levy, Y., Degani, D., Seginer, A., 1990. Graphical visualization of vortical flows by means of helicity. AIAA J. 28 (8), 1347–1352.

Leweke, T., Williamson, C.H.K., 1998. Cooperative elliptic instability. J. Fluid Mech. 360, 85–119.

Li, Q., Liu, C., 2010. LES for Supersonic Ramp Control Flow Using MVG at M= 2.5 and Re_θ= 1440. AIAA paper 2010-592.

Li, Q., Liu, C., 2011. Implicit LES for supersonic microramp vortex generator: new discoveries and new mechanisms. Model. Simul. Eng. 10 (1155), 934982.

Li, W.P., Liu, H., 2017. On the mechanism of turbulent darg reduction with riblets. In: 16th European Turbulence Conference. Stockholm, Sweden.

Li, L., Qiu, X.Y., Jin, S., Xiao, J., Gong, S., 2008. Weakly swirling turbulent flow in turbid water hydraulic separation device. J. Hydrodyn. 20 (3), 347–355.

Li, H., Wang, D., Xu, H., 2020. Numerical simulation of turbulent thermal boundary layer and generation mechanisms of hairpin vortex. Aero. Sci. Technol. 98, 105680.

Li, H., Yu, T., Wang, D., Xu, H., 2019. Heat-transfer enhancing mechanisms induced by the coherent structures of wall-bounded turbulence in channel with rib. Int. J. Heat Mass Transfer 137, 446–460.

Li, J., Zhang, W., 2016. The performance of proper orthogonal decomposition in discontinuous flows. Theor. App. Mech. Lett. 6 (5), 236–243.

Li, Z., Zhang, X., He, F., 2014. Evaluation of vortex criteria by virtue of the quadruple decomposition of velocity gradient tensor. Acta Phys. Sin. 63 (5), 054704.

Lian, Q.X., 1990. A visual study of the coherent structure of the turbulent boundary layer in flow with adverse pressure gradient. J. Fluid Mech. 215, 101–124.

Liang, Y., Tao, L., 2017. Interaction of vortex shedding processes on flow over a deep-draft semi-submersible. Ocean Eng. 141, 427–449.

Ligrani, P.M., Frendi, K., 2016. New Experimental Wind Tunnel Research Capabilities at UAH for Investigation of Shock-Wave-Boundary-Layer-Interactions. 9th Annual NASA/USAF Shock Wave Boundary Layer Interaction Technical Interchange Meeting, Cleveland, Ohio, USA.

Lin, J.C., Howard, F.G., Selby, G.V., 1989. Turbulent flow separation control through passive techniques. In: AIAA 2nd Shear Flow Conference.

Linnick, N.N., Fasel, H.H., 2005. A high-order immersed interface method for simulating unsteady incompressible flows on irregular domains. J. Comput. Phys. 204, 157–192.

Liu, C., Cai, X.S., 2017. New theory on turbulence generation and structure–DNS and experiment. Sci. China Phys. Mech. Astron. 60 (8), 084731.

Liu, C., Chen, L., 2011. Parallel DNS for vortex structure of late stages of flow transition. Comput. Fluids 45 (1), 129–137.

Liu, C., Gao, Y., 2020. Liutex-Based and Other Mathematical, Computational and Experimental Methods for Turbulence Structure. Bentham Science Publishers.

Liu, C., Liu, Z., 1995. Multigrid mapping and box relaxation for simulation of the whole process of flow transition in 3-D boundary layers. J. Comput. Phys. 119 (2), 325–341.

Liu, C., Liu, Z., 1997. Direct numerical simulation for flow transition around airfoils. In: Zakin, J.L., Patterson, G. (Eds.), Proceedings of First AFOSR International Conference on DNS/LES. Ruston, Louisiana.

Liu, J., Liu, C., 2019. Modified normalized Rortex/vortex identified method. Phys. Fluids 31, 061704.

Liu, C., Wang, Y., Yang, Y., Duan, Z., 2016. New omega vortex identification method. Sci. China Phys. Mech. Astron. 59 (8), 684711.

Liu, M., Xiao, L., Yang, J., Tian, X., 2017. Parametric study on the vortex-induced motions of semi-submersibles: effect of rounded ratios of the column and pontoon. Phys. Fluids 29 (5), 055101.

Liu, C., Yan, Y., Lu, P., 2014. Physics of turbulence generation and sustenance in a boundary layer. Comput. Fluids 102, 353–384.

Liu, C., Gao, Y., Tian, S., Dong, X., 2018a. Rortex—a new vortex vector definition and vorticity tensor and vector decompositions. Phys. Fluids 30 (3), 035103.

Liu, C., Li, Q., Yan, Y., Yan, Y., Yang, G., Dong, X., 2018b. High order large eddy simulation for shock-boundary layer interaction control by a micro-ramp vortex generation. Front. Aerospace Sci. 2.

Liu, C., Gao, Y., Dong, X., Wang, Y., Liu, J., Zhang, Y., Cai, X., Gui, N., 2019. Third generation of vortex identification methods: omega and Liutex/Rortex based systems. J. Hydrodyn. 31, 205–223.

Liu, J., Deng, Y., Gao, Y., Charkrit, S., Liu, C., 2019a. Mathematical foundation of turbulence generation-symmetric to asymmetric liutex/rortex. J. Hydrodyn. 31, 632–636.

Liu, J., Gao, Y., Liu, C., 2019b. An objective version of the Rortex vector for vortex identification. Phys. Fluids 31, 065112.

Lozano-Durán, A., Jiménez, J., 2014. Time-resolved evolution of coherent structures in turbulent channels: characterization of eddies and cascades. J. Fluid Mech. 759, 432–471.

Lu, P., Liu, C., 2012. DNS study on mechanism of small length scale generation in late boundary layer transition. Phys. Nonlinear Phenom. 241 (1), 11–24.

Lu, F.K., Pierce, A.J., Shih, Y., 2010. Experimental Study of Near Wake of Micro Vortex Generators in Supersonic Flow. AIAA Paper 2010-4623.

Lu, F., Li, Q., Liu, C., 2012a. Microvortex generators in high-speed flow. Prog. Aeosp. Sci. 53, 30–45.

Lu, P., Thapa, M., Liu, C., 2012b. Numerical Study on Randomization in Late Boundary Layer Transition. AIAA Paper 2012-0748.

Lugt, H.J., 1979. The dilemma of defining a vortex. In: Recent Developments in Theoretical and Experimental Fluid Mechanics. Springer-Verlag, Berlin Heidelberg, Germany.

Lugt, H.J., 1983. Vortex Flow in Nature and Technology. John Wiley & Sons, Inc, New York.

Lumley, J.L., 1967. The structure of inhomogeneous turbulent flows. In: Atmospheric Turbulence and Radio Wave Propagation, pp. 166–178.

Luo, W., Wei, Y., Dai, K., Zhu, J., You, Y., 2020. Spatiotemporal characterization and suppression mechanism of supersonic inlet buzz with proper orthogonal decomposition method. Energies 13 (1), 217.

Majda, A., Bertozzi, A., 2001. Vorticity and Incompressible Flow. Cambridge university press, Cambridge.

Martin, J.E., Meiburg, E., 1991. Numerical investigation of three-dimensionally evolving jets subject to axisymmetric and azimuthal perturbations. J. Fluid Mech. 230, 271–318.

Martins, R.S., Pereira, A.S., Mompean, G., Thais, L., Thompson, R.L., 2016. An objective perspective for classic flow classification criteria. Compt. Rendus Mec. 344 (1), 52–59.

Martins Afonso, M., Meneveau, C., 2010. Recent fluid deformation closure for velocity gradient tensor dynamics in turbulence: timescale effects and expansions. Phys. Nonlinear Phenom. 239, 1241–1250.

Marusic, I., McKeon, B., 2010. Wall-bounded turbulent flows at high Reynolds numbers: recent advances and key issues. Phys. Fluids 22, 065103.

Mathew, J., Ghosh, S., Friedrich, R., 2016. Changes to invariants of the velocity gradient tensor at the turbulent-nonturbulent interface of compressible mixing layers. Int. J. Heat Fluid Flow 59, 125–130.

Matsuura, K., 2018. DNS investigation into the effect of free-stream turbulence on hairpin-vortex evolution. WIT Trans. Eng. Sci. 120, 149–159.

Matsubara, K., Miura, T., Ohta, H., 2015. Transport dissimilarity in turbulent channel flow disturbed by rib protrusion with aspect ratio up to 64. Int. J. Heat Mass Tran. 86, 113–123.

McClure, J., Pavan, C., Yarusevych, S., 2019. Secondary vortex dynamics in the cylinder wake during laminar-to-turbulent transition. Phys. Rev. Fluids 4 (12), 124702.

Mendez, M.A., Raiola, M., Masullo, A., Discetti, S., Ianiro, A., Theunissen, R., Buchlin, J.M., 2017. POD-based background removal for particle image velocimetry. Exp. Therm. Fluid Sci. 181–192.

Meneghini, J.R., Saltara, F., Siqueira, C.L.R., Ferrari, J.A., 2001. Numerical simulation of flow interference between two circular cylinders in tandem and side-by-side arrangements. J. Fluid Struct. 15, 327–350.

Menter, F.R., 1994. Two equation eddy viscosity turbulence models for engineering applications. AIAA J. 32 (8), 1598–1605.

Meyer, K.E.E., Pedersen, J.M., Özcan, O., 2007. A turbulent jet in crossflow analysed with proper orthogonal decomposition. J. Fluid Mech. 583 (583), 199–227.

Miura, H., Kida, S., 1997. Identification of tubular vortices in turbulence. J. Phys. Soc. Jpn. 66 (5), 1331–1334.

Miura, T., Matsubara, K., Sakurai, A., 2010. Heat transfer characteristics and Reynolds stress budgets in single-rib mounting channel. J. Therm. Sci. Technol. 5 (5), 135–150.

Miura, T., Matsubar, a K., Sakurai, A., 2012. Turbulent-Heat-Flux and temperature-variance budgets in a single-rib mounting channel. J. Therm. Sci. Technol. 7 (1), 120–134.

Mizushima, J., Suehiro, N., 2005. Instability and transition of flow past two tandem circular cylinders. Phys. Fluids 17 (10), 104–107.

Moehlis, J., Smith, T.R., Holmes, P., Faisst, H., 2002. Models for turbulent plane Couette flow using the proper orthogonal decomposition. Phys. Fluids 14 (7), 2493–2507.

Moin, P., Leonard, A., Kim, J., 1986. Evolution of curved vortex filament into a vortex ring. Phys. Fluids 29 (4), 955–963.

Morgan, B., Kawai, S., Lele, S.K., 2010. Large-eddy Simulation of an Oblique Shock Impinging on a Turbulent Boundary Layer. AIAA Paper, 2010-4467.

Natrajan, V.K., Christensen, K.T., 2006. The role of coherent structures in subgrid-scale energy transfer within the log layer of wall turbulence. Phys. Fluids 18, 065104.

Nitsche, M., 2006. Vortex dynamics. In: Encyclopedia of Mathematics and Physics. Academic Press, New York.

NTNU, 2016. Francis-99 Workshops 2. Experimental Data for a High Head Francis Turbine Model at Several Operating Points.

Ooi, A., Martin, J., Soria, J., Chong, M.S., 1999. A study of the evolution and characteristics of the invariants of the velocity-gradient tensor in isotropic turbulence. J. Fluid Mech. 381, 141–174.

Pandit, J., Thompson, M., Ekkad, S.V., Huxtable, S.T., 2014. Effect of pin fin to channel height ratio and pin fin geometry on heat transfer performance for flow in rectangular channels. Int. J. Heat Mass Tran. 77, 359–368.

Papaioannou, G.V., Yue, D.K.P., Triantafyllou, M.S., Karniadakis, G.E., 2006. Three dimensionality effects in flow around two tandem cylinders. J. Fluid Mech. 558, 387–413.

Perry, A.E., Chong, M.S., 1982. On the mechanism of wall turbulence. J. Fluid Mech. 119, 173–217.

Piponniau, S., Dussauge, J.P., Debieve, J.F., Dupont, P., 2009. A simple model for low-frequency unsteadiness in shock-induced separation. J. Fluid Mech. 629, 87–108.

Pirozzoli, S., Bernardini, M., 2013. Probing high-Reynolds-number effects in numerical boundary layers. Phys. Fluids 25 (2), 021704.

Pirozzoli, S., Bernardini, M., Grasso, F., 2008. Characterization of coherent vortical structures in a supersonic turbulent boundary layer. J. Fluid Mech. 613, 205–231.

Pirozzoli, S., Grasso, F., 2006. Direct numerical simulation of impinging shock wave/turbulent boundary layer interaction at M= 2.25. Phys. Fluids 18 (6), 065113.

Priebe, S., Martín, M.P., 2012. Low-frequency unsteadiness in shock wave-turbulent boundary layer interaction. J. Fluid Mech. 699, 1–49.

Prothin, S., Djeridi, H., Billard, J.Y., 2014. Coherent and turbulent process analysis of the effects of a longitudinal vortex on boundary layer detachment on a NACA0015 foil. J. Fluid Struct. 47, 2–20.

Quadrio, M., Ricco, P., 2003. Initial response of a turbulent channel flow to spanwise oscillation of the walls. J. Turbul. 4, N7.

Quadrio, M., Ricco, P., 2004. Critical assessment of turbulent drag reduction through spanwise wall oscillations. J. Fluid Mech. 521, 251–271.

Quadrio, M., Ricco, P., Viotti, C., 2009. Streamwise-travelling waves of spanwise wall velocity for turbulent drag reduction. J. Fluid Mech. 627, 161–178.

Richardson, L.F., 1920. The supply of energy from and to atmospheric eddies. Proc. R. Soc. A 97, 354.

Richardson, L.F., 1922. Weather Prediction by Numerical Process. Cambridge University Press, Cambridge, UK.

Rehimi, F., Aloui, F., Nasrallah, S.B., Doubliez, L., Legrand, J., 2008. Experimental investigation of a confined flow downstream of a circular cylinder centred between two parallel walls. J. Fluid Struct. 24, 885.

Ricco, P., Wu, S., 2004. On the effects of lateral wall oscillations on a turbulent boundary layer. Exp. Therm. Fluid Sci. 29 (1), 41–52.

Robinson, S.K., 1990. Quasi-coherent Structures in the Turbulent Boundary Layer. II - Verification and New Information from a Numerically Simulated Flat-Plate Layer. Hemisphere, New York.

Robinson, S.K., 1991. Coherent motions in the turbulent boundary layer. Annu. Rev. Fluid Mech. 23, 601–639.

Robinson, S., Kline, S., Spalart, P., 1989. A review of quasi-coherent structures in a numerically simulated turbulent boundary layer. Tech. rep. NASA TM-102191.

Robinson, S.K., Spalart, P., Kline, S., 1990. A review of vortex structures and associated coherent motions in turbulent boundary layers. In: Structure of Turbulence and Drag Reduction. Springer.

Rogers, M.M., Moser, R.D., 1993. Direct simulation of a self-similar turbulent mixing layer. Phys. Fluids 6 (2), 903−923.

Roshko, A., 1954. On the Drag and Shedding Frequency of Two-Dimensional Bluff Bodies. NACA Technical Note No. 3169.

Roth, M., 2000. Automatic Extraction of Vortex Core Lines and Other Line-type Features for Scientific Visualization. ETH Zürich, Zürich, Switzerland.

Saffman, P.G., 1992. Vortex Dynamics. Cambridge University Press, Cambridge.

Sakamoto, N., Carrica, P.M., Stern, F., 2012. URANS simulations of static and dynamic maneuvering for surface combatant: Part 2. Analysis and validation for local flow characteristics. J. Mar. Sci. Technol. 17 (4), 446−468.

Sanada, Y., Tanimoto, K., Takagi, K., Gui, L., Toda, Y., Stern, F., 2013. Trajectories for ONR Tumblehome maneuvering in calm water and waves. Ocean Eng. 72, 45−65.

Scarano, F., 2007. Overview of PIV in supersonic flows. Part. Image Velocim. Top. Appl. Phys. 112 (1), 445−463.

Schlatter, P., Bagheri, S., Henningson, D.S., 2011. Self-sustained global oscillations in a jet in crossflow. Theor. Comput. Fluid Dynam. 25 (1−4), 129−146.

Schlatter, P., Örlü, R., 2010. Assessment of direct numerical simulation data of turbulent boundary layers. J. Fluid Mech. 659, 116−126.

Schlichting, H., Gersten, K., 2010. Boundary Layer Theory, 8th Revised and Enlarged Edition. Springer, Berlin, German.

Schoppa, W., Hussain, F., 2002. Coherent structure generation in near-wall turbulence. J. Fluid Mech. 453, 57−108.

Schmid, P.J., 2013. Advanced Post-Processing of Experimental and Numerical Data, VKI Lecture Series 2014-01.

Schroder, A., Geisler, R., Staack, K., Elsinga, G.E., Scarano, F., Wieneke, B., Henning, A., Poelma, C., Westerweel, J., 2011. Eulerian and Lagrangian views of a turbulent boundary layer flow using time-resolved tomographic PIV. Exp. Fluid 50 (4), 1071−1091.

Shu, C., 1988. Total-variation-diminishing time discretizations. SIAM J. Sci. Stat. Comput. 9 (6), 1073−1084.

Sen, M., Bhaganagar, K., Juttijudata, V., 2007. Application of proper orthogonal decomposition (POD) to investigate a turbulent boundary layer in a channel with rough walls. J. Turbul. 8 (41), N41.

Shen, Z., Wan, D.C., Carrica, P., 2014. RANS simulations of free maneuvers with moving rudders and propellers using overset grids in OpenFOAM. In: SIMMAN Workshop on Verification and Validation of Ship Maneuvering Simulation Methods, Lyngby, Denmark.

Shen, Z., Wan, D., Carrica, P.M., 2015. Dynamic overset grids in OpenFOAM with application to KCS self-propulsion and maneuvering. Ocean Eng. 108, 287−306.

Sirovich, L., 1987. Turbulence and the dynamics of coherent structures. Part I: coherent structures. Q. Appl. Math. 45 (3), 561−571.

Skote, M., 2012. Temporal and spatial transients in turbulent boundary layer flow over an oscillating wall. Int. J. Heat Fluid Flow 38, 1−12.

Slaouti, A., Stansby, P.K., 1992. Flow around two circular cylinders by the random-vortex method. J. Fluid Struct. 6 (6), 641−670.

Smith, C.R., Walker, J.D.A., Haidari, A.H., Sobrun, U., 1991. On the dynamics of near-wall Turbulence. Philos. Trans. R. Soc. Lond. A. 336, 131−175.

Souverein, L.J., Dupont, P., Debiève, J.F., Van Oudheusden, B.W., Scarano, F., 2010. Effect of interaction strength on unsteadiness in shock-wave-induced separations. AIAA J. 48 (7), 1480−1493.

Spivack, H.M., 1946. Vortex frequency and flow pattern in the wake of two parallel cylinders at varied spacing normal to an air stream. J. Aeronaut. Sci. 13 (6), 289–301.

Stanislas, M., 2017. Near wall turbulence: an experimental view. Phys. Rev. Fluids 2 (10), 100506.

Stewart, G.W., Sun, J., 1990. Matrix Perturbation Theory. Academic Press.

Strawn, R.C., Kenwright, D.N., Ahmad, J., 1999. Computer visualization of vortex wake systems. AIAA J. 37 (4), 511–512.

Sujudi, D., Haimes, R., 1995. Identification of Swirling Flow in 3D Vector Fields. AIAA Paper 95-1715.

Sullivan, R.D., 1959. A two-cell vortex solution of the Navier-Stokes equations. J. Aero. Sci. 26 (11), 767–768.

Suman, S., Girimaji, S.S., 2009. Homogenized Euler equation: a model for compressible velocity gradient dynamics. J. Fluid Mech. 620, 177.

Suman, S., Girimaji, S.S., 2010. Velocity gradient invariants and local flow-field topology in compressible turbulence. J. Turbul. 11, N2.

Sun, Z., 2014. Micro Ramps in Supersonic Turbulent Boundary Layers: An Experimental and Numerical Study.

Sun, Z., Schrijer, F.F.J., Oudheusden, B.W.V., 2012. The three-dimensional flow organization past a micro-ramp in a supersonic boundary layer. Phys. Fluids 24, 055105.

Sun, Z., Scarano, F., Oudheusden, B.W.V., Schrijer, F.F.J., Yan, Y., Liu, C., 2014a. Numerical and experimental investigations of the supersonic microramp wake. AIAA J. 52, 1518–1527.

Sun, Z., Schrijer, F.F.J., Scarano, F., Oudheusden, B.W.V., 2014b. Decay of the supersonic turbulent wakes from micro-ramps. Phys. Fluids 26 (2), 389–420.

Sun, Z., 2015. Micro vortex generators for boundary layer control: principles and applications. Int. J. Flow Contr. 7, 67–86.

Taylor, G.I., 1938. The spectrum of turbulence. Proc. R Soc. Lond. Ser. A 164, 476–490.

Tennekes, H., Lumley, J.L., 1972. A First Course in Turbulence. MIT Press, Massachusetts.

Theodorsen, T., 1952. Mechanism of turbulence. In: Proceeding of Second Midwestern Conference of Fluid Mechanics. Ohio State University, Columbus, Ohio, pp. 1–19.

Thomas, D.G., Kraus, K.A., 1964. Interaction of vortex streets. J. Appl. Phys. 35 (12), 3458–3459.

Tomkins, C.D., Adrian, R.J., 2003. Spanwise structure and scale growth in turbulent boundary layers. J. Fluid Mech. 490, 37–74.

Touber, E., Sandham, N.D., 2009. Large-eddy simulation of low-frequency unsteadiness in a turbulent shock-induced separation bubble. Theor. Comput. Fluid Dynam. 23 (2), 79–107.

Townsend, A.A., 1956. The Structure of Turbulent Shear Flow. Cambridge University Press, Cambridge.

Townsend, A.A., 1976. The Structure of Turbulent Shear Flow, second ed. Cambridge University Press, Cambridge.

Tran, C.T., Ji, B., Long, X., 2019. Simulation and analysis of cavitating flow in the draft tube of the Francis turbine with splitter blades at off-design condition. Teh. Vjesn. 26 (6), 1650–1657.

Tran, C.T., Long, X., Ji, B., Liu, C., 2013. Prediction of the precessing vortex core in the Francis-99 draft tube under off-design conditions by using Liutex/Rortex method. J. Hydrodyn. 32, 623–628.

Trivedi, C., Cervantes, M.J., Gandhi, B.K., Dahlhaug, O.G., 2013. Experimental and numerical studies for a high head Francis turbine at several operating points. J. Fluid Eng. 135 (11), 111102.

Truesdell, C., 1954. The Kinematics of Vorticity. Indiana University Press, Bloomington.

Trujillo, S., Bogard, D., Ball, K., 1997. Turbulent boundary layer drag reduction using an oscillating wall. In: 4th Shear Flow Control Conference, Snowmass Village, Colorado, USA.

Vollmers, H., Kreplin, H.P., Meier, H.U., 1983. Separation and vortical-type flow around a prolate spheroid-evaluation of relevant parameters. In: Proceedings of the AGARD Symposium on Aerodynamics of Vortical Type Flows in Three Dimensions AGARD-CP-342, Rotterdam, Netherlands.

Wallace, J.M., 2013. Highlights from 50 years of turbulent boundary layer research. J. Turbul. 13, 1.

Wang, Z., Chen, S., 2009. Structures of confined vortex breakdown in constant diameter pipe flow. J. Hydrodyn. 21 (3), 341–346.

Wang, Y., Gui, N., 2019. A review of the third-generation vortex identification method and its applications. Chinese J. Hydrodyn. 34 (4), 413–429 (in Chinese).

Wang, Y., Liu, C., 2019. Liutex (vorex) cores in transitional boundary layer with spanwise-wall oscillation. J. Hydrodyn. 31, 1178–1189.

Wang, J., Wan, D., 2019. Numerical simulations of viscous flows around jbc ship using different turbulence models. In: Proceedings of the 11th International Workshop on Ship and Marine Hydrodynamics, Hamburg, Germany.

Wang, B., Weidong, L., Yuxin, Z., Xiaoqiang, F., Chao, W., 2012. Experimental investigation of the micro-ramp based shock wave and turbulent boundary layer interaction control. Phys. Fluids 24 (5), 055110.

Wang, B., Weidong, L., Mingbo, S., Zhao, Y., 2014a. Energy and fluid transportation in turbulent boundary-layer under the micro-ramp control. In: 7th AIAA Flow Control Conference, p. 2650.

Wang, H., Cao, H., Zhou, Y., 2014b. POD analysis of a finite-length cylinder near wake. Exp. Fluid 55 (8), 1790.

Wang, W., Guan, X., Jiang, N., 2014c. TRPIV investigation of space-time correlation in turbulent flows over flat and wavy walls. Acta Mech. Sin. 30 (4), 468–479.

Wang, Y., Huang, W., Xu, C., 2015. On hairpin vortex generation from near-wall streamwise vortices. Acta Mech. Sin. 31 (2), 139–152.

Wang, J., Zhao, W., Wan, D., 2016a. Free maneuvering simu- lation of ONR tumblehome using overset grid method in naoe-FOAM-SJTU solver. In: Proceedings of the 31st Symposium on Naval Hydrodynamics, Monterey, California, USA.

Wang, Y., Al-Dujaly, H., Yan, Y., Zhao, N., Liu, C., 2016b. Physics of multiple level hairpin vortex structures in turbulence. Sci. China Phys. Mech. Astron. 59 (2), 624703.

Wang, Y., Yang, Y., Yang, G., Liu, C., 2017a. DNS study on vortex and vorticity in late boundary layer transition. Commun. Comput. Phys. 22 (02), 441–459.

Wang, J., Zou, L., Wan, D., 2017b. CFD simulations of free running ship under course keeping control. Ocean Eng. 141, 450–464.

Wang, Y., Gao, Y., Liu, C., 2018a. Letter: galilean invariance of rortex. Phys. Fluids 30 (11), 111701.

Wang, L., Guo, C., Su, Y., Wu, T., 2018b. A numerical study on the correlation between the evolution of propeller trailing vortex wake and skew of propellers. Int. J. Naval Architect. Ocean Eng. 10 (2), 212–224.

Wang, J., Zou, L., Wan, D., 2018c. Numerical simulations of zigzag maneuver of free running ship in waves by RANS-overset grid method. Ocean Eng. 162, 55–79.

Wang, Y., Gao, Y., Liu, J., Liu, C., 2019a. Explicit formula for the Liutex vector and physical meaning of vorticity based on the Liutex-Shear decomposition. J. Hydrodyn. 31 (3), 464–474.

Wang, D., Li, H., Li, Y., Yu, T., Xu, H., 2019b. Direct numerical simulation and in-depth analysis of thermal turbulence in square annular duct. Int. J. Heat Mass Tran. 144, 118590.

Wang, C., Liu, Y., Chen, J., Zhang, F., Huang, B., Wang, G., 2019c. Cavitation vortex dynamics of unsteady sheet/cloud cavitating flows with shock wave using different vortex identification methods. J. Hydrodyn. 31 (3), 475–494.

Wang, J., Zhao, W., Wan, D., 2019d. Simulations of self-propelled fully appended ship model at different speeds. Int. J. Comput. Methods 16 (5), 1840015.

Wang, Y., Zhang, W., Cao, X., Yang, H., 2019e. The applicability of vortex identification methods for complex vortex structures in axial turbine rotor passages. J. Hydrodyn. 31, 700–707.

Wang, D., 2020. Direct Numerical Simulations and In-Depth Analysis of Fully-Developed Thermal Turbulence in Square Annular Duct and Transition Flow Around Two Cylinders in Tandem (Ph.D. thesis). Department of Aeronautics and Astronautics, Fudan University.

Wang, Y., Gao, Y., Xu, H., Dong, X., Liu, J., Xu, W., Chen, M., Liu, C., 2020a. Liutex theoretical system and six core elements of vortex identification. J. Hydrodyn. 32 (2), 197–211.

Wang, D., Li, H., Cao, B., Xu, H., 2020b. Law-of-the-wall analytical formulations for type-A turbulent boundary layers. J. Hydrodyn. 32 (2), 296–313.

Warsi, Z.U.A., 2005. Fluid Dynamics: Theoretical and Computational Approaches. CRC Press, pp. 20–23. ISBN 0-8493-4436-0.

Wedgewood, L.E., 1999. An objective rotation tensor applied to non-Newtonian fluid mechanics. Rheol. Acta 38 (2), 91–99.

White, F.M., 2012. Viscous Fluid Flow. McGraw Hill, New York.

Wille, R., 1960. Karman vortex streets. Adv. Appl. Mech. 6, 273–287.

Williamson, C.H.K., 1988. The existence of two stages in the transition to three-dimensionality of a cylinder wake. Phys. Fluids 31, 3165.

Williamson, C.H.K., 1992. The natural and forced formation of spot-like vortex dislocations in the transition of a wake. J. Fluid Mech. 243, 393–441.

Williamson, C.H.K., 1996. Three-dimensional wake transition. J. Fluid Mech. 328, 345–407.

Wu, M., Martin, M.P., 2008. Analysis of shock motion in shockwave and turbulent boundary layer interaction using direct numerical simulation data. J. Fluid Mech. 594, 71–83.

Wu, X., Moin, P., 2009. Direct numerical simulation of turbulence in a nominally zero-pressure-gradient flat-plate boundary layer. J. Fluid Mech. 630, 5–41.

Wu, J., Xiong, A., Yang, Y., 2005. Axial stretching and vortex definition. Phys. Fluids 17 (3), 0381108.

Wu, J., Ma, H., Zhou, M., 2006. Vorticity and Vortex Dynamics. Springer, Berlin Heidelberg.

Wu, N., Sakai, Y., Nagata, K., Ito, Y., 2015a. Dynamics and geometry of developing planar jets based on the invariants of the velocity gradient tensor. J. Hydrodyn. 27, 894–906.

Wu, X., Moin, P., Adrian, R., Baltzer, J., 2015b. Osborne Reynolds pipe flow: direct computation from laminar through bypass transition to fully-developed turbulence. Proc. Natl. Acad. Sci. U. S. A. 112 (26), 7920–7924.

Wu, Y., An, G., Wang, B., 2019. Numerical investigation into the underlying mechanism connecting the vortex breakdown to the flow unsteadiness in a transonic compressor rotor. Aero. Sci. Technol. 86, 106–118.

Wu, Y., Zhang, W., Wang, Y., Zou, Z., Chen, J., 2020. Energy dissipation analysis based on velocity gradient tensor decomposition. Phys. Fluids 32, 035114.

Xing, T., 2014. Direct numerical simulation of open von Kármán swirling flow. J. Hydrodyn. 26 (2), 165–177.

Xing, T., Bhushan, S., Stern, F., 2012. Vortical and turbulent structures for KVLCC2 at drift angle 0, 12, and 30 degrees. Ocean Eng. 55, 23–43.

Xu, H., 2009. Direct numerical simulation of turbulence in a square annular duct. J. Fluid Mech. 621, 23–57.

Xu, C., Huang, W., 2005. Transient response of Reynolds stress transport to spanwise wall oscillation in a turbulent channel flow. Phys. Fluids 17, 018101.

Xu, W., Gao, Y., Deng, Y., Liu, J., Liu, C., 2019a. An explicit expression for the calculation of the Rortex vector. Phys. Fluids 31, 095102.

Xu, W., Wang, Y., Gao, Y., Liu, J., Dou, H., Liu, C., 2019b. Liutex similarity in turbulent boundary layer. J. Hydrodyn. 31, 1259–1262.

Xu, H., Cai, X., Liu, C., 2019c. Liutex (vortex) core definition and automatic identification for turbulence vortex Structures. J. Hydrodyn. 31, 857–863.

Yan, Y., Liu, C., 2014. Study on the ring-like vortical structure in MVG controlled supersonic ramp flow with different inflow conditions. Aero. Sci. Technol. 35, 106–115.

Yan, Y., Li, Q., Liu, C., Pierce, A., Lu, F., Lu, P., 2012. Numerical discovery and experimental confirmation of vortex ring generation by microramp vortex generator. Appl. Math. Model. 36 (11), 5700–5708.

Yan, Y., Chen, C., Lu, P., Liu, C., 2013. Study on shock wave-vortex ring interaction by the micro vortex generator controlled ramp flow with turbulent inflow. Aero. Sci. Technol. 30 (1), 226–231.

Yan, Y., Chen, C., Fu, H., Liu, C., 2014a. DNS study on Λ-vortex and vortex ring formation in flow transition at mach number 0.5. J. Turbul. 15, 1–21.

Yan, Y., Chen, C., Wang, X., Liu, C., 2014b. LES and analyses on the vortex structure behind supersonic MVG with turbulent inflow. Appl. Math. Model. 38 (1), 196–211.

Yang, Q., Fu, S., 2008. Analysis of flow structures in supersonic plane mixing layers using the POD methodSci. China Phys. Mech. Astron. 05, 93–110.

Yang, Y., Yan, Y., Liu, C., 2016. ILES for mechanism of ramp-type MVG reducing shock induced flow separation. Sci. China Phys. Mech. Astron. 59, 124711.

Yang, B., Xiang, Y., Cai, X., Zhou, W., Liu, H., Li, S., Gao, W., 2017. Simultaneous measurements of fine and coarse droplets of wet steam in a 330 MW steam turbine by using imaging method. Proc. IME J. Power Energy 231 (3), 161–172.

Yao, J., Zhao, Y., Fairweather, M., 2015. Numerical simulation of turbulent flow through a straight square duct. Appl. Therm. Eng. 91, 800–811.

Ye, J., Zou, Z., 2007. Large-eddy simulation periodic wake/laminar separation bubble interaction under low Reynolds number conditions. J. Eng. Thermophys. 28, 215–218.

Ye, Q., Schrijer, F.F.J., Scarano, F., 2016. Boundary layer transition mechanisms behind a micro-ramp. J. Fluid Mech. 793, 132–161.

Ye, Q., Schrijer, F.F.J., Scarano, F., 2018. On Reynolds number dependence of micro-ramp-induced transition. J. Fluid Mech. 837, 597–626.

Yoon, D.H., Yang, K.S., Choi, C.B., 2009. Heat transfer enhancement in channel flow using an inclined square cylinder. J. Heat Tran. 131 (7), 074503.

Yu, Y., Shrestha, P., Nottage, C., Liu, C., 2020. Principal coordinates and principal velocity gradient tensor decomposition. J. Hydrodyn. 32, 441–453.

Yudhistira, I., Skote, M., 2011. Direct numerical simulation of a turbulent boundary layer over an oscillating wall. J. Turbul. 12, N9.

Zdravkovich, M.M., 1972. Smoke observations of wakes of tandem cylinders at low Reynolds numbers. Aeronaut. J. 76, 108–114.

Zhang, S., Choudhury, D., 2006. Eigen helicity density: a new vortex identification scheme and its application in accelerated inhomogeneous flows. Phys. Fluids 18 (5), 058104.

Zhang, W., Zou, Z., Ye, J., 2012. Leading-edge redesign of a turbomachinery blade and its effect on aerodynamic performance. Appl. Energy 93 (5), 655–667.

Zhang, Q., Liu, Y., Wang, S., 2014. The identification of coherent structures using proper orthogonal decomposition and dynamic mode decomposition. J. Fluid Struct. 49, 53–72.

Zhang, W., Zou, Z., Qi, L., Ye, J., Wang, L., 2015. Effects of freestream turbulence on separated boundary layer in a low-Re high-lift LP turbine blade. Comput. Fluids 109, 1–12.

Zhang, Y., Liu, K., Xian, H., Du, X., 2018a. A review of methods for vortex identification in hydroturbines. Renew. Sustain. Energy Rev. 81 (1), 1269–1285.

Zhang, Y., Qiu, X., Chen, F., Liu, K., Dong, X., Liu, C., 2018b. A selected review of vortex identification methods with applications. J. Hydrodyn. 30, 767–779.

Zhang, Y., Wang, X., Zhang, Y., Liu, C., 2019. Comparisons and analyses of vortex identification between Omega method and Q criterion. J. Hydrodyn. 31 (2), 224–230.

Zhao, Y., Liao, W., Ruan, H., Luo, X., 2015. Performance study for Francis-99 by using different turbulence models. J. Phys. Conf. 579, 012012.

Zhao, H., Wei, A., Luo, K., Fan, J., 2016. Direct numerical simulation of turbulent boundary layer with heat transfer. Int. J. Heat Mass Tran. 99, 10–19.

Zhao, W., Zou, L., Wan, D., Hu, Z., 2018. Numerical investigation of vortex-induced motions of a paired-column semi- submersible in currents. Ocean Eng. 164, 272–283.

Zhao, W., Wang, J., Wan, D., 2020. Vortex identification methods in marine hydrodynamics. J. Hydrodyn. 32, 286–295.

Zhou, J., Adrian, R.J., Balachandar, S., 1996. Autogeneration of near wall vortical structure in channel flow. Phys. Fluids 8 (1), 288–291.

Zhou, J., Adrian, R.J., Balachandar, S., Kendall, T.M., 1999. Mechanisms for generating coherent packets of hairpin vortices in channel flow. J. Fluid Mech. 387, 353–396.

Zou, Z., Wang, S., Liu, H., Zhang, W., 2014. Turbine Aerodynamics for Aero-Engine: Flow Analysis and Aerodynamics Design. Shanghai Jiao Tong University press.

Zou, Z., Liu, J., Zhang, W., Wang, P., 2016. Shroud leakage flow models and a multi-dimensional coupling CFD (computational fluid dynamic) method for shrouded turbines. Energy 103, 410–429.

Zou, Z., Shao, F., Li, Y., Zhang, W., Berglund, A., 2017. Dominant flow structure in the squealer tip gap and its impact on turbine aerodynamic performance. Energy 138, 167–184.

Zwart, P., Gerber, A., Belamri, T., 2004. A two-phase flow model for predicting cavitation dynamics. In: Fifth International Conference on Multiphase Flow, Yokohama, Japan, p. 152.

Index

Note: Page numbers followed by "f" indicate figures and "t" indicate tables.

Printed in the United States
By Bookmasters